PAVEMENT LIFE-CYCLE ASSESSMENT

PROCEEDINGS OF THE PAVEMENT LIFE-CYCLE ASSESSMENT SYMPOSIUM, CHAMPAIGN ILLINOIS, USA, 12–13 APRIL 2017

Pavement Life-Cycle Assessment

Editors

Imad L. Al-Qadi & Hasan Ozer
University of Illinois at Urbana-Champaign, IL, USA

John Harvey
University of California at Davis, CA, USA

CRC Press is an imprint of the
Taylor & Francis Group, an **informa** business

A BALKEMA BOOK

CRC Press/Balkema is an imprint of the Taylor & Francis Group, an informa business

© 2017 Taylor & Francis Group, London, UK

Typeset by V Publishing Solutions Pvt Ltd., Chennai, India
Printed and bound in the USA by Edwards Brothers, Inc, Lillington, NC

All rights reserved. No part of this publication or the information contained herein may be reproduced, stored in a retrieval system, or transmitted in any form or by any means, electronic, mechanical, by photocopying, recording or otherwise, without written prior permission from the publisher.

Although all care is taken to ensure integrity and the quality of this publication and the information herein, no responsibility is assumed by the publishers nor the author for any damage to the property or persons as a result of operation or use of this publication and/or the information contained herein.

Published by: CRC Press/Balkema
P.O. Box 11320, 2301 EH Leiden, The Netherlands
e-mail: Pub.NL@taylorandfrancis.com
www.crcpress.com – www.taylorandfrancis.com

ISBN: 978-1-138-06605-2 (Hbk)
ISBN: 978-1-315-15932-4 (eBook)

Pavement Life-Cycle Assessment – Al-Qadi, Ozer & Harvey (Eds)
© 2017 Taylor & Francis Group, London, ISBN 978-1-138-06605-2

Table of contents

Foreword	vii
Organization	ix
Pavement life cycle assessment: A comparison of American and European tools J. Santos, S. Thyagarajan, E. Keijzer, R. Flores & G. Flintsch	1
Lessons learned in developing an environmental product declaration program for the asphalt industry in North America A. Mukherjee & H. Dylla	11
Current difficulties with creation of standardized digital climate calculations for infrastructural projects L. Strömberg	23
Development of an environmental Life-Cycle Assessment (LCA) protocol for flexible pavements that integrates life-cycle components to a proprietary software F. Osmani, M. Hettiwatte, S. Kshirsagar, S. Senadheera & H.C. Zhang	31
Review and comparison of freely-available tools for pavement carbon footprinting in Europe D.L. Presti & G. D'Angelo	41
Route level analysis of road pavement surface condition and truck fleet fuel consumption F. Perrotta, L. Trupia, T. Parry & L.C. Neves	51
Impact of PCC pavement structural response on rolling resistance and vehicle fuel economy D. Balzarini, I. Zaabar & K. Chatti	59
Cool pavement LCA tool: Inputs and recommendations for integration H. Li, J.T. Harvey & A. Saboori	69
Rolling resistance and traffic delay impact on a road pavement life cycle carbon footprint analysis L. Trupia, T. Parry, L.C. Neves & D.L. Presti	79
LCA case study for O'Hare International Airport taxiway A&B rehabilitation J. Kulikowski	89
Exploring alternative methods of environmental analysis A.F. Braham	103
An uncoupled pavement-urban canyon model for heat islands S. Sen & J.R. Roesler	111
The importance of incorporating uncertainty into pavement life cycle cost and environmental impact analyses J. Gregory, A. Noshadravan, O. Swei, X. Xu & R. Kirchain	121

Functional unit choice for comparative pavement LCA involving use-stage
with pavement roughness Uncertainty Quantification (UQ) 133
M. Ziyadi, H. Ozer & I.L. Al-Qadi

Role of uncertainty assessment in LCA of pavements 145
S. Inti, M. Sharma & V. Tandon

Calculation method of stockpiling and use phase in road LCA: Case study
of steel slag recycling 157
O. Yazoghli-Marzouk, M. Dauvergne, W. Chebbi & A. Jullien

Concrete pavement life cycle environmental assessment & economic analysis:
A manitoba case study 165
M. Alauddin Ahammed, S. Sullivan, G. Finlayson, C. Goemans, J. Meil & M. Akbarian

Life-cycle assessment tool development for flexible pavement in-place recycling techniques 179
M.K. Senhaji, H. Ozer & I.L. Al-Qadi

Life-cycle assessment of road pavements containing marginal materials: Comparative
analysis based on a real case study 189
M. Pasetto, E. Pasquini, G. Giacomello & A. Baliello

Integrated sustainability assessment of asphalt rubber pavement based
on life cycle analysis 199
R. Cao, Z. Leng, M. Shu-Chien Hsu, H. Yu & Y. Wang

Environmental assessment and economic analysis of porous pavement at sidewalk 211
X. Chen, H. Wang & H. Najm

LCCA for silent surfaces 221
F.G. Praticò

Life cycle assessment and benchmarking of end of life treatments of flexible
pavements in California 231
A. Saboori, J.T. Harvey, A.A. Butt & D. Jones

Life cycle assessment of pavements under a changing climate 241
O. Valle, Y. Qiao, E. Dave & W. Mo

Capitalizing green pavement: A method and valuation 251
X. Liu, D. Choy, Q. Cui & C.W. Schwartz

A methodology for sustainable mechanistic-empirical pavement design 261
N. Soliman & M. Hassan

Energy consumption and Greenhouse Gas Emissions of high RAP central plant hot
recycling technology using Life Cycle Assessment: Case study 271
Y. Lu, H. Wu, A. Liu, W. Ding & H. Zhu

Implementation of life cycle thinking in planning and procurement at the Swedish
Transport Administration 281
S. Toller & M. Larsson

Emission-controlled pavement management scheduling 289
U.D. Tursun, R. Yang & I.L. Al-Qadi

Author index 299

Pavement Life-Cycle Assessment – Al-Qadi, Ozer & Harvey (Eds)
© 2017 Taylor & Francis Group, London, ISBN 978-1-138-06605-2

Foreword

An increasing number of agencies, academic institutes, and governmental and industrial bodies are embracing the principles of sustainability in managing their activities and conducting business. As a sustainability quantitative measurement tool, life-cycle assessment (LCA) has been increasingly used in Europe in the construction industry and is being integrated into green construction regulations in a number of countries. Similarly, North American has recently witnessed the development of life-cycle assessment tools such as the first regional Tollway pavement LCA tool developed by the Illinois Center for Transportation of the University of Illinois at Urbana-Champaign.

Organized by the University of Illinois at Urbana-Champaign (UIUC), the Pavement Life-Cycle Assessment Symposium-2017 will provide a forum for assessing the status of pavement LCA development (including all aspects from goal and scope definition, to inventory databases, impact assessment and interpretation), plans for implementation for the project- and network-levels, and identifying the extent of consensus. This symposium is a follow-on to the 2010 Pavement LCA Workshop, Davis, California; 2012 RILEM Symposium on LCA for Construction Materials, Nantes, France; and 2014 Pavement LCA Symposium, Davis, California.

The symposium is preceded by a meeting of the Federal Highway Administration (FWHA) Sustainable Pavement Technical Working Group (SPTWG). Presentations and panel discussions summarize the current status, challenges, and pressing issues related to pavement LCA, thus paving the way for global knowledge sharing of the latest developments and advances in the field of pavement sustainability.

The conference brought together academic and industrial leaders from around the world. Selected papers are included in this proceedings. The papers published in these proceedings are *fully refereed*. Each paper submitted to this conference was peer-reviewed by at least three professionals in this field. Based on the reviewers' recommendations, those papers which are of high technical value and best suited the conference goals and objectives were chosen for inclusion in the proceedings.

The proceedings include papers on various research and practical issues related to pavement LCA, including regional inventory database and modeling; review of LCA tools and new LCA tool development; the importance of incorporating uncertainty into pavement LCA data; sustainable mechanistic-empirical pavement design; case studies highlighting LCA implementation for various pavement applications; impact of pavement in-place recycling techniques; development of use-stage models including rolling resistance and heat island effect; among other diverse topics.

The technical program of the symposium consisted of approximately 30 oral sessions, three panel discussions—focusing on building sustainable and cost-efficient pavements, discussing current and future pavement life cycle assessment and environmental product declarations (EPD) from an international perspective, and steps to be taken moving forward—and a poster session. In addition to the FHWA SPTWG pre-symposium meeting, the symposium included two receptions and a dinner to allow discussion and networking.

The symposium was developed in consultation with an advisory committee and a scientific committee of over 40 members. The organizers of the Pavement Life-Cycle Assessment Symposium 2017 acknowledge the efforts of all members of the scientific committee whose help has vastly contributed to the success of the symposium. We are thankful for all those who volunteered their time to thoroughly review the submitted papers and offer constructive comments to authors.

Special thanks must go to Illinois Center for Transportation, the Illinois Department of Transportation, and the University of California Pavement Research Center for sponsoring this event. Thanks are also due to the Federal Highway Administration for being a strong advocate of the development of sustainable transportation practices.

Imad L. Al-Qadi, PhD, PE, Dist.M.ASCE
University of Illinois at Urbana-Champaign

Hasan Ozer, PhD
University of Illinois at Urbana-Champaign

John Harvey, PhD, PE, M. ASCE
University of California, Davis

Organization

CHAIRMEN

Imad L. Al-Qadi, Chairman, *University of Illinois at Urbana-Champaign*
Hasan Ozer, Co-Chairman, *University of Illinois at Urbana-Champaign*
John Harvey, Co-Chairman, *University of California-Davis*

SCIENTIFIC COMMITTEE

Loizos Andreas, *National Technical University of Athens, Greece*
Amit Bhasin, *The University of Texas at Austin, USA*
Bjorn Birgisson, *Aston University, UK*
Alexander Brown, *Asphalt Institute, USA*
Karim Chatti, *Michigan State University, USA*
Jo Sias Daniel, *University of New Hampshire, USA*
Heather Dylla, *Federal Highway Association, USA*
Mostafa Elseifi, *Louisiana State University, USA*
Jon Epps, *Texas A&M University, USA*
Gonzalo Fernández-Sánchez, *Universidad Politécnica de Madrid, Spain*
Gerardo Flintsch, *Virginia Polytechnic Institute and State University, USA*
Navneet Garg, *Federal Aviation Administration, USA*
Steve Gillen, *Illinois State Toll Highway Authority, USA*
Elie Hajj, *University of Nevada, Reno, USA*
Kevin Hall, *University of Arkansas, USA*
Baoshan Huang, *University of Tennessee, USA*
Said Jalali, *Universidade do Minho, Portugal*
Agnès Jullien, *Institut Français des Sciences et Technologies des Transports, De L'Amenagement et des Reseaux, France*
Kamil Kaloush, *Arizona State University, USA*
Lev Khazanovich, *University of Minnesota, USA*
Sébastien Lasvaux, *Haute école spécialisée de Suisse occidentale (HES-SO), Switzerland*
James Mack, *CEMEX, USA*
Rajib Mallick, *Worcester Polytechnic Institute, USA*
Joep Meijer, *The Right Environment Ltd, USA/Netherlands*
Amlan Mukherjee, *Michigan Technological University, USA*
Jorge Pais, *University of Minho, Portugal*
Tony Parry, *University of Nottingham, UK*
Brian Pfeifer, *Illinois Department of Transportation, USA*
Jeff Roesler, *University of Illinois at Urbana-Champaign, USA*
Luis Loria-Salazar, *Universidad de Costa Rica, Costa Rica*
Charles Schwartz, *University of Maryland, USA*
Nadarajah Sivaneswaran, *Federal Highway Administration, USA*
Mike Southern, *Eurobitume, Belgium*
Silvia Caro, *Universidad de los Andes, Colombia*
Wynand Steyn, *University of Pretoria, South Africa*

Nina Štirmer, *University of Zagreb, Faculty of Civil Engineering, Croatia*
Erol Tutumluer, *University of Illinois at Urbana-Champaign, USA*
Tom Van Dam, *Nichols Consulting Engineers, USA*
Anne Ventura, *Université de Nantes, France*
Hao Wang, *Rutgers University, USA*
Kelvin Wang, *Oklahoma State University, USA*
Linbing Wang, *Virginia Polytechnic Institute and State University, USA*
Leif Wathne, *American Concrete Pavement Association, USA*
Matthew Wayman, *TRL Ltd, UK*
Kua Harn Wei, *National University of Singapore, Singapore*

ADVISORY COMMITTEE

Gina Ahlstrom, *Federal Highway Administration, USA*
Heather Dylla, *National Asphalt Pavement Association, USA*
Larry Galehouse, *Michigan State University, USA*
Kurt Smith, *Applied Pavement Technology, USA*
Leif Wathne, *American Concrete Pavement Association, USA*

LOCAL ORGANIZING COMMITTEE

Waad Ayoub, *University of Illinois at Urbana-Champaign*
Lori Heinz, *University of Illinois at Urbana-Champaign*
Mouna Krami Senhaji, *University of Illinois at Urbana-Champaign*
Kang Seung Gu, *University of Illinois at Urbana-Champaign*
Rebekah Yang, *University of Illinois at Urbana-Champaign*
Mojtaba Ziyadi, *University of Illinois at Urbana-Champaign*

Pavement life cycle assessment: A comparison of American and European tools

J. Santos
IFSTTAR, AME-EASE, Bouguenais, France

S. Thyagarajan
Turner Fairbank Highway Research Center, VA, USA

E. Keijzer
TNO, Utrecht, The Netherlands

R. Flores
ACCIONA S.A. Corporate Division, Sustainability and CSR Area, Madrid, Spain

G. Flintsch
Center for Sustainable Transportation Infrastructure, Virginia Tech Transportation Institute, USA

ABSTRACT: Road pavements have considerable environmental burdens associated with their initial construction, maintenance and usage, which has led the pavement stakeholder community make congregate efforts to better understand and mitigate these negative effects. Life Cycle Assessment (LCA) is a versatile methodology to quantify the effect of decisions regarding the selection of resources and processes. However, there is a considerable variety of tools for conducting pavement LCAs. The objective of this paper is to provide the pavement stakeholder community with insights on the potential differences in the life cycle impact assessment results of a pavement by applying American and European LCA tools, namely PaLATE V2.2, VTTI/UC asphalt pavement LCA model, GaBi, DuboCalc and ECORCE-M, to a Spanish pavement reconstruction project. Construction and maintenance life cycle stages were considered in the comparison. Based on the impact assessment methods adopted by the different tools, the following indicators and impact categories were analyzed: energy consumption, climate change, acidification, eutrophication and photochemical ozone creation. The results of the case study showed that it is of crucial importance to develop (1) a more standardized framework for performing a LCA of road pavement that can be adapted to various tools and (2) local databases of materials and processes, which follow national and international standards.

1 INTRODUCTION

Road pavements have considerable environmental burdens associated with their construction, maintenance, and use. Concurrently, the environmental issues are becoming more relevant in social and political contexts. This has led the pavement stakeholder community to congregate efforts to better understand and mitigate these negative effects.

A "twining" activity was initiated in 2014 between the LCE4ROADS consortium (FP7 European Union—funded project Grant Agreement nº 605748), led by ACCIONA, and the U.S. National Sustainable Pavements Consortium pooled fund effort, led by the Virginia Department of Transportation (VDOT), supported by the U.S. Federal Highway Administration (FHWA) and three other State Departments of Transportation (DOTs), and managed by the Virginia Tech Transportation Institute (VTTI). This cooperative initiative resulted from an arrangement signed in Washington on February 12, 2013, by the European Commission (EC) and the U.S. DOT that aims to foster collaboration on research, development, and technology transfer activities that are of mutual benefit. In particular, the main objective of this twining activity is to foster the exchange of knowledge across the Atlantic, finding synergies in research aimed at enhancing sustainability in pavements. The agreement focuses on the following aspects: (1) Life Cycle Assessment (LCA) methodologies and their applications to roads pavement construction and maintenance practices; (2) Life Cycle Cost Analysis (LCCA) for pavements and integration of use phase models, including analysis of the influence of pavement deterioration on vehicle fuel consumption and emissions and the interaction between pavement, environment, and humans; (3) Climate Change (CC) adaptation measures for road infrastructures; (4) Product Category Rules (PCRs) and Environmental Product Declarations (EPDs); and (5) implementation of strategies in terms of Green Public Procurement for road infrastructures.

To improve the sustainability of road pavement infrastructure, road agencies and construction companies are adopting appropriate methodologies and tools to identify priority areas for improvement. Thus, it is necessary to know the impact of pavements on the environment to develop and implement approaches and procedures that can produce the greatest gains in all aspects and dimensions of the system. LCA is a versatile methodology capable of informing decisions on resource and process selection to better understand, measure, and reduce the environmental impacts of a system Glass et al. (2013).

However, there is a considerable variety of tools for conducting pavement LCA, and there are notable differences between them. Available tools cover different phases and processes of the pavement's life cycle, take different environmental issues into account, and model with distinct levels of accuracy within chosen functional units and system boundaries. They can be global, national, or even regional or local. They have also been developed for different purposes, (e.g., research, consulting, and decision making), and their domain of applicability is tailored for different phases of a project's life cycle, (e.g., planning, designing, construction and maintenance). Furthermore, they use different foreground and background generic or industry data. Also distinct is the level of interaction they allow with the user. While some of the tools are "black-boxes" in the sense that only the default processes and data can be used, others allow users to use their own data, to choose the database that best match the features of the case study, or even to modify the existing datasets.

2 BACKGROUND

Over the last few years, many LCA tools have been developed for assisting decision makers in evaluating the environmental performance of their pavement-related decisions. The set of pavement-specific LCA tools includes, among others, PaLATE V2.2 (PaLATE V2.2 2011), UK asphalt pavement LCA model (Huang et al. 2009), PE-2 (Mukherjee & Cass 2012), ECORCE-M (Dauvergne et al. 2014), DuboCalc (Rijkswaterstaat 2015), CO_2NSTRUCT (Fernández-Sánchez et al. 2015), VTTI/UC asphalt pavement LCA model (Santos et al. 2015a,b) and Athena Impact Estimator for Highways (ASMI 2012). Commercial LCA tools, such as SimaPro (PRé Consultants 2016) and GaBi (PE International 2012), despite being not specifically designed for pavement-specific LCAs, have been used for that purpose (Blankendaal et al. 2014) since they are quite complete in terms of the elementary flows inventoried and unit processes taken into account, some of which are particularly applicable to the pavement domain (e.g., raw materials and equipment fuel combustion).

Moreover, LCA is a data-intensive method and thus the LCA tools are provided with databases which commonly present distinctive features in terms of data sources, elementary flows inventoried and unit processes taken into account, technical, temporal and geographical representativeness. For further information on the impacts of using different databases on the final results of infrastructure LCA studies the reader is referred to (Takano et al. 2014).

3 OBJECTIVES

The main objective of this paper is to provide the pavement stakeholder community with insights on the potential differences in the Life Cycle Impact Assessment (LCIA) results of a pavement LCA by applying American and European LCA tools to a Spanish case study. As a consequence of comparing the features of different tools and potential life cycle environmental impacts, the differences in datasets and life cycle inventory will be analyzed as well. In order to avoid an excessive level of complexity, the number of tools considered in the study had to be controlled. The tools selected were: (1) PaLATE V2.2, (2) VTTI/UC asphalt pavement LCA model, (3) GaBi; (4) DuboCalc and, (5) ECORCE-M.

4 OVERVIEW OF THE TOOLS COMPARED

An overview of the features of each tool is given in Table 1.

5 METHODOLOGY

5.1 *Goal and scope definition*

This paper presents and compares the results of an LCA of a pavement reconstruction project on a Spanish road section, N-340, located in Elche (Alicante), performed through the application of several LCA tools.

Table 1. Overview of the different LCA tools.

Feature		GaBi	PaLATE V2.2	DuboCalc	VTTI/UC	ECORCE-M
Country		Germany	USA	Netherlands	USA	France
Primary data source		Literature and industrial data; other databases (US LCI, ELCD, ecoinvent, etc.)	Carnegie Mellon University EIO-LCA software; Transportation Energy Data Book	National data	Literature data	Literature and industrial data
Impact category	AD	–*	–	Y	–	–
	CC	Y	Y	Y	Y	Y
	OD	Y	–	Y	–	–
	POC	Y	–	Y	Y	Y
	AC	Y	–	Y	Y	Y
	EU	Y	–	Y	Y	Y
	HT	Y	–	Y	–	–
	FAE	Y	–	Y	–	–
	MAE	Y	–	Y	–	–
	TE	Y	–	Y	–	–
	EC	Y	Y	–	Y	Y
	HHCP	–	–	–	Y	–
	CE	–	–	–	–	Y**
	CT	–	–	–	–	Y

Legend: AD—abiotic depletion; CC—climate change; OD—ozone depletion; POC—photochemical ozone creation; AC—acidification; EU—eutrophication; HT—human toxicity; FAE—freshwater aquatic ecotoxicity; MAE—marine aquatic ecotoxicity; TE—terrestrial ecotoxicity; EC—energy consumption; HHCP—Human health criteria pollutants; CE—chronic ecotoxicity; CT—chronic toxicity; Notes: *"–"means impact category not measured; **Beyond the toxicity specific to humans, which has been treated separately in ECORCE-M (chronic toxicity), all other toxicity indicators for the various ecosystems (i.e., freshwater aquatic, marine aquatic and terrestrial) have been aggregated into this single ecotoxicity indicator.

5.2 *Functional unit*

The function of the product system is to provide safe, comfortable, economical, and durable driving conditions over the project analysis period. The functional unit considered as a reference basis is the quantified function provided by the product system. In this case study, it is defined as *1 km of mainline pavement and year*. The analysis period is 20 years and comprises the maintenance of the top pavement structure layer at year 10. The assessed road section is 1,568 m long and has four lanes, divided into two roadways separated by a central separator. The inputs (raw materials and energy consumption) were collected and quantified from the ACCIONA work site in 2012.

5.3 *System boundaries and general assumptions*

The N-340 road received an EPD in December 2013 (Fernandez 2013). EPD is an Eco-label type III that aims to communicate transparently the environmental performance of a product, process, or system. It follows the rules established both in the International Organization for Standardization (ISO) 14025 (ISO 2006) and in the PCR guideline. In this project, the PCR named "highways, streets and roads" (EPD 2013) was used.

To compare the different LCA tools in a fair way, only the pavement life cycle phases and sub-phases that can be assessed with all five LCA tools were included in the analysis: (1) materials extraction and production, (2) transportation of materials, and (3) construction and Maintenance and Rehabilitation (M&R). The environmental impacts related to the usage phase and the traffic disruption caused by the performance of M&R activity were not assessed because not all of the tools evaluated in this study are capable of assessing these phases. Finally, the EOL was not taken into account because of its negligible contribution to the environmental life cycle impacts (<1%) (EPD 2013). As far as the materials extraction and production phase is concerned, it must be noted that at least 99% of material and energy requirements during the pavement life cycle were considered. The construction stages accounted for in this study are as follows: (1) demolition of the old pavement and fence; (2) soil excavation and movement; (3) pavement structure construction; (4) road sub-structure construction (e.g., drainage system); (5) M&R of the top layer.

Other stages, such as the production of traffic control devices (for signposting and for diverting traffic) and the construction of tunnels and bridges, were not included in the analysis due to their residual existence in this specific case study. When modeling the transportation of materials phase, an average distance of 20 km was considered for all concrete-based materials as there is a concrete plant near the road. For the borrowed soil and aggregates/gravel materials, an average distance of 15 km was assumed. With regard to the transportation of the soil removed from the work site, a 3-km long hauling movement was adopted.

Finally, the environmental impacts stemming from the construction of the infrastructure associated with intermodal activities, the operation of vehicles for loading and uploading at terminals, the production of manufacturing equipment and personnel activities were also disregarded.

6 LIFE CYCLE INVENTORY

The LCI stage of an LCA aims to identify and quantify the environmentally significant inputs, such as material and energy, and outputs, such as air emissions, water effluents and solid waste disposal, of a system by means of mass and energy balances. The elementary output flows were inventoried according to the methodology of each tool and the databases that feed them. Table 2 summarizes the type of materials applied in each construction stage considered in the case study. Because the tools have different ways of modeling the energy consumption, it was converted and expressed in terms of electricity.

Table 2. Input materials as modeled in different tools.

Stage	Baseline*	Quantity	DuboCalc	ECORCE-M	GaBi	PaLATE V2.2	VTTI/UC
1	Total energy consumption	32779 kWh	Grey electricity, NL	Energy consumption of construction equipment & hauling trucks	Electricity grid mix (ES)	Default construction & transportation equipment	US diesel for non-road engines
2	Total energy consumption	250686 kWh	Included in soil processing data	Energy consumption of construction equipment & hauling trucks	Electricity grid mix (ES)	Default construction & transportation equipment	US diesel for non-road engines
	General fill (soil)	1797 m³	Soil movement (3 km)	_**	Gravel, grain size 2/32	Accounts for soil, impact similar to aggregate	–
	Water	627 m³	–	Water (no environmental impacts)	Water	Water	Tap water
	Selected material (soil)	13256 m³	Soil movement (3 km)	–	Gravel, grain size 2/32	Accounts for soil, impact similar to aggregate	–
3	Total energy consumption	415021 kWh	Included in soil processing data	Energy consumption of construction equipment & hauling trucks	Electricity grid mix (ES)	Default construction & transportation equipment	US diesel for non-road engines
	Soil from borrowed site	398 m³	Soil from local project (15 km)	–	Gravel, grain size 2/32	Accounts for soil, impact similar to aggregate	–
	Water	124 m³	–	Water (no environmental impacts)	Water	Water	Tap water
	Graded aggregates	3187 m³	Gravel from rivers (15 km)	Aggregates, in quarry	Limestone, crushed gravel, grain size 2/16	Graded aggregates	Limestone; quartzite
	Bitumen emulsion	24 t	Bituminous emulsion	Bituminous emulsion	Bitumen emulsion	Bitumen emulsion; does not differentiate emulsion & PG bitumen	Bitumen emulsion 65%
	Asphalt concrete AC 32 Base G	5395 t	Stone mastic asphalt, 0% recycled content	Hot asphalt mixes, in gas plant	Asphalt supporting layer	Modeled as individual materials	Asphalt concrete AC 32 Base G
	Asphalt concrete AC 22 Bin S	2324 t	AC Surf, dense asphalt concrete	Hot asphalt mixes, in gas plant	Asphalt pavement	Modeled as individual materials	Asphalt concrete AC 22 Bin S
	Penetration grade (PG) bitumen	320 t	Bituminous emulsion (proxy)	Bitumen, 20 to 220 grade	Bitumen (grade)	PG bitumen; does not differentiate emulsion & PG bitumen	PG 70–22 binder
	Concrete (brick)	310 m³	Concrete C20/25 (CEM I)	Concrete, at mixing plant	Concrete (stones, bricks)	Modeled as individual materials	Concrete (brick)
	Glass fibers filaments	92 m³	Plastic fibers (sub process from "fiber reinforced concrete")	–	Glass fibers mesh	Glass fibers filaments	–

(Continued)

Table 2. (Continued).

Stage	Baseline*	Quantity	DuboCalc	ECORCE-M	GaBi	PaLATE V2.2	VTTI/UC
4	Total energy consumption	3820 kWh	Included in soil processing data	Energy consumption of construction equipment & hauling trucks	Electricity grid mix (ES)	Default construction & transportation equipment;	US diesel for non-road engines
	Concrete C20	510 m³	Concrete C20/25 (CEM I)	Concrete, at mixing plant	Concrete C20/25	Modeled as individual materials	Concrete HM20
	Soil	200893 m³	Soil movement (3 km)	–	Gravel	Accounts for soil, impact similar to aggregate	–
	Formwork	0.04 m³	Traditional formwork (converted to m²)	–	Laminated wood	–	–
	Concrete C15	41 m³	Concrete C12/15 (CEM I)	Concrete, at mixing plant	Concrete C12/15	Modeled as individual materials	Concrete HM15
5	Total energy consumption	106562 kWh	Included in soil processing data	Energy consumption of construction equipment & hauling trucks	Electricity grid mix (ES)	Default construction & transportation equipment; specific details not available	US diesel for non-road engines
	Asphalt concrete AC 22 Bin S	2324 t	AC Surf, dense asphalt concrete	Hot asphalt mixes, in gas plant	Asphalt pavement	Modeled as individual materials	Asphalt concrete AC 22 Bin S
	Prime coat	13 t	Bituminous emulsion (proxy)	Bituminous emulsion	Asphalt binder	Prime coat; does not differentiate emulsion & PG bitumen	Prime coat (bituminous emulsion)

* The "baseline" was the starting point of all tools; ** "–" means that this process was not available in the specific tool and was excluded in calculations.

7 LIFE CYCLE IMPACT ASSESSMENT

In the LCIA stage of an LCA, the LCI results are assigned to different impact categories based on the expected types of impacts on the environment. In this study the Center for Environmental Studies of the University of Leiden's "CML 2001" impact assessment method (Guinée 2002) is implemented by several tools, either in a direct way (i.e., GaBi) or by adapting the original indicators (i.e., ECORCE-M and DuboCalc). Alternatively, the VTTI/UC pavement LCA tool adopts the Tool for the Reduction and Assessment of Chemical and other environmental Impacts 2.0 (TRACI 2.0) method (Bare 2011). In the case of PaLATE V2.2 only the CC impact category is considered, taking the CO_2 emissions exclusively into account. The LCIA indicators were calculated at the mid-point level from (1) resource consumption flows, e.g., energy, (2) air emission flows, e.g., the 100-year horizon CC, etc., and (3) air, soil, and water pollutant flows, i.e. toxicity indicators.

8 RESULTS AND DISCUSSION

8.1 *Impacts on a material level*

Figure 1 shows the potential environmental impacts of the materials used in this case study, per kilogram of material, and calculated by the different tools. All axes are cut-off and the scores greater than the cut-off threshold are displayed in boxes. On x-axis, the percentages (%) next to each material show the Coefficient of Variation (CV) of the values per material in each impact category. Furthermore, because not all tools cover the same impact categories (as shown in Table 1), the graphs present only the results obtained with the tools which are able to consider the impact category under evaluation. At first glance, it is clear that the impacts per kilogram of material differ largely among the tools for some of the materials, while other materials have rather comparable impacts. Taking the CC impact category as an example, Figure 1a shows that the CV

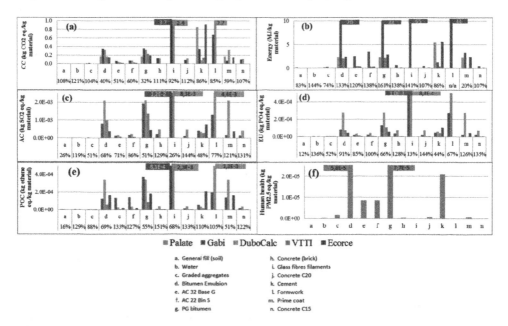

Figure 1. Environmental impacts per kilogram of product, calculated with the five different tools for six environmental impact categories (a) CC; (b) energy consumption; (c) AC; (d) EU, (e) POC and (f) human health.

values range approximately from 32 to 121%. Water and concrete C20 present the highest variability (121% and 112%, respectively). Cement, concrete brick, and concrete C15 also exhibit high CV values (86 to 111%), though the LCIs associated with these materials are well defined and quantified by researchers for many years. On the contrary, asphalt concrete, bitumen and bitumen emulsion denote the lowest variability (30 to 60%). In general, the scores for the remaining materials have a much lower impact than the materials that present high levels of variability. One can conclude that the generality of the most common and bulk materials are well researched and represented by the tools, while the LCIs of more specific materials like water, formwork and glass fibers are more difficult to quantify in accurate fashion, and, thus, have been disregarded by several tools or based on proxy elements.

When comparing the CV values of the same material across the several impact categories, the CC impact category was found to exhibit the lowest levels of variability for the generality of the materials. This result is explained by the fact that all the LCIA methods adopted by the tool use the characterization factors based on the Intergovernmental Panel on Climate Change model. On the other hand, the energy consumption indicator generates the highest CV values. To a great extent this outcome can be explained by the fact that the impact category scores calculated with the GaBi tool are extremely high for the majority of the materials in comparison to the scores calculated with the other tools. Such a result suggests that GaBi might have other definitions for these materials or consider different system boundaries, which might influence the conclusions drawn on this case study. Furthermore, this discrepancy also illustrates the importance of using consistent sources and local databases as different materials may have different sources or may be produced using different processes with significantly different environmental loads. For example, different form materials may be used in different regions.

8.2 *Life cycle impact assessment comparison*

Figure 2 presents the environmental impacts associated with each construction stage considered in the case study (represented in log-scale) and the relative contributions to the total score on various impact categories that are computed by the majority of the LCA tools. In general, considerable variation was observed within each impact category computed by the different LCA tools. The GaBi tool was found to yield the lowest impact scores when compared with the other evaluated tools. Interestingly, the POC scores associated with stages 1 and 2 possess a negative value. This result indicates that in these stages there is a mitigation effect on the POC impact category. To the contrary, PaLATE V2.2 was found to produce the highest scores for the two impact categories that it is able to account for (i.e., CC and energy consumption). The only exception to this general trend was observed in the case of the energy consumed during stage 3. The VTTI/UC and ECORCE-M tools denoted similar CC and energy consumption scores. This result contrasts with those observed for the remaining impact categories, as they were found to vary considerably. Also, the scores obtained with VTTI/UC and ECORCE-M tools were higher than those generated by GaBi for all impact categories. The DuboCalc tool produced intermediate scores relative to those generated by GaBi and VTTI/UC for the CC and energy consumption indicators. However, the AC, EU, and POC scores computed by DuboCalc for stages 3, 4, and 5 were the highest among those calculated by all the compared tools.

Regarding the relative contributions of each construction stage to the total scores, Figure 2 (right) shows that construction stage 1 is the smallest contributor for each impact category, regardless of the LCA tool considered. Construction stage 3 was found to be the main contributor. The only exception to this uniform outcome was obtained with PaLATE V2.2. As explained before, the quantity of soil used in stage 4 combined with the LCA approach adopted by PaLATE V2.2 led to a higher relative contribution from this stage in both CC and energy consumption indicators.

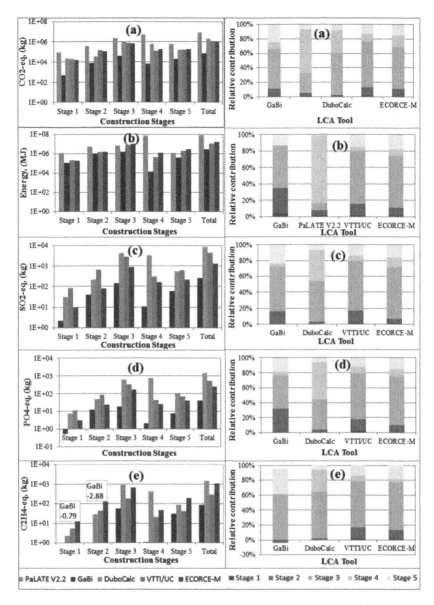

Figure 2. Comparison of the impact categories scores (left) and percentage contribution (right) from different LCA tools for the following impact categories and indicators: (a) CC; (b) energy consumption; (c) AC; (d) EU, and (e) POC.

9 SUMMARY AND CONCLUSIONS

The research work described in this paper investigates the extent to which the choice of an LCA tool may influence the LCA results for road pavement infrastructure. Several tools with different functionalities and geographic contexts were compared by applying them to a Spanish case study.

The results indicate that there is a considerable variability in the environmental impact scores computed with the different LCA tools for each impact category. In particular, this case study demonstrates that the impacts of the most common materials are less sensitive to the choice of the LCA tool, in contract with less-common materials.

Based on the findings of this case study, the following recommendations can be made to improve LCA tools, the databases connect to them, and LCA practices in general: (1) there is a need for a formal consensus framework and PCR specific for pavements so that a standardized framework can be adapted to the various tools; (2) local databases of materials and processes should be developed that, for the sake of consistency, comply with national and international standards regarding technical, geographical, and temporal representativeness requirements. Those databases should be built based primarily on tight, international cooperation between academia and industry, and updated on a regular basis. The availability of such a database would improve the reliability of LCA and thereby stimulate its application; (3) the accuracy and comprehensiveness level of the datasets should be tailored to the impact category and impact assessment method; and (4) a sensitivity analysis is necessary to ascertain the uncertainty and, thus, the credibility and value of the final results.

Finally, it is important to mention that this paper focused only on the construction and maintenance phases of the pavement LCA, leaving the use phase outside of its scope as most of the tools can not include it. Therefore, it is recommended that similar studies be conducted using the use phase.

REFERENCES

Athena Sustainable Materials Institute (ASMI). 2012. *Impact estimator for highways: user guide*.
Bare, J. 2011. TRACI 2.0: the tool for the reduction and assessment of chemical and other environmental impacts 2.0. *Clean Technologies and Environmental Policy* 13: 687–696.
Blankendaal, T., Schuur, P. & Voordijk, H. 2014. Reducing the environmental impact of concrete and asphalt: a scenario approach. *Journal of Cleaner Production* 66(1): 27–36.
Dauvergne, M., Jullien, A., Boussafir, Y., Tamagny, P. & Proust, C. 2014. *Logiciel ECORCE-M, version multilingues*.
Environmental Product Declaration (EPD). 2013. *PCR CPC 53211 highways (except elevated highways), streets and roads*.
Fernández, R. 2013. *Environmental Product Declaration of "N-340" road (Elche, Alicante). Reg. no. S-EP-00516*.
Fernández-Sánchez, G., Berzosa, A., Barandica, J.M., Cornejo, E. & Serrano, J. 2015. Opportunities for GHG emissions reduction in road projects: a comparative evaluation of emissions scenarios using CO_2NSTRUCT. *Journal of Cleaner Production* 104: 156–167.
Glass, J., Dyer, T., Georgopoulos, C., Goodier, C., Paine, K., Parry, T., Baumann, H. & Gluch, P. 2013. Future use of life-cycle assessment in civil engineering. *Proceeding of the ICE: Construction Materials* 106(4): 204–212.
Guinée, J. (ed.) 2002. *Handbook on life cycle assessment: operational guide to the ISO standards*. Series: eco-efficiency in industry and science. Dordrecht: Kluwer Academic Publishers.
Huang, Y., Bird, R. & Heidrich, O. 2009. Development of a life cycle assessment tool for construction and maintenance of asphalt pavements. *Journal of Cleaner Production* 7: 283–296.
International Standard Organization (ISO), 2006. *ISO 14025: Environmental labels and declarations-Type III Environmental Declarations-Principles and procedures*.
Mukherjee, A. & Cass, D. 2012. Project emission estimator: implementation of a project-based framework for monitoring pavement greenhouse gas emissions. *Transportation Research Record: Journal of the Transportation Research Board* 2282: 91–99.
PaLATE v2.2 for Greenroads 2011. Software and user guide modified by Civil & Environmental Engineering Department, University of Washington (UW).
PE International. 2012. *Gabi Manual*.
PRé Consultants. 2016. *SimaPro Tutorial*.
Rijkswaterstaat, 2015. *DuboCalc version 4.01.1*.
Santos, J., Bryce, J., Flintsch, G., Ferreira, A. & Diefenderfer, B. 2015b. A life cycle assessment of in-place recycling and conventional pavement construction and maintenance practices. *Structure and Infrastructure Engineering: Maintenance, Management, Life-Cycle Design and Performance* 11(9): 1199–1217.
Santos, J., Ferreira, A. & Flintsch, G. 2015a. A life cycle assessment model for pavement management: methodology and computational framework. *International Journal of Pavement Engineering* 16(3): 268–286.
Takano, A., Winter, S., Hughes, M. & Linkosalmi, L. 2014. Comparison of life cycle assessment databases: a case study on building assessment. *Building and Environment* 79: 20–30.

Lessons learned in developing an environmental product declaration program for the asphalt industry in North America

Amlan Mukherjee
Department of Civil and Environmental Engineering, Michigan Tech., Houghton, MI, USA

Heather Dylla
Director of Sustainable Engineering, National Asphalt Pavement Association, Lanham, MD, USA

ABSTRACT: The objective of this paper is to report the technical and organizational challenges involved in the development of the North American Environmental Product Declaration program for asphalt mixtures. Developing a Life Cycle Assessment (LCA) for asphalt mixtures presents the challenge of coordinating consistent assumptions across industries and stakeholders, and requiring a harmonized decision-making process that accounts for the impacts of materials across the supply and value chain. While, the methods of LCA are rational and well defined, the decisions defining the various assumptions are often arrived at through a negotiation process shaped by stakeholder relationships and priorities. There is much discussion in the literature regarding the technical challenges of conducting an LCA involving choice of system boundary, functional unit, and allocation procedures used for co-products and recycled products. However, the formulation process of these technical questions within the context of stakeholder biases and heuristics are seldom explicitly discussed, even though they play an important role in how the technical challenges are resolved. Hence, the paper explores how differences in stakeholder priorities and perspectives, in the pavement construction industry, directly shape the Product Category Rules (PCR) defining the program by drawing attention to specific LCA related technical questions and highlights how the solutions were negotiated. The primary challenge identified is how to ensure technical rigor of the underlying LCA, while recognizing the interests of the stakeholders and ensuring the delivery of a program that is effective. The paper discusses how technical issues regarding system boundary choice, data use and allocation presented challenges for the PCR Development Working Group accounting for different stakeholder interests. Within this context the paper will highlight the developed PCR and present relevant results from the underlying LCA.

Keywords: Life Cycle Assessment, Sustainability, Environmental Product Declaration, Innovation Adoption, Change Management

1 INTRODUCTION

The principles and framework for conducting an attributional Life Cycle Assessment (LCA) for products and processes are outlined in ISO standard 14040. An LCA accounts for cradle-to-grave environmental impacts of a product or process, including the impacts incurred during the mining, extraction, manufacturing and production phases of all the raw materials involved in the process; the distances travelled in transporting them through the supply chain, and the impacts involved in the use and eventual end-of-life disposal of the materials.

For a complex system such as a pavement system, developing an LCA poses the challenge of coordinating consistent assumptions across industries and stakeholders, requiring a harmonized decision-making process that accounts for the impacts of materials across the supply and value chain. While, the methods of LCA are rational and well defined, the decisions defining the various assumptions are often arrived at through a negotiation process defined by stakeholder relationships and priorities. There is evidence from decision science that human decision-making is influenced by individual biases and heuristics and is sensitive to formulation, context and procedure (Kahneman and Lovallo 1993; Kahneman and Tversky 2000). While there is much discussion in the literature regarding the technical challenges of conducting an LCA involving choice of system boundary, functional unit, and allocation procedures used for co-products and recycled products; the formulation process of the technical LCA questions within the context of stakeholder biases and heuristics are seldom explicitly discussed, even though they play an important role in how the technical challenges are resolved. In addition, an understanding of the collaborative and competitive forces shaping the stakeholder interactions and the incentive mechanisms in place can be crucial to this discussion.

Hence, the objective of this paper is to explore the extent to which technical decisions, involved in an LCA underlying the development of an Environmental Product Declaration (EPD) program, are shaped by negotiations between stakeholders with competing objectives. The development of North American EPD program for asphalt mixtures presented the opportunity to observe direct stakeholder involvement in shaping LCA questions within the context of the technical and organizational challenges in the pavement construction industry.

The paper starts with an introduction to the EPD program development process. Next it establishes the role played by stakeholders in the decision-making process given their affiliations and relationship to the asphalt materials industry. Finally, the paper discusses the technical and organizational challenges that were negotiated through the decision-making process and their impact on the LCA supporting the EPD program.

2 THE STRUCTURE OF EPD PROGRAMS

The EPD program was developed with the goal of standardizing industry specific LCA assumptions, allowing for credible and transparent reporting. The EPD is a Type III Environmental Label as defined in *ISO Standard ISO 14025:2006, Environmental Labels and Declarations – General principles* (International Orginization for Standardization 2006). It communicates the environmental impacts of a product or service using methods in Life Cycle Assessment (LCA). The process used to develop an EPD ensures consistent data collection, analysis and reporting requirements, supported by third party verification. This ensures the reliability of the information communicated through an EPD. Typically an EPD development process adheres to various international standards; chief among them is the ISO 14025 standard.

EPDs are developed on the basis of Product Category Rules (PCR)—a consensus document that defines the rules, requirements and guidelines for conducting the LCA that supports an EPD. Specifically, it provides guidelines for:

1. Defining the functional and declared units of comparison for a product.
2. Establishing the goal and scope, and defines the system boundaries for conducting the LCA that supports the EPD. It specifies the modules and processes in each of the life cycle stages and that would have to be considered in an LCA for developing an EPD.
3. Outlining data collection and specification when developing the life cycle inventory, requiring reporting across a twelve-month period, reflecting technology in current use and ensuring the use of geographically pertinent data.
4. Ensuring the quality of the data collected, (including tolerances) for conducting the underlying LCA.
5. Reporting the environmental impacts across relevant product impact categories using appropriate characterization factors. Example categories are: environmental impact indicators (such as Global Warming Potential), total primary energy consumption and material resource consumption.

Based on a given PCR, an EPD can be developed to convey information from Business-to-Business (B-to-B: cradle to gate) or from Business-to-Consumer (B-to-C: cradle to grave).

In order to maintain the accuracy, reliability, and the unbiased integrity of an EPD, multiple stakeholders are involved with the peer-review and third party verification in place. Typically three different types of organizations can play the role of program operator: professional organizations such as American Society for Testing and Materials (ASTM) International, environmental compliance consultants and industry trade associations. In the United States state agencies have not played the role of a program operator, but they are qualified to. Program operators develop industry specific PCR in compliance with ISO 14025. The PCR development process involves participation from various stakeholders including customers, producers and upstream manufacturers. In addition, an independent review panel provides a peer-review of the developed PCR. Using the PCR developed by the program operators, manufacturers and producers of the product can develop a specific EPD, based on an LCA conducted using input data, on materials and energy use, specific to their operations and processes. A third party (or the program operator) must certify the EPD as being compliant with the PCR. In the construction materials industry, currently, there is a growing demand for directly involving industries in developing PCR that are reflective of realities within the industry.

For this program, the operator is the National Asphalt Pavement Association (NAPA), an industry association with significant expertise in asphalt materials, rather than a central regulatory agency. Organizationally, this is akin to a bottom-up consensus driven effort that directly engages stakeholders with voluntary adoption, driven by market forces; as different from a top-down regulatory process.

An ISO compliant EPD program must go through multiple stakeholder review processes. The PCR must be prepared in consensus with a PCR Development Working Group (DWG) representing a consensus from various industry, agency and academic stakeholders. Before publication the PCR has to be available for public comment for a period of one month and in order to be ISO compliant the program operator must address all comments and an independent external committee of three reviewers representing expertise in the specific field and LCA methodology must verify the final document. This clearly presents a situation where multiple conflicting perspectives require attention, and the PCR becomes a document that is socially negotiated. The next section outlines the stakeholders who were part of the PCR DWG and their relationships to the asphalt industry.

3 STAKEHOLDERS ON THE PCR DWG

Within the context of the asphalt pavement industry, the stakeholders included customers, representatives from other industries that collaborate with, or are part of the upstream product supply chain, state agencies that may play a regulatory role and academic advisors. Non-profit associations that are usually funded by industry constituents (typically for-profit contractors and producers) represent industry interests by providing technical support and advocacy. Examples of such organizations are: National Asphalt Pavement Association (NAPA), the Asphalt Institute (AI) and the National Stone, Sand and Gravel Association (NSSGA), each representing the interests of their respective industries. State agencies that may play a regulatory role include Departments of Transportation (DOT), Tollway Authorities, Cities and Metropolitan Planning Organizations (MPO) at the state level, and the US Department of Transportation, specifically, the Federal Highway Administration (FHWA). They serve as public owners directly responsible for planning and design decisions and contracting authority. Academic organizations and think tanks who contribute to furthering knowledge and practice in the field. The relationships between all the different stakeholders can be classified primarily as regulatory, collaborative, competitive, and advisory.

The relationship between the industries can vary from collaborative to cooperative, depending on their position in the product supply chain. For example, concrete and asphalt mixtures are products that can each be used to design and construct pavement sections of comparable functionality, thus making the respective industries competitors. As a result, agencies must intentionally keep pavement type selection fair so that neither industry is disadvantaged inadvertently due to practice guidelines. The stone crushing and aggregate industry on the other hand, tends to have a collaborative relationship with both the concrete and asphalt industries as they provide crushed stone, and coarse aggregate that is used in both asphalt and concrete mixtures. However, all industries tend to cooperate in their governmental advocacy for improved highway and infrastructure funding.

State Departments of Transportation (DOT) and municipal agencies at the federal, state and local level play the role of setting design and construction specifications. In addition, as the owner, DOTs can call for alternative bids based on life cycle costing methods when selecting between competing asphalt and concrete pavement designs. This ensures the most optimal use of taxpayer dollars, while selecting designs that promise the best long-term performance.

At the federal level, under the US DOT umbrella, of specific interest is FHWA's role in supporting state and local governments through financial and technical assistance. Notably, FHWA supports research and collaboration efforts and considers its mission to improve mobility through national leadership, innovation, and program delivery. Specifically, the emphasis on "innovation and program delivery" often positions FHWA to help further best practices and develop guidelines for adopting cutting edge technologies in addition to providing educational briefs. In addition, they have methods of incentivizing the adoption of practices by states through funding mechanisms.

Academic organizations including universities and think tanks are a part of this stakeholder ecosystem though their position is less participatory and more advisory. University faculty often led research into questions that shed light on best practice. Strictly speaking, such positions are considered neutral, with an emphasis on supporting best practices.

Interactions between the stakeholders have led to a combination of forces that have been crucial towards the development and adoption of EPD programs.

1. *Incentive:* Public agencies such as Illinois Tollway with their aggressive embrace of LCA in their overall decision-making process, as well as the implications of legislation in California requiring reduction of greenhouse gas emissions, have created incentives for industry to evolve towards low emission processes. The inclusion of EPDs within the International Green Construction Codes, various rating systems and consumer demand for greener infrastructure is creating demand and incentives for industry to evolve.
2. *Competitive:* Given that concrete and asphalt pavements compete in pavement selection decisions, there has been a movement towards developing "greener" mixtures that are also more durable and cost effective in the long run. This has fuelled the use of recycled materials in both concrete and asphalt pavements. Investments by both industries' research concerning pavement-vehicle interactions have been spurred by competition to establish the relative benefits of each material type from a life cycle perspective, with an emphasis on vehicle fuel consumption performance during the use phase. In each case, the competitive forces have nudged the industry into accepting life cycle thinking.
3. *Collaborative:* The Sustainable Pavements Technology Working Group effort that has brought agency, industry, and academia to the table has been critical in furthering standards and consensus driven protocols that support the adoption of life cycle thinking in practice.

These three forces have created a greater demand for pavement related EPDs. Broadly, the forces motivating EPD programs are the promise of competitive advantage in the market place, fueled by incentive opportunities provided by regulatory agencies, and often as a collaborative necessity when it is necessary to synchronize best practice with partner organizations and allied industries. In acknowledgement of these forces, the PCR DWG was selected as follows: Asphalt plant owners/producers (three members), representative from Asphalt Institute, and the petroleum industry, both allied collaborative industry (one member each), pavement construction contractor (one member), public agency owners (one member from a state DOT, one from a city municipality), representative from a public regulatory agency (one member) and one asphalt industry expert from academia. The authors of this paper facilitated this committee of ten members.

4 TECHNICAL AND ORGANIZATIONAL CHALLENGES IN EPD PROGRAM DEVELOPMENT

The process followed the EPD program development guidelines as per ISO 14025:2006. The discussion examines how technical issues regarding system boundary choice, data use and allocation presented challenges for the PCR Development Working Group accounting for different stakeholder interests.

4.1 *System boundary and declared units*

The purpose of the PCR is to accommodate the use and implementation of EPDs that will provide the basis for comparing cradle-to-gate environmental impacts for the production of asphalt mixtures. As per the recommendations of ISO 14025:2006, the environmental impacts of all asphalt mixtures that have an EPD compliant with this program can be compared. Therefore, EPDs compliant with this PCR will only reflect differences in plant energy use, material use, and plant emissions, thus providing an effective approach to comparing the environmental impacts of the process used in producing asphalt mixtures. Some of the challenges encountered in defining the system boundaries were related to identifying justifications for cut-off exclusion criteria, as listed below:

1. As this EPD program functions at the B-to-B interface (or cradle-to-gate), a declared unit of 1 short ton of asphalt mixture is used. This is not the same as a functional unit, as the unit is not associated with a functionality of the asphalt mixture.
2. The impact of plant infrastructure is discounted as these impacts are similar across all plants and can be considered as a common overhead. This is justifiable because, differences in plant energy use due to age and/or maintenance requirements are already accounted for as part of the process energy calculations. Hence, capital goods are being omitted in this study, and being considered non-essential to the comparison and not relevant with regard to the decisions that will be supported by the EPDs.
3. No differentiation is being made between a hot asphalt mixture and a warm asphalt mixture, instead for each asphalt mixture the plant production temperature will be declared in the EPD. Reduced production temperature can reduce the energy requirements and thus lower the environmental impacts of asphalt production. Different plants achieve temperature reduction in different ways, however the use of Reclaimed Asphalt Pavement (RAP) and/or polymer-modified asphalts can place a limit to how far the temperature can be reduced. This creates significant variability in the actual temperatures at which so-called lower temperature asphalt mixtures are produced. Therefore, it is preferable if each mixture explicitly declares the production temperature, and the use of any pertinent warm mix technology that has been used to do so.
4. In case of insufficient input data or data gaps for a unit process, the cut-off criteria is being limited to 1% of renewable and non-renewable primary energy usage, and 1% of the total mass input of that unit process, unless a material has the potential of causing significant emissions into the air, water or soil, or is known to be resource intensive. The total sum of neglected input flows is limited and shall not exceed 5% of energy usage and mass.
5. Materials that are less than 1% of the total mass input, but are considered environmentally relevant are chemical additives and polymers.
6. Upstream impacts of extraction, production and manufacturing of any material that is not consumed in the production of the asphalt mixture, and is considered to be "part" of the plant infrastructure, is being excluded. Consumables such as conveyor belts and lubricants are being excluded, as the total consumption of these materials that can be ascribed to a mixture is negligible.

The system boundaries defined by the PCR are strictly cradle-to-gate for asphalt mixtures.

4.2 *Accuracy, availability and use of datasets*

A common impulse among program reviewers is to look for "accuracy" in the data that is used or the outcomes that are measured using the indicators. As the word "accuracy" is best applied to measurements it applies to the reporting of primary data (i.e. the process specific data within the asphalt plant that is seeking an EPD including definite measurable quantities such as annual energy use, total quantities of materials used and observed stack emissions data). Primary data is reported directly at the plants and includes the following items reported over a 12 month period, in the last 5 years:

1. Total asphalt produced at the plant, reported in US short tons
2. Total electricity in kWh.
3. Generator energy—Diesel fuel in gallons or in liters.
4. Plant burner energy (primary and secondary)—used in one or more of the following:
   a. Natural gas use in MCF or MMBTu
   b. Propane used in gallons or in liters
   c. Diesel fuel in gallons or in liters
   d. Recycle fuel oil in gallons or in liters
   e. Biofuels in gallons or in liters
5. Hot-oil heater energy
6. Mobile equipment energy—Diesel fuel use in gallons or in liters
7. Aggregate used in production in US short tons
8. Asphalt binder used in production in US short tons
9. One-way distances travelled to plant for asphalt binder and aggregate (both virgin and recycled), expressed in US short ton-miles.
10. Water used in gallons or in liters.
11. Stack emissions from plant in Lbs. or in Kg.

Pre-determined scenarios: For the parameters that may be difficult to estimate or collect primary data for, the following has been used.

1. Default energy requirements for processing of RAP/RAS is 0.1 gal/short ton or 0.4 kWh/short ton.
2. Distance travelled by RAP/RAS to plant is 50 miles.

The following principles have supported the primary data collection design and process for this LCA study.

1. Ease of Collection: Ensure that the primary data collection process is practical and can be conducted by plant managers. This will reduce the data collection burden in the long run thus improving the possibility of adoption of the EPD program.
2. Data Aggregation: The total annual (12 month period) use of primary energy and material use data will be collected. Daily average data for consumption is also being collected, to provide a validity check for the annual reported data. This allows assessment of differentials in impact categories due to (i) energy use (electricity, natural gas etc.), (ii) the mixture design, and (iii) the distances travelled by the raw materials to the plant.
3. Primary Data Analysis: An analysis of the primary data is provided to examine trends in energy use and their relative sensitivity to moisture and aggregate type in different regions. The results from this analysis can help plants identify ways of improving plant operating efficiencies while also providing a method to identify possible errors in data reporting.
4. Data Quality Assurance: As plant managers are directly reporting all data, it is very important to create checks and balances to insure data quality, and identify possible errors or anomalies in reporting. Hence the following criteria have to be met:
   a. Time period: All data reported must be reflective of plant production over a period of 12 uninterrupted months, within the last 5 years, or the most recent data available.
   b. Documents on file: Primary data reported should be based on utility and energy bills, sales records and other similar documents all of which should be on file and easily accessible.

c. Correctness Check: Data reported by plants that do not fall within the error margins based on these trends should be checked for reporting errors.
d. Geography: All data reported for a plant must be specific to that plant. Company averages should not be used.
5. Data Gaps: Efforts should be made to ensure gaps in primary data collection are limited to only those items for which a predetermined scenario has been provided.

When the assessment involves data from upstream processes the notion of accuracy becomes difficult to enforce, as the data is derived from estimated upstream life cycle inventories. For example, the upstream impacts due to the extraction and refining of crude oil. These data sets are referred to as secondary data (i.e. data from processes that are within the system boundary but not immediate to the process being studied). Currently life cycle inventories for upstream processes are based on industry wide estimates. Some upstream datasets that come from other chemical industries, such as chemical additives in warm mix asphalt or anti-strip agents, can be difficult to acquire.

Some of these datasets are publicly available while others are proprietary. The publicly available datasets, while most accessible are often presented with limited guarantee or review. The proprietary datasets are available for a fee, but their guarantee of quality is often based on professional reputation and internal review processes that are not always transparent to stakeholders.

The interests of the asphalt industry producers and contractors are to develop the EPD program using free and public datasets. This choice is motivated by two reasons: (i) public datasets are transparent and verifiable, and (ii) being free they reduce the cost of the EPD program. Transparency is important to the industry as it helps establish credibility. This is particularly important as most customers in the asphalt pavement industry are public. Keeping the cost low is also very important as a high price point will deter industry members from getting their mixtures certified through the EPD program. The challenge here is to choose between the following choices along with the accompanying trade-offs:

1. *Use proprietary data:* The risk of data quality is no longer owned by the program, but rather transferred to data providers like EcoInvent or Gabi (Thinkstep, Inc). The trade-off: the cost of obtaining an EPD will go up as the significant cost of the proprietary data will get transferred to the producers and contractors. This may prove to be a deterrent for the adoption of EPD programs in the industry.
2. *Use public data:* In this case the data sets being used in the EPD are verifiable and transparent. This works well when the data sets are from specific public agencies, such as the eGRID electricity production datasets, or the data maintained by the US Department of Energy. However, the US life cycle inventories provided by National Renewable Energy Laboratory, explicitly come without review or guarantee. Therefore, this leaves the EPD program open to criticism. The cost of the EPD remains low and the transparency is attractive.

In addressing the above trade-off, the following principles have supported the secondary data selection process.

1. Uniformity in Use of Life Cycle Inventories: The scope of the PCR supported by this LCA requires asphalt mixtures with EPD from this program be comparable. Therefore, it is critical *all* LCA supporting EPD certified by this program use recommended upstream inventories. Previous work has shown that even with the same primary data a choice of different upstream inventories can create significant differences in the final LCA results. Therefore, it is of critical importance, that upstream inventories specified by the program operator be used in any LCA conducted to support an EPD certified by this program. If this uniformity is not maintained, EPDs provided by this program will not be comparable, thus defeating its goal.
2. Transparency of Life Cycle Inventories: This EPD program intends to respect the spirit of transparency in environmental performance reporting. Therefore, it is of critical importance to this program all upstream data sources be available *publicly* and *economically* to anybody who wishes to reproduce the results of the impact assessment. The program intends to remove barriers to providing third parties access to the process and calculations supporting the underlying LCA.

3. <u>Geography and Regionalization:</u> This effort discussed use of upstream data specific to the United States. US average data is used for electricity. However, for LCA supporting EPDs, it is critical regional energy mixes from eGRID be used to reflect regional differences. Similarly, at this time, the inventory and allocation for asphalt binder is based on the US average, however EPDs for specific mixes should reflect regional allocation factors based on their Petroleum Administration for Defense Districts (PADD) region.
4. <u>Data Gaps:</u> Given the emphasis on transparency and uniform use of the same upstream inventories, a trade-off is public datasets are not easily available for all mixture components – particularly chemical additives and polymers. While these chemical additives are usually less than 1% by mass of the mixture, they cannot be ignored, as they tend to have a disproportionately large environmental impact. Excluding them reduces the scope of the EPD, even though it comes with the recognition that this is a first step towards expansion of scope in future, as and when the data for the chemicals become available. Therefore, the PCR suggested the use of similar chemicals to reasonably approximate the impacts. This also may compromise the EPD, as the choice of an approximate chemical can be subjective and prone to error and potential over estimation of impacts.
5. <u>Dependence on LCI Data from Allied Industries:</u> The life cycle inventory of asphalt mixtures is dependent on upstream data from various other industries, most importantly the petroleum refining industry. At this time reasonable placeholder data is being used for the binder. NAPA is in conversation with Asphalt Institute and as they develop a detailed LCI for the asphalt binder, it will be harmonized as an input to this EPD. Therefore, it is important to recognize this is currently not a data gap, but rather work in progress that will ensure harmony between asphalt mixtures and various critical upstream products.

The above values were arrived at in discussion with the stakeholders in the PCR DWG, and were driven by a desire to balance two goals: (i) develop an EPD program that is accessible and easily adoptable by the industry, and (ii) ensure that the technical rigor of the EPD program is maintained within the limitations of cost and available data. Both are laudable goals, while the latter is a technical challenge requiring a careful LCA, the former is a question of easing adoption of tools such as EPDs that further life cycle thinking.

A trade-off point may prove to be based in a gradual introduction of the program, with each version expanding the scope of mixtures that can be certified while ensuring that the burden of adopting the program is low. However, that is based on two assumptions: (i) the quality and completeness of public datasets will improve over time, and (ii) in future datasets from other industries (like additives) will become available. Both these issues remain outside the control of the stakeholders.

4.3 *Allocation factors*

The question of allocation in LCA provides an opportunity to examine the role of social collaboration within and across industries. Allocation refers to the process involved in dividing the environmental impacts of a multifunction process between each of the co-products that are produced from it. For example, the crude oil refining process produces multiple products that are used by many different downstream industries. Another situation where the issue of allocation becomes pertinent is when a product serves multiple industrial processes, some of which may be disparate, through the course of its life cycle, i.e. it gets recycled after its first use in one process to be used in a different process one or more times thereafter. The question revolves around how the life cycle environmental impacts of the product are to be assigned across the different processes that utilize the product.

A consistent allocation procedure is necessary to account for the impacts in each of these cases, so that the same impacts are not accounted for twice, and/or that no impacts have been excluded from the accounting process. This procedure is not simply a technical question, as each stakeholder is motivated to be held accountable for as little of the burden for a co-product or recycled product. Hence, ISO 14041 recommends that when it comes to co-products, allocation should be avoided whenever possible, in favor of system boundary expansion (International Orginization for Standardization 1998). System boundary expansion allows for the system boundary of the multifunction process to be expanded to include processes that benefit from one or more of the co-products. This is particularly useful when one of the co-products is replacing a virgin material, allowing the producer of the multifunction process to argue that the co-product produced by them is helping avoid the burden of producing the virgin material it is replacing.

While this technique attempts to view the question of allocation purely as one of mass and energy balance, it is difficult to apply in all cases. The social negotiation of allocation is difficult to avoid. The different perspectives can be examined as follows:

1. *The producer of a multifunction process:* The producer's priorities are shaped by the economic value of each co-product. The fate of a co-product that is not the primary source of economic value is dictated by its demand in other markets. The producer has an interest in profiting from it by selling it or stockpiling it for future use.
2. *The downstream user of a co-product originating in a multi-function process:* The user will want to inherit only the portion of the environmental burden that is associated with the co-product that they are using. The allocation of the impacts can be done either based on the relative economic value of the co-product or based on its relative mass/volume fraction.
3. *The recycler:* The recycler expects to gain credit for creating a re-use opportunity for a waste material from their process, rather than landfilling it. This credit is reflected in the incentives provided to customers for recycling glass and paper.
4. *The producer who utilizes a recycled product:* When the use of recycled or post use material can be used to reduce the use of virgin material, the producer expects credit for reducing the impacts associated with use of virgin material.

The above perspectives present a potential for inconsistent allocation, based in how a material is valued differently by a producer and user. These issues become apparent within the asphalt pavement materials industry as each of the four perspectives are represented.

Asphalt binder is a fractional distillation column bottom product, a co-product of the petroleum refining—a multi-function process. It is a critical component of the asphalt mixture that is produced in an asphalt plant and used as a pavement material. Further, after its first application, the asphalt pavement can be milled and crush and converted into Reclaimed Asphalt Pavement (RAP). RAP can be used to replace virgin aggregate (a quarry product of stone crushing), and also a small, but not negligible portion, of asphalt binder to the new mixture. The following questions of allocation arise:

1. How should the impacts of crude oil extraction, transportation to refinery and refining be allocated appropriately to the asphalt binder?
2. How should the upstream impacts of allocation of recycled material such as RAP be allocated to the mixture?

The response to the first question requires a negotiation between the asphalt mixture production industry and the petroleum refining industry with the objective of arriving at a consistent allocation procedure across all co-products. The challenges framing such a discussion are as follows:

1. A mass based allocation is not entirely fair because, as a heavy fraction, the density for asphalt places it at a disadvantage compared to the lighter fractions that are the primary consumers of energy during distillation. Conversely, it could be argued that as the asphalt does not use any energy during distillation, the energy consumed should be allocated only to the distillates. Both these situations are difficult to defend as they, respectively, either over estimate or underestimate the impacts associated with the binder.
2. An economic allocation based on the relative economic values of each of the co-products can be considered, but this comes with accompanying challenges. All refineries do not produce liquid asphalt binder. Sometimes the heaviest fraction is further refined into lighter fractions. Depending on the demand for asphalt binder, refineries can, from time to time, selectively choose to sell or further refine the asphalt binder. The process energy in refining the binder is significantly higher.

Currently a combination of mass based and economic allocation are applied on a case by case basis while the asphalt industry and the petroleum industry are negotiating the best path forward.

When it comes to the use of RAP, the question of allocation is based on who takes on the credits associated with the use of the reclaimed pavement. An agency may seek to gain credit for using RAP—indeed, green-rating systems such as FHWA INVEST or Greenroads, credit the use of RAP in projects. Even though this is not a closed loop recycling process (i.e., the product is not being recycled into the same pavement), it is within the same industry and hence the cut-off allocation method is used. In this method, the stakeholder recycling the pavement does not get any credit, but gets to avoid the end-of-life impacts (including landfill), while the stakeholder using the RAP to substitute virgin aggregate takes on the burden of preparing the RAP for reuse (fractionating) receives getting credit for the reduced impacts associated with the reduced use of virgin aggregate. As RAP could be generated multiple times, from the same material, this approach ensures that the impacts for each use are allocated appropriately.

The allocation of recycled material becomes more complicated when the industry creating the recycled product is directly in competition with the industry that is using the recycled product. For example, Recycled Concrete Aggregate (RCA) could be used in both asphalt and concrete pavements, although given the competition between the industries; the price of the recycled product varies. The possibility exists where the downstream industry and the upstream industry inappropriately avoid the burdens of first production and manufacturing, each expecting the other to account for it. This will require harmonization between competing industries in future.

5 ASPHALT MIXTURE FAMILIES

The objective of defining mix design families is to illustrate the sensitivity of LCA indicators to marginal changes in asphalt mixture designs. It is likely to serve the following purposes:

1. As an asphalt mixture design support tool to meet LCA targets (e.g., equivalent CO_2 emissions in global warming potential), while considering alternative designs.
2. As a method to cluster mixtures that have minor variations in design and environmental impacts.

The asphalt binder has the highest relative impact of all the constituent materials in the mixture. The asphalt binder content in the mixture is a combination of the virgin asphalt binder content and the contributions from recycled materials used such as RAP and Recycled Asphalt Shingles (RAS). As the RAP and RAS contents of a mixture increase, the percentage of virgin asphalt binder replaced also increases. Hence, as a mixture design varies starting with a virgin asphalt mix (with no RAP or RAS), the impacts are varied as follows: there is a reduction in use of virgin asphalt binder and virgin aggregate, there is an added impact due the processing and transport of the RAP and RAS from their respective sources. Over all, they present a reduction in the estimated impacts compared to the baseline impacts for a virgin mix. Of course, the baseline is specific to the primary data inputs for a particular plant, and may vary by region and plant energy use trends.

Therefore, a reasonable way to cluster asphalt mixes by environmental impacts is to rank them by reduction in virgin asphalt binder content, with respect to a baseline virgin mixture. For example, starting with a mixture that has 5% virgin asphalt binder, considering a family of mixes (specific to the plant) that can be designed by introducing RAP and/or RAS till the asphalt binder has been replaced by 50%, i.e. the mixture has a 2.5% virgin asphalt binder. Needless to say, the members of such a family can vary in design for different percentages of RAP and RAS. A range for the GWP for the family can be established, allowing mix designers to understand how the GWP changes as they vary their mix designs. Figure 1, illustrates the sensitivity of GWP to the change in mixture through asphalt binder replacement by adding RAP and RAS, starting from two base mixes one with 5% virgin binder and the other with 8% virgin binder for primary data specific to a plant. It is worth reiterating that the trends across a family are dependent on a baseline mix design and plant specific energy use data.

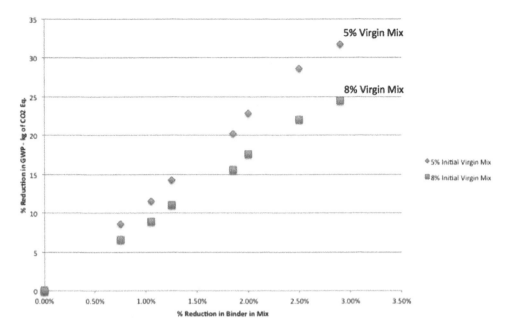

Figure 1. Percentage change in GWP as asphalt binder is reduced.[1]

6 CONCLUSION

This paper presents the challenges that were addressed by asphalt industry stakeholders who participated in the PCR development process for the NAPA EPD program. The paper discusses the ongoing technical and organizational challenges in addressing the following specific LCA related technical issues: (i) challenges in defining system boundaries, (ii) challenges in identifying sources of transparent and high quality inventory data, (iii) challenges in estimating missing inventory data, (iv) challenges in harmonization with allied upstream material supplier industries; namely the petroleum refining industry, (iv) challenges in allocating impacts to co-products and recycled materials such as asphalt binder, and reclaimed asphalt pavements among others. Each of these factors is discussed, both from the perspective of the technical requirements of the program as well as the stakeholder concerns regarding product economics and production viability.

While these issues are yet to be resolved through accepted long term industry based protocols; at their heart is the tension between implementing rigorously verified, transparent programs that make accurate estimates of environmental and process improvements versus, easing the adoption of life cycle based approaches through a step-by-step approach based in incremental improvement in the estimation process. The LCA process is critical to quantifying environmental impacts of products and processes. However, a lack of standardized LCA rules has been a stumbling block in its adoption in the decision-making process. While the FHWA SPTWG is in the process of identifying pavement industry LCA standards, through a collaborative process, the negotiations of that process closely reflect the industry forces discussed in this paper. As such efforts are likely to have direct impacts on stakeholder economic bottom lines, implementation must be carefully negotiated without compromising the underlying rigor, or deterring long-term adoption.

---

[1] This plot is plant specific and based on working data that has not been finalized yet as this work is in progress. The intention of this diagram is to highlight the trend.

REFERENCES

International Orginization for Standardization. (1998). "1998 Environmental management—Life cycle assessment—Goal and scope definition and inventory analysis." *ISO 14041*.

International Orginization for Standardization. (2006). "2006 Environmental labeling and declarations for Type III EPD—principles and procedures." *ISO 14025*.

Kahneman, D., and Lovallo, D. (1993). "Timid Choices and Bold Forecasts: A Cognitive Perspective on Risk Taking." *Management Science*, 39(1), 17–31.

Kahneman, D., and Tversky, A. (2000). *Choices, values and frames*. Russell Sage Foundation.

Current difficulties with creation of standardized digital climate calculations for infrastructural projects

L. Strömberg
NCC (Nordic Construction Company), Solna, Sweden

ABSTRACT: The aim of this paper is to illustrate current difficulties and potential driving forces for creation of standardized digital climate calculations in existing computer-based LCA software. The standardized climate calculation is suggested to perform according to the requirements in international and European standards for making and verification of Environmental Product Declarations (EPD). An industry-wide standardized process for making digital climate calculations would make it possible to create an unbroken information flow with environmental information between material suppliers, contractors and clients. Several Swedish and Norwegian stakeholders were engaged in this joint project to perform a sharp test of making some climate declarations for bridge and roads projects.

1 INTRODUCTION

Sweden faces major challenges in the climate field. Where the government and parliament now have taken the initiative with the goal of Sweden as a nation to be climate neutral by 2050 (The Swedish government proposition, 2012). An active involvement of various market players is necessary to manage with the adoption of this goal to the practical work in infrastructural projects.

With this background the major Swedish client of infrastructural projects, the Swedish Transportation Administration, Trafikverket (STA) has established requirements for suppliers and contractors regarding energy and climate gas emissions accounting and reporting (Trafikverket, 2016). In Sweden, the STA starts to encourage contractors to use project-specific EPDs as a verification of project's actual climate impact since spring 2016 (Trafikverket, 2016). Product-specific EPDs are also required for public procurement in some major projects in Norway (Johannessen, 2016), by a few municipalities in Sweden and in the international building assessment schemes (LEED, BREEAM etc.).

At the same time a lot of Swedish major clients of infrastructural projects have established requirements for suppliers and contractors regarding the reporting of project documentation in a digital form. All project' changes as e.g. optimization of project' design solutions should be performed in a digital Building Information Modelling (BIM) (Trafikverket, 2016).

There is a need for a technical solution to include climate efficiency of infrastructural projects into the overall design and optimization process in the BIM. This requires that the environmental properties, in the form of emission factors for building materials, construction parts etc. are available directly in the BIM software for design engineers. Some main challenges are to extract a material list (Bill of Materials, BoM) from a project's BIM and match this list with emission factors from a secure data source, e.g. a commercial LCA database or a verified EPD.

Another challenge is to use the EPD format for the calculations of environmental and climate performance of a project's various stages and in the final report to the client. To do so requires the same system boundaries regarding detailed construction parts and life cycle stages are applied during the construction process. It demanded a system where entrepreneurs at an early stage of the project starts with a rough calculation of climate impact and then gradually improves the project's climate performance throughout the project. One challenge for such a system is to be able to plan for a reduced climate impact at an early planning stage when detailed information about technical solutions and purchased material is missing.

2 METHODS

2.1 The joint project

The joint project organization has been formed by several contractor companies, building material manufacturers, branch organizations and clients. Different stakeholders from some major Nordic building companies, some public clients and municipalities, industry joint organizations have participated at this joint project.

2.2 Goal and scope

The aim of this joint project is to illustrate current difficulties and potential driving forces for creation of standardized digital climate calculations in existing computer-based LCA software. The standardized climate calculation is suggested to perform according to the requirements in international and European standards for making and verification of Environmental Product Declarations (EPD). An industry-wide standardized process for making digital climate calculations would make it possible to create an unbroken information flow with environmental information between material suppliers, contractors and clients. Several Swedish and Norwegian stakeholders were engaged in this joint project to perform a sharp test of making some climate declarations for bridge and roads projects.

2.3 Verified LCA-calculation

In scope of this joint project was defined a need for an industry-wide LCA-calculation process, which can support entrepreneurs and other actors to perform a rough calculation of the actual climate impact in the early project stage and gradually improve climate performance throughout the project. One challenge for that is to be able to plan for the reduction of climate impact already in the early planning stage where detailed knowledge of solutions and purchased material is missing.

To make comparison of different project alternatives, it is crucial to develop definitions and adoption of an industry joint LCA-calculation process as a basis for comparison. The same system boundaries and data quality should be applied into such industry joint LCA-calculation process to allow the using different commercial LCA-software. To achieve a comparable LCA-data collection for an infrastructural project, the data should to be broken down to various building components and material according to the EN 15978 (EN 15978:2011). The different input and output flows in a performed LCA should be allocated to related project' stages, e.g. the material production, the use phase etc. according to EN 15804 (EN 15804:2014–07). The same method for assessment of impact categories e.g. the CML-method need to be used.

An important question is how to include the project specific climate impact in the LCAs, e.g. climate impact from operation and maintenance. It isn't enough to calculate the climate impact only from the material production (Modules A1-A3), since maintenance and operation actions (Module B1–7) are planned for the entire construction or the construction part and seldom at the material level. Contractors can control this part of the construction process and can contribute with more environmentally-friendly solutions for A4-C4, e.g. by optimization of planning, design and production processes. Using environmental assessment only for A1-A3 in the evaluation of alternative contractor' designs may lead to the sub-optimization.

2.4 Pre-verification of a LCA-tool

An industry common LCA-calculation process would allow usage of the same environmental information for various purposes such as a company's internal monitoring of the climate performance in infrastructural projects, internal product development, purchasing of building materials, monitoring for the environmental management system, public procurement, creation of business strategies, reporting according to building assessment schemes (BREEAM, LEED, DGNB etc.) and reporting to client. This can be achieved by pre-verification of some existing commercial LCA-tools in order to produce comparable verified EPDs and climate declarations for infrastructural projects. EPDs are the only LCA-format, which is allowed to be used in procurement to evaluate and compare alternative designs according to the European standards (ISO/TR 14025: 2000).

A pre-verification of existing digital LCA-tools to produce EPDs doesn't replace the single EPD-verification, but it makes the verification procedure simpler as the tool has been pre-verified to conduct comparable LCAs in accordance with a specified PCR. This is different from EPD Process Certification, which replaces a single EPD verification. EPD Process Certification means that the internal procedures and processes of a company to generate verified EPDs are checked yearly by an accredited certification body.

A key component of a LCA in an EPD-format is the method description document, referred to as Product Category Rules (PCR). A PCR defines the LCA-model, the calculation method and data requirements (The International EPD System, 2016). For pre-verification of a digital LCA-tool, based on a certain LCA-calculation process, according to a specific PCR, e.g. for bridges or roads, following pre-verification steps in the LCA tool and outside of the tool are required by the Swedish operator for EPD-system, Environdec (The International EPD®System, 2016):

1. Checking of few examples with input data for climate calculations (a material recipe or so-called Bill of Materials, BoM)
2. Used LCA-databases must be approved, verified and structured according to EN 15804
3. Verification of links between building materials, machinery, etc. in the BoM and emission factors in the used LCA database
4. Control of used system boundaries for the LCA-calculation process (according to PCR and EN 15804)
5. Monitoring of a final EPD-report
6. A LCA-tool should be pre-verified by the EPD-operator, Environdec.

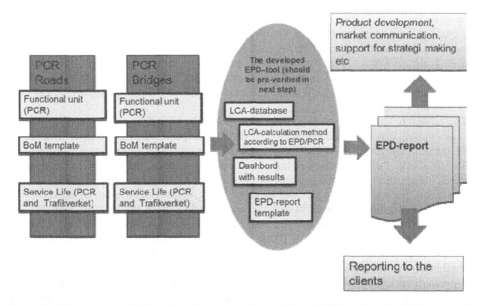

Figure 1. The conceptual LCA-calculation process for making of EPDs and verified climate calculations in a digital pre-verified LCA-tool.

The test of these verification steps for the LCA-calculation process has been done on some infrastructural projects and with help of thinkstep's LCA-software, LCA Service (thinkstep, 2016). These steps can be applied to other digital LCA-tools as well, see Figure 1. The project participants tested to calculate climate impact for four infrastructural projects.

3 RESULTS

3.1 *Test of a conceptual LCA-calculation process*

The project has tested the conceptual LCA-calculation process for climate calculations based on the Swedish-based, the International EPD® System, see §2.4. The verification rules require using a "locked" standardized LCA-calculation process, which can be applied to different LCA-tools in order to produce comparable calculation results. The LCA-calculation process to produce EPD or climate calculations in an EPD-format starts with the BoM-template preparation. The pre-verification of a digital commercial LCA-tool requires the checking of the input data collection for a LCA. In this joint project an Excel-based template for collection of data on materials; operations etc. with a pre-configured input structure were created (Step 1). The project information from contractors' existing cost calculation systems was used in this joint project, because of the lack of a complete BIM with comprehensive description of all used materials and building parts for an infrastructural project.

Several participated contractors tested the data collection process and the LCA-calculation method and usage of verified LCA databases (Step 2 and 4), which is recommended by PCRs (PCR: Bridges and elevated highways, 2013 and PCR: Highways (except elevated highways), streets and roads, 2013). The project participants created a detailed specification for pre-verification of existing digital LCA-tools (Step 6) in order to produce comparable climate declarations and EPDs for infrastructural constructions and projects.

Next, it was created the mapping (Step 3) of the BoM' items to emissions factors in a LCA-database (Gabi database). Then the calculation results were presented in the pre-defined Word-template for an EPD-report (Step 5).

3.2 *System boundaries*

Each participant collected life cycle inventory data for the selected infrastructure in the pre-defined Excel-based template (BoM template). The functional unit for the pilot project was one piece of infrastructure. For this reason, the collected data has to be entered in the corresponding BoM for one piece of infrastructure (either one bridge or one road). All life cycle phases (materials preparation, construction, operation/maintenance and end-of-life) can be covered in this BoM template depending on what is specified in the infrastructural project. The system boundaries considered as default in the delivered BoM template are described in the following Figure 2.

The main intention with the Excel-based BoM template for data collection was that each user can modify this BoM according to their needs following some guidelines and can adapt it for the different project phases. For example if the user wants to analyze a project in the tender phase, the BoM template will have to be adapted accordingly. This means that not all life cycle stages and/or material positions will be covered or given a value in the early stage. For example, the construction site and therefore the groundworks might not be known yet and will not be included in the BoM (e.g. soil and gravel volumes will be set to zero).

The created BoM template in this branch-joint project was based on the rules for comparable EPDs for bridges and roads. PCRs must be followed in order to compare the EPDs for infrastructural constructions with a similar function. The completed BoM template can then be used in a pre-verified LCA-tool. All calculations on climate impact can then be performed in a pre-verified LCA-tool based on the standardized LCA-calculation model. The considered system boundaries have to be documented in a standardized layout for the EPD-report with the final results, which may be delivered to a client.

Upstream Module	Core Module	Downstream Module							Other environmental information
Construction			Operation	Maintenance	End-of-Life				
Raw material supply (extraction, processing, recycled material) Transport to manufacturer Manufacturing	Transport to construction site	Construction of the bridge	Use/application Operational energy use Operational water use	Maintenance, Repair Replacement Refurbishment	Deconstruction / demolition	Transport to end-of-life	Waste processing for reuse Recovery or recycling	Disposal	Benefits and loads beyond the system boundaries (BLBSB)
A1-A3	A4	A5	B1, B6, B7	B2-B5	C1	C2	C3	C4	D
X	X	X	MND	X	MND	X	X	X	X

Figure 2. System boundaries of the BoM template for the tested infrastructural projects (X = declared module; MND = module not declared).

3.3 Data collection

Swedish construction companies use various IT-based internal systems to store the project-specific information. In one of these systems, cost calculation systems entrepreneurs use to create a spreadsheet with the planned purchase of building materials and machine hours for each infrastructural project. It is important to note that these cost calculation systems have been developed to be able to make a very rough estimation of project costs. A cost calculation is usually updated during the various projects' stages: tender, design and production.

In some isolated cases the spreadsheet with the project cost can be updated after the commissioning of a project. In this case this cost calculation will contain the total final cost for the whole infrastructural project: purchased building materials, cost for all work performed and additional costs. However this follow-up calculation isn't the established practice in the construction industry today.

Participants provided information and data collection about the construction stage of their infrastructural projects (including materials, transport to construction site and construction site works). Even the material and energy flows for the operation and the maintenance were added in the BoM template. The material flows have been collected for A1-A3 and the energy and operating materials' flows for B2 according to the PCR' requirements for bridges. Since no specific information about the allocation of materials in the maintenance phase is given in the PCR for Roads (PCR: Highways (except elevated highways), streets and roads, 2013), the same approach as for bridges was used. Energy carriers and water use are allocated to B2.

It was made some general assumptions on the end-of-life scenarios, e.g. the amount of material to landfill, incineration and recycling. Module C1 (demolition, deconstruction) wasn't considered in the used BoM template, since no detailed data (e.g. energy, water, etc.) was available in the current project planning IT-systems. The project participants had to perform a lot of manual modifications of the existed project data from their cost calculation systems to match the requirements by EN 15804, EN 15978 and PCRs. Another important aspect is that the selection and comparison of design options is currently conducted in the BIM design software such as AutoCAD, Revit etc. Some construction companies try to link the analysis of design options in the CAD/BIM software and the cost calculation system in "real time". This is far from a standardized workable routine in the industry.

4 DISCUSSIONS

4.1 *Automatization of EPDs' development*

Contractors' internal IT-systems for design and planning of infrastructural projects (e.g. BIM, Cost Planning System) have a different data structure to store project information, which isn't compatible with EPD/PCR' requirements. There is a need for a technical solution to automate the collection of project-specific data, according to the requirements for verified EPDs. At the moment the data collection for climate calculations requires a manual handling. Some standardized routines and support to collect data for the entire life cycle of engineering works should be established in the construction companies. This will support an unbroken information flow with LCA-data between different market players.

The material list in the existing cost calculation and BIM systems usually contain fragmented input data for the Modules A1 to A3. Data collection for other Modules is even more difficult to collect from the current project design and planning IT-systems. Data for the operation and the maintenance, percentage of recycled content in building materials, the transportation of building materials to the construction site are examples of input data for LCA-calculations, which aren't stored in these IT-systems. There are various proposals on how this would be done in the future, for example as part of the information in the Building Information Model (BIM). The automation of an EPD process creation can be achieved by pre-verification of existing commercial LCA-tools. This will reduce the cost for creation of project-specific EPDs and climate declarations.

4.2 *Using a EPD-format for verification process*

EPD is an environmental declaration of the final product. Adoption of the EPD-format to set up, measure and follow-up operational goals for an infrastructural project should be developed. There are also some uncertainties with using EPDs as a verification of the environmental performance for a project design in the early stages. For early planning stage or tender phase there should be an industry-wide agreement with listed assumptions regarding the system boundaries (e.g. level of detail on the bill of materials) and generic emission factors to be used for different materials/construction parts.

It's also unclear in the current PCR/EPD-system how to aggregate single EPDs for some main construction elements (tunnel, bridge) to calculate the total environmental impact of the entire infrastructural project. Harmonization of the various functional units, assumptions etc. in different PCRs for civil engineering structures (bridges, tunnels, etc.) must be standardized by ISO/EN. It would be a great practical benefit to include bridges, tunnels and even a few more engineering structures in the scope of the same PCR.

PCRs are available for engineering constructions (e.g. bridges, highways etc.) and construction materials and services. However each PCR is linked to a particular EPD program operator (e.g. EPD-system Norway or the International EPD® System in Sweden). EPDs from different operators aren't comparable yet. The calculation rules for e.g. carbon dioxide emissions differ between the current EPD-systems in Europe and USA, which can be misleading when comparing EPDs for products and processes. However this is already addressed within the framework of the EPD/PCR and also within the ongoing work in the EU-project (Eco-Platform, 2016).

Reporting according to the EPD-format includes a mix of accounting and consequence LCAs. For example, Module D of an EPD is such a "consequence LCA", where all potential missing emissions and "credits" to other product systems should be reported. Even for the use phase, various potential usage scenarios far ahead in time, for example a particular source of energy, should be disclosed. The verification rules for these future potential scenarios for the use/maintenance and the end-of-life phase are still unclear in the PCRs.

4.3 *More tests are needed*

During this joint project it was foreseen that the participants test some testing examples for one design scenario. More testing scenarios should be developed in the next step (early design, tender, planning, production, additional works and reporting to the client). This project was the very first step in establishment of LCA-calculation process for standardized climate calculation.

Benefits for the industry from this joint project is more knowledge on the existing gap between the long-term overall planning by the EU, the Swedish government, the ongoing standardization work and the contractor's practical work on climate issues in engineering projects. The project has also identified and clarified the requirements for the continued industry-wide cooperation in the development of verified comparable calculations of the climate impact for civil engineering structures.

ACKNOWLEDGEMENTS

This industry joint project was initiated in early 2013. The project is partly financed by The Development Fund of the Swedish Construction Industry (SBUF), Trafikverket and several Nordic contractors, material suppliers, industry joint associations and clients.

Participants in the project express gratitude for the opportunity that SBUF has given the construction industry by funding the project. Special thanks to the Infrastructural committee at SBUF.

REFERENCES

Eco-platform. 2016. www.eco-platform.org.
EN 15804:2014-07: Sustainability of construction works—Environmental Product Declarations—Core rules for the product category of construction products.
EN 15978:2011: Sustainability of construction work—Assessment of environmental performance of buildings—Calculation method.
ISO/TR 14025: 2000 Environmental labels and declarations—Type III environmental declarations.
Johannessen, A.K. 2016. Customer demands on LCA in the asphalt industry in Norway. Presentation at the workshop at the joint project (in Norwegian).
PCR: Bridges and elevated highways. 2013. Version 1.0: 2013:23. UN CPC 53221.
PCR: Highways (except elevated highways), streets and roads. 2013. Version 1.02: 2013:20. UN CPC 53211.
The International EPD®System. 2016. http://www.environdec.com/
The Swedish government proposition Fossiloberoende fordonsflotta - ett steg på vägen mot nettonollutsläpp av växthusgaser, Dir. 2012:78 (in Swedish).
thinkstep. 2016. https://www.thinkstep.com.
Trafikverket, 2016, http://www.trafikverket.se/for-dig-i-branschen/teknik/ny-teknik-i-transportsystemet/informationsmodellering-bim/bim-trappan/ (in Swedish).
Trafikverket. 2016. http://www.trafikverket.se/klimatkalkyl (in Swedish).

Development of an environmental Life-Cycle Assessment (LCA) protocol for flexible pavements that integrates life-cycle components to a proprietary software

F. Osmani, M. Hettiwatte, S. Kshirsagar, S. Senadheera & H.C. Zhang
Texas Tech University, Lubbock, TX, USA

ABSTRACT: Significant progress has been made in research on environmental Life-Cycle Assessment (LCA) of pavements. Use of this new knowledge has been slowed by a lack of standardized datasets and analysis protocols. Several software packages have been developed to address this by providing datasets with broader coverage and better quality, but practitioners need application-specific data input interfaces for easy use. The authors report work on developing a protocol for environmental LCA of flexible pavements that uses a proprietary software system with data input interface plug-ins they developed for each of the five flexible pavement life-cycle phases. This protocol allows designers to integrate environmental LCA with Life-Cycle Cost Analysis (LCCA) to select the optimum design from several alternatives. With looming transformative changes in pavement materials and technology landscapes, this protocol also has the potential to allow a more effectively holistic assessment of such novel systems.

1 INTRODUCTION

1.1 Background

Environmental LCA of products and processes has been a topic of interest for the last couple of decades, particularly since the introduction of the International Organization for Standardization (ISO) 14001 in 2004 (ISO14001, 2004). Many agencies have adopted this framework since then. In 2006, the ISO 14001 was replaced with ISO 14040 and ISO 14044 (ISO14040, 2006; ISO14044, 2006), and both government agencies and private sector began assessing sustainability impacts of their products and services. More and more agencies are now moving towards incorporating holistic sustainability concepts to make decisions based not only on LCCA but also on impacts to their organization, surrounding community and the society in general.

1.2 Literature review

The following literature review takes a broader look at LCA combined with some focused studies that relate to flexible pavements. The review showed a wide variation in approaches and results that are sometimes contradictory. However, many of these variations are the result of a diverse array of goals and objectives in these studies.

1.2.1 Environmental LCA

The concepts of environmental LCA were first introduced in the 1960s to monitor and estimate impacts due to solid waste as well as emissions to air, land and water (Harvey et al., 2016). Later it was broadened to include emissions from chemicals, energy production and use of resources. The first international standard was introduced in 2004, when ISO 14001 provided a framework to estimate these impacts in order to make important decisions on products and services (ISO14001, 2004). The life-cycle of a product comprises of consecutive and interlinked phases from raw material acquisition or generation to product end-of-life and disposal. LCA for a product is the compilation and evaluation of inputs, outputs and their potential environmental impacts throughout its life-cycle. This systematic approach will ensure accurate assignment of a potential environmental burden to the appropriate life-cycle phase or individual process (ISO14040, 2006).

The ISO 14040 LCA has the following four phases: goal and scope definition, inventory analysis, impact assessment, and interpretation. According to ISO 14040, the system boundary for an LCA must be framed to highlight key phases of the product life-cycle. This typically includes unit process for raw materials, inputs and outputs of the primary production/manufacturing/processing sequence, distribution, product use and maintenance, disposal of process wastes and used products, recovery of used products (including reuse, recycling and energy recovery), production and use of fuels, electricity and heat, manufacture of ancillary materials, manufacture, maintenance and decommissioning of capital equipment, lighting and heating (ISO14040:2006).

1.2.2 *Flexible pavement life-cycle*

In general, a pavement life-cycle consists of six phases, all of which have significant impacts on the environment: material production, pavement design, construction, service period, maintenance/rehabilitation, and end-of-life (Harvey et al., 2016). It can be considered as a closed loop for a particular project depending on the system boundaries for which the LCA is to be conducted. Design is not identified as a separate phase in most instances, and its impact is considered to be included among the other five phases.

1.2.3 *Pavement LCA*

Pavement LCA is an area of active interest among both researchers and practitioners, and both groups generally follow the ISO 14040 guidelines (AzariJafari et al., 2016). However, efforts of these groups are constrained by the lack of reliable data with appropriate levels of accuracy. Most sustainability studies focus on quantifying the environmental impacts and only a few address other important aspects such as design, material selection, constructability, maintenance and development of policies that promote achieving sustainability goals (Santero et al., 2011a). This situation can be addressed by standardizing the way functional units and system boundaries are established, thus paving the way for unification of LCA protocols around the world. LCA is still at its infancy in areas such as pavements, and many gaps in data and methodologies that still exist need to be addressed in order to accurately quantify their environmental impacts (Santero et al., 2011b). These gaps include misinterpretation of data due to lack of standardization and differences in analysis methods. If a common set of guidelines for data collection and analysis are agreed upon, these gaps can be further reduced and the data made more valuable to the pavement engineering community.

1.2.4 *Pavement LCA with commercial databases*

A number of pavement sustainability LCA studies have used commercial software platforms primarily to get access to accurate data. However, literature suggest that the user should recognize the strengths and limitations before using them. One such limitation for US users is that many of the leading LCA software have been developed in Europe and their data and methodologies cater to conditions there (Kang et al., 2014). In other instances, some software and databases have been reported to have double counting of sustainability impacts among unit process components (Vidal et al., 2013). Furthermore, the proprietary nature of commercial LCA software makes it very difficult to automate sustainability evaluations for clients' unique requirements (Gopalakrishnan et al., 2014).

1.2.5 *Introduction of paper*

A review of technical literature suggests a strong need for standardized protocols to conduct LCA of pavements. The purpose of this paper is to present a software-based framework for flexible pavements in which all five phases of a pavement life-cycle (excluding the design phase) are incorporated. These phases are materials processing, construction, pavement in service, maintenance/rehabilitation, and end-of-life. Unit process flow charts were developed for each life-cycle phase and a software plug-in was developed to seamlessly integrate customer design data with the proprietary LCA software and the database to allow the calculation of sustainability metrics. This method will allow the user to easily identify key process components in each phase that show high sustainability impacts and focus on ways to minimize those impacts to make the product (i.e. the pavement) more sustainable. The software-based plug-in interfaces for the five life-cycle phases have been developed in such a way that the pavement designer can evaluate each alternative using both the LCCA and environmental LCA at the time of design, so both criteria can be used to select the optimum design.

2 LCA FRAMEWORK FOR FLEXIBLE PAVEMENTS

The LCA approach is gaining ground as a viable concept to make pavements more sustainable (Huang et al., 2009). For this study, a process-based modeling approach was selected since it facilitates the assessment of every minute detail of the pavement life-cycle to assess sustainability impacts so remediation measures can be identified. The five phases of the pavement life-cycle considered here are material production, construction, pavement in service, maintenance/rehabilitation, and end-of-life. Performing all inventory analyses manually can be very time-consuming, so modeling for individual life-cycles phases was done using the commercial software platform GaBi™ (thinkstep, 2016). Many software platforms are available for inventory analysis, including GaBi™, open LCA, Simapro, Umberto, and PaLATE. A free pavement LCA software tool, Athena Pavement LCA (Athena, 2016) is also currently available with limited data availability. Athena Pavement LCA includes data specific to Canada and selected US regions; data includes materials manufacturing, roadway construction, and maintenance life-cycle stages. Data does not include demolition and disposal of the pavement. This tool calculates environmental impacts in accordance with the US Environmental Protection Agency's (EPA) methodology for the Reduction and Assessment of Chemical and other Environmental Impacts (TRACI).

This paper is the result of a broader research study to develop and assess novel material systems for flexible pavements. The commercial software platform GaBi™ was found to offer capabilities convenient to developing process flow charts, and it also provides extensive datasets for United States to go with its robust European database. GaBi™ offers twenty-three vendor-developed databases for a wide array of industries, materials and continents, and a number of these databases are relevant for flexible pavements. The databases include Professional, Energy, End-of-Life, Manufacturing Processes, Renewable Raw Materials, Construction Materials, National Renewable Energy Laboratory (NREL) US Life-Cycle Inventory (LCI) Integrated, and Full US databases. The Full US Database allowed this research team to conduct LCA of flexible pavement systems by entering candidate pavement designs short-listed from LCCA. The schematic in Figure 1 illustrates this analysis framework. The GaBi™

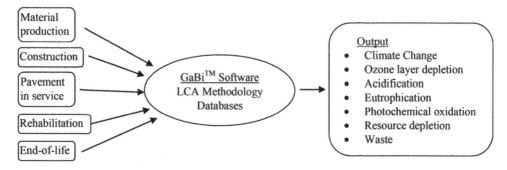

Figure 1. Illustration of LCA method.

software can also identify process hotspots for more detailed analysis and conduct what-if analyses for situations such as materials and process replacement candidates. The LCA for this study used a number of sustainability metrics (balances) available in GaBi™ including climate change, ozone layer depletion, acidification, eutrophication, formation of photo-chemical oxidants, depletion of fossil and mineral resources, and hazardous/nonhazardous waste.

The effects of these metrics can be assessed by using Institute of Environmental Sciences (CML) 2001 and TRACI methodologies commonly used for characterization of environ-mental impact assessment. In this study, the whole life-cycle of the pavement was considered, from material production to end-of-life scenarios, processes such as production of asphalt cement, asphalt concrete and base materials, fuel and electricity, construction and rehabilita-tion, pavement in service, and recycling. The Full US database in GaBi™ software (think-step, 2016) was used in this study because it has many complete US life-cycle inventory data sets. The GaBi™ software and the database allow the quantification of inputs (material, fuel, and electricity) required, and outputs (air, water, and soil emissions) released during the pave-ment life-cycle. Figure 2 illustrates the components of the flexible pavement life-cycle and its components, along with the system boundary used for LCA.

To conduct this study, we defined a functional unit (a comparative medium with all the properties of a system) and reference flows supporting the information for the establishment of functional units. Functional unit represents the system's function and provides an equiva-lent level of function or service for comparison. The functional unit of the conventional pave-ment system for this study is listed in Table 1. It was defined for a principle arterial highway located in Lubbock County, Texas. The LCA software and the data entry plug-in allows the use of the corresponding electricity grid mix from the Southwest Power Pool (SPP) (US EIA, 2016).

A reference flow is a quantified amount of product(s), including product parts, necessary for a specific product system to deliver the performance described by the functional unit. For this study the reference flow quantities according to the life-cycle phases are: *Structural design*

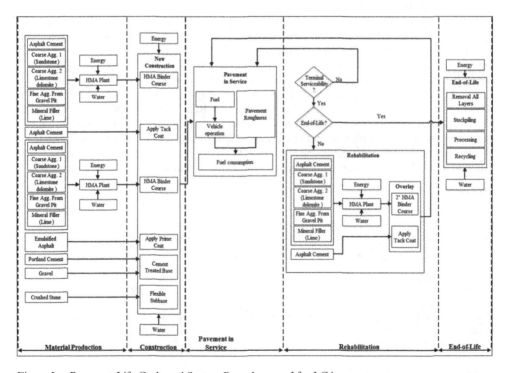

Figure 2. Pavement Life-Cycle and System Boundary used for LCA.

Table 1. Functional unit.

Parameters	Values
AADT	9,146
Truck%	35%
Traffic growth rate	2%
Design Life	25 years
Surface Area	1 yard × 1 yard

Table 2. Flexible pavement layer configurations.

Design	Layer 1	Layer 2	Layer 3	Layer 4	Rehabilitations	Total Cost/SY
Design 1	HMA 4.5"	HMA 2"	CTB 7"	FB 8"	8 and 16	$ 23.73
Design 2	HMA 2"	HMA 5"	CTB 7"	FB 8"	8 and 17	$ 24.36
Design 3	HMA 6"	HMA 2"	CTB 5"	FB 8"	8 and 16	$ 24.46

Note: HMA = Hot mix asphalt; CTB = Cement stabilized base; FB = Flexible base; SY = Square yard.

of the pavement (number of layers and their composition): *Construction* (activity type): *Rehabilitation of the pavement* (when-after construction, expected design life-how many years will pass before the activity needs to happen again, percentage of element affected by activity, material types and quantities).

3 INTEGRATION OF LIFE-CYCLE COMPONENTS

The Life-Cycle Impact Assessment (LCIA) phase of an LCA is the estimation of potential human health and environmental impacts of the environmental resources and emissions during the life-cycle of a process or product (US EPA, 2006). Impact assessment should include ecological and human health effects; it also includes the resource depletion. An LCIA attempts to establish a connection between the product or process and its impacts on the environment. In this study, three layered configurations were used as shown in Table 2. These configurations were obtained for the same functional units as mentioned in Table 1, using the Flexible Pavement Design System (FPS-21) software used by Texas Department of Transportation (TxDOT) to design flexible pavements. All three configurations have four layers with the same material composition, but with different layer thickness values.

3.1 *Material production*

Material production consists of the processes involved to produce pavement materials. This process includes the entire upstream of the supply chain required to produce each material and mixing at the asphalt plant (FHWA, 2014). Traffic level and the soil characteristics at a particular location are the most important variables regarding pavement structural design and material composition. Using the functional units indicated in Table 1, a layered configuration was designed using FPS-21 software for one square yard of pavement surface. Based on the design output from FPS-21, the Hot-Mix Asphalt (HMA) layers consist of 40.1% sandstone, 34.7% limestone-dolomite, 9.4% aggregate from a gravel pit, 9.4% mineral filler (lime) and 6.3% asphalt cement. The Cement-Treated Base (CTB) consists of 5% Portland cement treated limestone-dolomite compacted to 120 lb./ft^3. The Flexible Base (FB) also consists of limestone-dolomite compacted to 115 lb./ft^3. These layer material compositions are unique for this study, and using the software plug-in interface for the material production phase, designers have the flexibility of changing these materials and compositions to suit their pavement design.

3.2 Construction

As shown in Figure 2, the environmental impacts of the construction phase involve the impacts from the diesel fuel combusted in the project equipment and the production of the fuel. Information like the daily output of each activity and the Horse Power (HP) of every piece of equipment was used, and an eight-hour workday was assumed. The construction phase includes a series of paving processes. In this model, the system boundary of the asphalt pavement construction phase is defined in generic form for any situation. This system boundary includes stabilized subgrade, unbound granular base, bound granular base tack coat/prime coat application and HMA/Warm Mix Asphalt (WMA) placement/compaction. The environmental burdens associated with the construction phase mainly come from the energy consumption, emissions from equipment used in these activities and from diesel fuel production, which is a function of the thickness of different layers and their material compositions.

3.3 Pavement in service

Figure 2 illustrates the system boundary of the pavement in service phase consisting of the vehicles using the pavement, fuel production, and combustion. In this study, the environmental impact inventory covers both gasoline and diesel impacts for the vehicle mix on the roadway. Vehicles operating on diesel fuel include all types of vehicles from medium car to articulated truck, while vehicles operating on gasoline cover only the smaller vehicles such as medium car, van and SUV. The pavement in service phase is mainly composed of vehicular operation on the pavement. For the vehicle to stay in movement, it has to overcome many forces, such as inertia, gravity, internal friction, aerodynamic drag, and rolling resistance. All these factors, with the exception of rolling resistance, are vehicle-related. Rolling resistance is affected primarily by pavement texture, roughness, and the pavement type. These properties may significantly increase energy consumed by the vehicles, thus increasing the fuel consumption (Willis et al., 2015). The widely accepted International Roughness Index (IRI) standard was introduced by the World Bank in 1982. The IRI is a scale for roughness based on the simulated response of a generic motor vehicle to the roughness in a single wheel path of the road surface (NCHRP, 1978). IRI values are measured in inches/mile or m/km (Al-Rousan et al, 2010). As pavement roughness increases with time, it increases the vehicle fuel consumption rate. Authors have used results from previous research by Chatti & Zaabar (2012) for the base fuel consumption and how it changes based on travel speed and pavement roughness.

The effect of maintenance activities, such as patching and crack sealing, on the pavement IRI is not clear (Yang, 2014). Because of that, this study only considered the effect of rehabilitation (overlays) on the pavement IRI. It is assumed that the rehabilitation of a pavement restores its condition, but not to the point when the pavement was initially constructed. Rehabilitation will be done after a fixed number of years and will be repeated again, until it reaches its end-of-life.

The pavement in service phase is the most complex section when analyzing the environmental impact of a pavement system (Yang, 2014). The interaction between vehicles and the pavement surface has a significant effect on vehicle fuel consumption (Santero et al., 2011b). The vehicle fuel consumption is influenced by the load and pavement roughness. In this study, the flexible pavement roughness is represented using one of two parameters: pavement serviceability expressed by Present Serviceability Index (PSI) and pavement roughness expressed by IRI. The IRI is incorporated into the sustainability assessment using the model provided in equation 1, 2 and 3 (Hall & Muñoz, 1999). The IRI values will change during the life-cycle of the pavement depending on its performance cycle, and it can be calculated for a certain year using equations 4 and 5. All these processes are incorporated in the GaBi™ plug-in to calculate the vehicle emissions.

$$PSI = 5 - 0.2937x^4 + 1.1771x^3 - 1.4045x^2 - 1.5803x \qquad (1)$$

$$x = \log(1+SV) \quad (2)$$

$$SV = 2.2704 * IRI^2 \quad (3)$$

$$IRI_n = IRI_{Initial} e^{bn} \quad (4)$$

$$b = \frac{\ln\left(\dfrac{IRI_{Terminal}}{IRI_{Initial}}\right)}{Time\ between\ rehabilitation} \quad (5)$$

where, PSI = present serviceability index; SV = slope variance used to characterize the pavement roughness; IRI = International Roughness Index; n = number of year; b = site constant for IRI.

The authors used this approach to obtain the IRI value for a certain year of the life-cycle, using the initial PSI, final PSI, and time between rehabilitations calculated during the pavement design process. Hence, we calculated the varying fuel consumption depending on the IRI for a given time of the life-cycle. For this study, a traffic mix of 10% medium cars (gasoline), 20% vans (gasoline), 30% SUVs (gasoline), 20% light trucks (diesel) and 20% articulated trucks (diesel) was considered. The impacts were calculated based on the extra fuel consumption due to increasing IRI for an average speed of 55 mph. The traffic mix used in this study is unique for the selected design, but using the software plug-in interface for the pavement in service phase, designers can change the traffic mix and average speed to suit their pavement design.

3.4 Rehabilitation

As shown in Figure 2, the rehabilitation process considered in this study is the overlay of the 2" HMA surface course. In this study, it is assumed that the overlay uses the same type of materials as the new construction. The environmental impacts considered in the rehabilitation process include the impacts released by the production of asphalt cement, aggregates, asphalt concrete, diesel fuel combusted by the equipment during the rehabilitation process, and the production of the diesel fuel used to run the equipment during the rehabilitation process. In order to calculate the amount of emission released, the model used for the material production phase was applied.

The data input plug-in interface for the rehabilitation phase provides flexibility to the pavement design engineer to enter the timing and overlay material information for each rehabilitation operation conducted during its life-cycle. The number of rehabilitation operations for a design life-cycle is limited to a maximum of four. The rehabilitation and preventive maintenance phase will generally include three alternatives: overlay, seal coat, slurry seal. But for this study we only looked at a 2" overlay alternative.

3.5 End-of-life

A pavement is expected to provide satisfactory service over its design life. It is rehabilitated a number of times before it is demolished and a new pavement is built at the same location. Figure 2 shows the system boundaries followed in the study for end-of-life of a pavement. As seen, it was assumed that at the end of its design life, the asphalt cement layer is removed completely from the road by milling machines. The other layers are reused and function as salvaged materials. For calculating the environmental impacts from the milling machine, the impacts from the diesel fuel combusted in the equipment and the production of the fuel are considered. The milled asphalt concrete is transported back to the HMA plant, where it is reintroduced with the raw materials as Reclaimed Asphalt Pavement (RAP). The process included after transporting the reclaimed asphalt is stockpiling, which involves the use of an excavating machine and loader, and crushing, a procedure that involves the use of a dozer machine (Yang et al., 2014). When calculating the environmental impacts for this equipment, the impacts from the diesel fuel combusted in the equipment and the production of the fuel are considered. For this study, it was assumed that all the waste material was transported by trucks 100 miles away from the pavement location. The authors have assumed that all material are either landfilled or transported to a recycling/processing facility, and the materials will be reused in another pavement system. The effect of landfilling is captured in this study, but to capture the effect of recycled materials this system has to be connected with another pavement system which is out of the system boundary of this study.

4 RESULTS AND DISCUSSION

The results for different impact categories determined by GaBi™ Software for the three design alternatives are listed in Table 3. The results for each impact category for a particular design alternative is the sum of impact categories for each of the five phases of the pavement life-cycle. As expected, for each impact category, the values are dominated by the 'pavement in service' phase of the life-cycle. The next high impact life-cycle phase is material production. Due to the high material intensity of pavement structures, the material depletion and waste impact categories show up significantly as expected. The important thing about this analysis is that it provides detailed data by impact category and by life-cycle phase of the pavement system for each design alternative. This form of LCA, which integrates both conventional life-cycle economics and environmental sustainability, will lay the groundwork to create a unified assessment of design alternatives.

As shown in Table 2, FPS-21 calculated different life-cycle cost values for the three designs being considered in this illustration. If a designer was to select the best design based only on the LCCA criteria, design 1 would be selected because it has the lowest cost ($23.73/SY) of the three designs. But if we considered the LCA criteria, according to the values in Table 3, the best design would have been design 2, which has the lowest footprint values for the overall pavement life-cycle. Therefore, this clearly indicates that considering only the LCCA criteria will overlook the sustainability aspect of a design, and end up in giving a less

Table 3. LCA results for three design alternatives, for one square yard of pavement surface area.

Design alternative	Life-cycle phase	Impact category						
		Climate change	Ozone layer depletion	Acidification	Eutrophication	Photochemical oxidation	Material depletion	Waste
Design 1	MP	1.70e2	1.62e-7	4.60e-1	1.14e0	3.50e-1	2.91e4	2.93e4
	C	4.02e0	2.59e-11	5.00e-2	3.63e-3	6.90e-2	1.68e2	1.76e2
	PIS	2.09e5	3.77e-6	1.91e3	1.43e4	3.62e3	6.17e6	6.53e6
	R	2.27e1	3.43e-8	1.50e-1	8.44e-3	1.53e-1	4.65e3	4.79e3
	EOL	1.64e1	1.82e-10	5.00e-2	4.19e-3	5.50e-2	3.51e3	3.85e3
Design 2	MP	1.17e2	1.94e-7	4.18e-1	9.5e-1	3.12e-1	2.14e4	2.14e4
	C	4.46e0	2.87e-11	5.50e-2	4.03e-3	7.70e-2	1.86e2	1.96e2
	PIS	2.00e5	3.60e-6	1.83e3	1.37e4	3.47e3	5.92e6	6.27e6
	R	2.27e1	3.43e-8	1.50e-1	8.44e-3	1.53e-1	4.65e3	4.79e3
	EOL	1.78e1	1.97e-10	5.00e-2	4.63e-3	6.20e-2	3.79e3	4.15e3
Design 3	MP	2.08e2	1.58e-7	5.22e-1	1.32e0	4.07e-1	3.68e4	3.72e4
	C	3.54e0	2.28e-11	4.40e-2	3.19e-3	6.10e-2	1.48e2	1.55e2
	PIS	2.09e5	3.77e-6	1.91e3	1.43e4	3.62e3	6.17e6	6.53e6
	R	2.27e1	3.43e-8	1.50e-1	8.44e-3	1.53e-1	4.65e3	4.79e3
	EOL	1.99e1	2.22e-10	6.00e-2	4.88e-3	6.30e-2	4.31e3	4.73e3

Note: MP = Material Production, C = Construction, PIS = Pavement in Service, R = Rehabilitation, EOL = End-of-Life, Climate change (Global warming potential) using kg of CO_2-eq.; Ozone layer depletion using kg of CFE 11-eq.; Acidification using kg of SO_2^--eq.; Eutrophication using kg of N-eq.; Photochemical oxidation using kg NMVOC Equiv.; Material depletion in kg; Waste in kg (Emissions to air, water, soil and deposited goods); SY = Square Yard.

sustainable design. Hence, we must consider both LCA and LCCA methods to achieve the best viable sustainable design. The software-based data entry plug-in interfaces for the five life-cycle phases developed here will allow the designers to achieve the best design with minimum effort, since the data required for the interfaces are typical pavement design parameters that are already known by the designers at the design stage.

5 CONCLUSIONS AND RECOMMENDATIONS

Pavement structures are highly material and energy intensive and as a result, their sustainability impacts are substantial. By making pavements more sustainable and environmentally friendly, a significant positive impact will benefit the society as a whole. There is a strong need to develop tools that will facilitate more accurate and practical assessment such that pavement designers can conduct LCA of pavement design alternatives and combine LCA results with results from conventional LCCA to select the optimum design alternative by considering both criteria. This paper presents software-based data entry plug-in interfaces for the five life-cycle phases of LCA of flexible pavements that were developed around a proprietary LCA software platform. The detailed unit process flow charts developed for each of the five phases of the flexible pavement life-cycle will allow the designers to assess the sustainability impacts of each component in each phase in the pavement life-cycle. This method will also allow the designers to change their designs to get the best-balanced design in terms of both LCA and LCCA criteria, where it will help the sustainable development of the future flexible pavement systems. The results of this study are encouraging and further work is currently underway in looking at incorporating novel material flexible pavement systems to this protocol.

ACKNOWLEDGEMENT

Authors would like to acknowledge the Texas Department of Transportation (TxDOT) for their support as sponsor of this research. Authors would also like to acknowledge the technical support received by Dr. Feri Afrinaldi and Mrs. Shahrima Maharubin during this study.

REFERENCES

Al-Rousan, T., Asi, I., & Baker, A. A. 2010. *Roughness evaluation of Jordan highway network*. In 24th ARRB Conference, Melbourne, Australia 41.
Athena. 2016. Athena Sustainable Materials Institute. http://www.athenasmi.org/our-software-data/pavement-lca/; visited: 11/26/2016.
AzariJafari, H., Yahia, A., & Amor, A. B. 2016. *Life cycle assessment of pavements: reviewing research challenges and opportunities*. Journal of Cleaner Production, 112, 2187–2197.
Chatti, K., & Zaabar, I. 2012. *Estimating the effects of pavement condition on vehicle operating costs* 720. Transportation Research Board.
Federal Highway Administration. 2014. *Life Cycle Assessment of Pavements*. Retrieved from http://www.fhwa.dot.gov/pavement/sustainability.
Gopalakrishnan, K., Steyn, W. J., & Harvey, J. 2014. *Climate Change, Energy, Sustainability and Pavements*. Green Energy and Technology. Springer. Heidelberg.
Hall, K., & Muñoz, C. 1999. *Estimation of present serviceability index from international roughness index*. Transportation Research Record: Journal of the Transportation Research Board, (1655), 93–99.
Harvey, J. T., Meijer, J., Ozer, H., Al-Qadi, I. L., Sabori, A. and Kendall, A. 2016. *Pavement Life Cycle Assessment Framework*. FHWA-HIF-16-014
Huang, Y., Bird, R., & Heidrich, O. 2009. *Development of a life cycle assessment tool for construction and maintenance of asphalt pavements*. Journal of Cleaner Production, 17(2), 283–296.
ISO 14001:2004(E). *Environmental management systems-requirements with guidance for use*
ISO 14040:2006(E). *Environmental management-life cycle assessment-principles and framework*
ISO 14044:2006(E). *Environmental management-life cycle assessment-requirements and guidelines*

Kang, S., Yang, R., Ozer, H., & Al-Qadi, I.L. 2014. *Life-Cycle Greenhouse Gases and Energy Consumption for Material and Construction Phases of Pavement with Traffic Delay*. Transportation Research Record: Journal of the Transportation Research Board, (2428), 27–34.

NCHRP. 1978. *Calibration of Response Type Road Roughness Measuring System*. D.C, U.S.A: Transportational Research Board.

Santero, N. J., Masanet, E., & Horvath, A. 2011a. *Life-cycle assessment of pavements Part I: Critical Review*, 55(9), 801–809.

Santero, N. J., Masanet, E., & Horvath, A. 2011b. *Life-cycle assessment of pavements Part II: Filling the research gaps*. Resources, Conservation and Recycling, 55(9), 810–818.

thinkstep. 2016. *GaBi Software*, v7.3 Leinfelden-Echterdingen, www.gabi-software.com

US EIA. 2016. United States Energy Information Administration. https://www.eia.gov/electricity/; visited: 06/25/2016

US EPA. 2006. *Life Cycle Assessment: Principle and Practice*. Cincinnati, OH: National Risk Management Research Laboratory and EPA. United States Environmental Protection Agency.

Vidal, R., Moliner, E., Martinez, G., & Rubio, M. C. 2013. *Life cycle assessment of hot mix asphalt and zeolite-based warm mix asphalt with reclaimed asphalt pavement*. Resources, Conservation and Recycling, 74, 101–114.

Willis, J. R., Robbins, M. M., & Thompson, M. 2015. *Effects of Pavement Properties on Vehicular Rolling Resistance*: A Literature Review.

Yang, R. Y. 2014. *Development of a Pavement Life Cycle Assessment Tool Utilizing Regional Data and Introducing an Asphalt Binder Model* (Doctoral dissertation, University of Illinois at Urbana-Champaign).

Yang, R., Ozer, H., Kang, S., & Al-Qadi, I.L. 2014. *Environmental Impacts of Producing Asphalt Mixtures with Varying Degrees of Recycled Asphalt Materials*. International Symposium on Pavement LCA 2014, Davis, CA, October 14–16, 2014.

Review and comparison of freely-available tools for pavement carbon footprinting in Europe

Davide Lo Presti & Giacomo D'Angelo
Nottingham Transportation Engineering Centre (NTEC), Faculty of Engineering, University of Nottingham, Nottingham, UK

ABSTRACT: This paper wants to raise awareness of LCA/CF tools that have been built within European projects in the last 5–6 years and that are freely-available on the internet: asPECT, ECORCE M and CARBON ROAD MAP. The tools will be compared in terms of their architecture and system boundaries considered, as well as on the basis of the results of a comparative carbon foot printing exercise of several high content reclaimed asphalt wearing courses. This will provide practitioners with clearer ideas on the benefits and limitations of the existing tools, will provide suggestions to perform the carbon footprinting analyses in Europe and through the comparison of the results will allow drawing some guidelines for the final user.

1 INTRODUCTION

Full process-based LCA is certainly the most advisable technique to assess the environmental impact of road components as well as road pavement services. Research and industry should cooperate to build tools that can allow an easy access to these techniques as well as a uniform interpretation of results. However, pavement LCA is still a very young subject and up-to-date still too many uncertainties and lack of primary data makes the full process LCA very difficult to undertake. Furthermore, in many cases, authorities charged with managing transportation infrastructure are concerned mainly with the impacts towards climate change, and are not as concerned with the additional information produced by an LCA. This may be driven by legislative actions that set goals towards the reduction of carbon emissions, such as when countries adopt the goals set forth in the Kyoto Protocol. In response to these goals, many countries also standardize the measurement of carbon emissions linked to the global warming process, such as the UK's PAS-2050 (Huang et al. 2013). Additionally, the process of calculating the Carbon Footprint (CF) is simplified from an LCA given that the reminder of the environmental impacts associated with LCA (e.g., acidification, eutrophication, etc.) do not need to be calculated in a carbon footprint analysis, leading to a much smaller required inventory. The CF may be seen as a proxy for environmental impacts, and thus displace the need for conducting a full LCA. However, the limitations of using the CF as a proxy for environmental sustainability was thoroughly evaluated in Laurent et al. (2012), and the authors found that limited conclusions can be drawn from carbon footprints regarding other pollutants. However, the authors note that in products, which are dominated by relatively few processes, the correlation between CF and other impact assessment results may become much stronger.

1.1 *Aim of the study*

This paper wants to raise awareness and analyze benefits and limitations of LCA/CF tools that have been built within European projects in the last 5–6 years and that are freely-available on the internet: asPECT (Wayman et al. 2014), ECORCE M (2014) and CARBON ROAD MAP (Spriensma et al. 2014). The study will perform a comparison of the tools in terms of their architecture and system boundaries considered, as well as comparing the results of environmental impact assessment exercise based on a CF study performed within another European project Allback2Pave (2016). This will provide engineers of Road Authorities and practitioners with an idea of which could be the more effective tool for performing environmental impact assessment of road pavement components (e.g. asphalt mixtures) in EU. A final comparison of the CF results will allow drawing some guidelines for the final user.

2 REVIEW OF THE LCA/CF TOOLS IN EUROPE

A recent comprehensive overview of the currently available pavement LCA and carbon footprinting tools allowing an estimate of the environmental impacts of road pavement technologies (i.e. asphalt mixtures) within their lifecycle, is provided by Spriensma et al. (2014). This section presents an overview of suggested tools that could be used from European Road Authorities to perform an environmental impact assessment exercise. These tools have the common characteristics of being:

- Freely available and accessible
- Based on full process LCA
- User-friendly and in any case accompanied by a user manual
- Able to perform at least a cradle-to-laid Carbon foot printing of road pavement technologies

Furthermore some of them also allow to:

- Perform a full pavement LCA, not only providing the carbon footprint
- Perform a cradle-to-grave analysis (up to end of life)
- Use references and databases developed in EU countries

Additionally, other more general, complex professional LCA tools can be used to assess/compare the environmental impact of road pavement technologies to be used in road pavements. These tools are not presented in this report and consist in softwares such as BEES, GaBi and SimaPro, which allows higher flexibility and possibly a more detailed estimation, but are not cost-free and need professional expertise.

2.1 *asPECT*

asPECT (asphalt Pavement Embodied Carbon Tool) is a pavement carbon footprint tool which was developed by the Transport Research Laboratory (TRL), and released in its current version in 2014 (Wayman et al 2014). The tool estimates CO_2e emissions from asphalt paving processes in a cradle to gate scenario, and has been designed to meet the specifications in the UK standard PAS 2050 (Huang et al 2014). Several thorough reviews have been conducted on the capabilities and limitations of asPECT, one of which is presented in Spriensma et al. (2014). One trade off when using asPECT that is noted in Spriensma et al. (2014) is the required input from the user. Other tools contain many default values within the database for processes, such as Hot Mix Asphalt (HMA) plant specifics, which limits the user input to material amounts, material types, construction processes and transportation types and distances. asPECT requires several user inputs, such as the characteristics of the HMA mixing plant (e.g., annual production values, the profile of the power consumed by the plant, etc.). This specificity of data has the benefit of increasing the reliability of the results, due to more accurate information, but has the drawback of considerably increasing the complexity over similar tools.

2.2 *ECORCE M*

ECORCE M (ECO-comparator applied to Road Construction and Maintenance) was developed by the French Institute of Science and Technology for Transport Development and Networks (IFSTTAR), and the international version was released in mid-2014 (ECORCE M. 2014). ECORCE M is a process based LCA tool with an extensive integrated default database used to complement the lifecycle inventory phase. The data used to populate the integrated database is drawn from multiple sources. However, in cases that an international consensus does not exist for data, ECORCE M draws upon research conducted in France to populate values. Inputs into ECORCE M include: volumetric data, equipment type (e.g. roller, paver, etc.), layer composition data, the type of mixing plant, and the transport distances and modes. Based on these inputs, several default values that are stored in the ECORCE M LCI database are used to populate an inventory of environmental data. The input data is then combined with inventory data to develop environmental impact values. The following indicators are presented as results from ECORCE M: Total Energy Consumption (MJ), Total Water Consumption (m^3), Greenhouse Effect (kg eq.CO_2), Acidification (kg eq.SO_2), Eutrophication (kg eq.PO_4), Tropospheric Ozone Formation (kg eq. ethylene), Ecotoxicity and Chronic Toxicity (both measured in terms of kg equivalent of 1,4 dichlorobenzene).

2.3 *Carbon Road Map (CEREAL)*

Royal Haskoning DHV BV, in a consortium with KOAC*NPC, The Netherlands and the Danish Road Directorate (DRD) in Denmark, initiated the research and development of a European Carbon footprinting tool for more sustainable road management and construction in the project called "CEREAL": CO_2 emission Reduction in road lifecycles (Spriensma et al. 2014). The consortium of CEREAL evaluated data of several popular national tools and finally created a tool, "Carbon Road Map", which concentrates on maintenance and rehabilitation of in service roads and was harmonized in the Western-European Countries that funded the project: Germany, Denmark, Ireland, The Netherlands, Norway, Sweden and The United Kingdom. When compared to asPECT, Carbon Road Map has the strong points of being as user-friendly but needs fewer data easily accessible by engineers of road authorities and it is able to include lifecycle maintenance strategies for road pavements allowing calculating a full lifecycle carbon footprint. The tool is programmed in Excel and already includes all CO_2 data for materials, transports and equipment for UK, Denmark and the Netherlands. Furthermore, the tool allows customization to other regions.

3 COMPARISON OF THE TOOLS: SYSTEM BOUNDARIES

The purpose of a pavement LCA is to quantify the total environmental impact of the pavement throughout the pavement's life, which is generally divided into the following five phases (Santero et al. 2011); (*1*) raw materials extraction and production, (*2*) construction, (*3*) use, (*4*) maintenance and (*5*) end of life. Ideally an LCA should be considered a cradle to grave analysis that accounts for the entire life of the materials, all the processes involved with the system, as well as other processes directly impacted by the system.

Nevertheless, a lack of information, such as the impact of infrastructure decisions on what are considered secondary systems, leads to a constraint on the system boundaries for a pavement. Thus, in the case of pavements, typical LCA boundaries are constricted to cover only the time period from material extraction through the end of the construction phase of the project also referred to as cradle to laid analysis. asPECT and ECORCE M are designed to perform CF/LCA considering this type of reduced lifecycle for road pavement which includes a cradle-to-laid + end-of-life scenario (Figure 1). These tools allow obtaining a much more detailed CF/LCA of new designed road pavement components, such as new asphalt mixtures, and a more accurate final outputs (e.g. $KgCO_2$/t of mix). However, this type of analysis can't be considered a comprehensive road pavement CF/LCA, because it does not allow taking in consideration the Use phase, which is a fundamental phase of the road pavement life cycle (Trupia et al 2016).

Furthermore, in order to use LCA/CF tools for decision-making, these must be more related to asset management rather than road pavement technologies. In other words, the CF of the pavement components (e.g. asphalt mixes) should be a mere input and the overall methodology should focus mainly on dealing with data such as road geometry, maintenance strategies, traffic, pavement conditions and statistical parameters to account for data changing over the

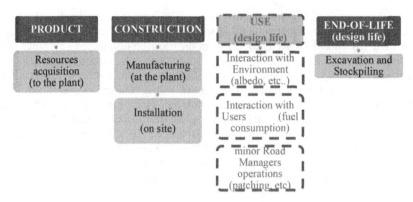

Figure 1. Proposed system boundaries for CF/LCA of road pavements components such as asphalt concrete. Use phase not included as in asPECT and ECORCE M.

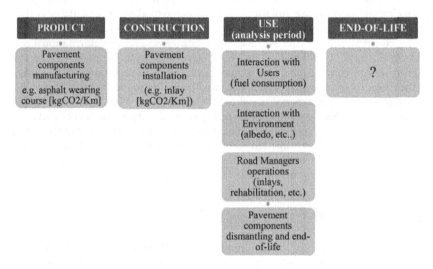

Figure 2. Proposed system boundaries for CF/LCA of maintenance of existing road pavement (e.g. CARBON ROAD MAP).

analysis period. An ideal tool to account for environmental impact within asset management should overcome these limitations. This would provide a methodology for decision-making which is composed by lifecycle stages that are different from those considered in the other two tools, so to allow accounting for the CF of maintenance of existing road pavements. This type of analysis is far more complex than the one related to only pavement components. Furthermore the standard on Sustainability of Construction Works EN 15804:2012, reports that the Use phase of LCA incorporates the maintenance, repair, replacement "…including provision and transport of all materials, products and related energy and water use, as well as waste processing up to the end-of-waste state or disposal of final residues during this part of the use stage". From this it can be interpreted that the Use phase is equal to the analysis period and should include pavement layers dismantling, replacement and also stockpiling, so that the End-of-life phase, as intended for pavement components as asphalt mixtures (Figure 2), might not occur or it occurs only when the whole road pavement is dismantled or changes functionality (this is the reason of the "?" on Figure 2). CARBON ROADMAP (CEREAL 2014) is built upon this philosophy and it was also conceived to be a tool able to account for CF of maintenance of existing road pavement at EU level.

4 COMPARISON OF THE TOOLS: CARBON FOOTPRINTING RESULTS

In order to compare the outcomes and user-friendliness of the selected CF/LCA software, a LCA study conducted within the AllBack2 Pave (AB2P) 2013 – 2015 project, funded by the Conference of European Road Directorate, was selected. More details of the LCA study will be published elsewhere and can be found on the project website (http://allback2pave.fehrl.org). The strategic aim of the AB2P project was to elevate asphalt recycling levels across Europe, on the basis that recycling surface into surface course would yield the greatest benefits, since this would best utilize the properties of the high grade aggregates used in the original surface course. This was funded on the premise that recycling asphalt into wearing courses was able to best utilize the preserved properties of the recycled material, both aggregate and binder. The Work Package (WP) 5 of the project aimed at providing results of the difference in the environmental performance over 60 years, between the currently maintained typical European major road asphalt pavements and scenarios in which the currently used asphalt mixtures are replaced by the AllBack2Pave (AB2P) asphalt mixes: eight asphalt mixtures for wearing course containing up to 90% Reclaimed Asphalt (RA). In order to take into account the European level of the project, three case studies were considered: Italy for South Europe, Germany for Central Europe and UK for North Europe.

4.1 CF baseline scenarios, maintenance strategies and general assumptions (LCI)

Each case study was crafted with the help of the interested Road Authorities and was intended to be representative of "typical" inter-urban roads of the selected countries (Table 1). Results from the other WPs of the project, direct data collection from the asphalt production plant in Germany and Italy and tailored questionnaires completed by the interested Road Authorities, were the sources of information on the key variables, such as the asphalt mixes recipes, energy and fuel consumption and transport distances. When needed, reports, standards and reputable data sources were utilized to provide emissions factors for fuels, transport and embodied carbon values for constituent materials. Gathered data including more detailed such as mix recipes, site locations and other information are presented elsewhere (AllBack2Pave 2016). Maintenance strategies for a 60 years analysis period and practices are reported in Table 2.

The philosophy behind the development of the AB2P technologies is to allow maintenance of existing roads through inlay treatments that re-use as much as possible of the milled material so to reduce RA stockpiling and downgrading to base or unbound layers, and avoid depletion of new natural resources. For this reason, in this exercise, in order to account for the potential lifecycle environmental benefits of the AB2P technologies, all the case studies will consider a similar maintenance strategy consisting of the repeated inlays of the wearing

Table 1. Asphalt pavement geometry, structures, traffic and durability.

	South EU—IT	Central EU—D	North EU—UK
Pavement course	(Anas 2015)	(Bast 2015)	–10
Section Width	9.50 m	11.80 m	11.00 m
Section Length	2000 m	800 m	720 m
Wearing	Asphalt 30 mm	Asphalt 30 mm	Asphalt 40 mm
Binder	Asphalt 40 mm	Asphalt 80 mm	Asphalt 50 mm
Base	Asphalt 100 mm	Asphalt 140 mm	Asphalt 100 mm
Foundation	Cement treated sand 300 mm	Unbound gravel +frost blanket 350 mm	Cement treated limestone 258 mm
Traffic levels	High Traffic	Medium Traffic	Low Traffic
Durability of wearing c.	5 years (AC)	16 years (SMA)	10 years (SMA)
Durability of binder c.	25–30 years		
Durability of base c.	50 years		

Table 2. Maintenance treatments, mix durability and assumptions.

Maintenance treatment	• Surface treatments with periodic inlay of wearing course and occasional inlay of binder and base course • Maintenance is undertaken in one carriageway (two lane), or one lane (single lane road) at a time, with the traffic diverted onto the other carriage/lane. • Workzones are extended for the whole length and the width of the full carriage • In the case studies with dual carriageway, maintenance event is considered only in one direction.
Materials	Current asphalt mixtures for each case study will be compared with the following asphalt mixes for wearing course and occasionally binder and base course: AB2P mixes technologies 1. AC16 30%RA+additives 4. SMA8S 30%RA 2. AC16 60%RA+additives 5. SMA8S 30%RA 3. AC16 90%RA+additives 6. SMA8S 60%RA+additives
Analysis period	60 years

courses, complemented with binder course inlays and rehabilitation over an analysis period of 60 years. Of course this will consider as a common assumption that the road pavement foundation and sub-layers will not deteriorate and the difference between the maintenance plans will be strictly linked to the estimated service lives of the wearing courses provided by primary data (ANAS 2015, BAST 2015; Spray 2014) or extrapolated from other similar investigation (EARN 2015; Re-Road 2013). Furthermore, for the purpose of modelling, it is assumed that the asphalt will only last for its full estimated lifetime with no under/over performance. The equipment, schedule and production rate are provided elsewhere (AllBack2 Pave 2016).

4.2 *Modeling carbon footprinting over 60 years*

The selected tools were all used to calculate the environmental performance of the AB2P technologies through a carbon foot printing exercise carried out for each case study. This exercise focuses mainly on comparing the CF generated from using the AB2P technologies in place of the currently used asphalt mixes over the entire analysis period of 60 years The total amount of tonnes of CO_2e over 60 years is the final outcome considered to evaluate each design alternative and is calculated as explained in Figure 3.

It has to be highlighted that whenever the maintenance intervention involves also binder and base courses, it is assumed that these asphalt mixes are the same of the considered wearing course. In fact, it looks not realistic that future interventions will make use of RA only in wearing courses. The obtained CF was then multiplied for the number of interventions included within the current maintenance scenarios of existing road sections in South, Central and North Europe (Table 3).

4.3 *asPECT 4.0*

asPECT is in the authors' opinion the most flexible and customizable free tool for road pavement components CF. Its main benefits come from its extreme flexibility: it allows inserting and/or changing almost all inputs, from the energy and resources consumed at the plant to the constants related to the grid electricity, fuels, transport, etc. and this increases the reliability of the results. asPECT allows implementing new design of the mixtures in the plant and understanding CF in each operation. Although requiring very detailed inputs, it also allows higher level of customization of CF.

At the same time this has the drawback of considerably increasing the complexity of the analysis over similar tools, especially for new users. However, other parts of the tool are not as detailed as those above discussed. In particular for the part concerning the laying and compaction on site, it would be more rigorous to import the specific CO_2e value from other

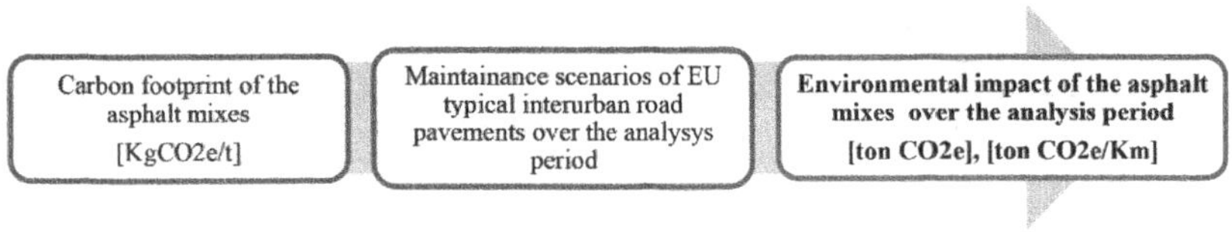

```
┌─────────────────┐   ┌─────────────────────┐   ┌─────────────────────────┐
│ Carbon footprint│   │ Maintainance scenarios│ │ Environmental impact of │
│ of the asphalt  │──▶│ of EU typical interurban│▶│ the asphalt mixes over │
│ mixes           │   │ road pavements over  │   │ the analysis period     │
│ [KgCO2e/t]      │   │ the analysys period  │   │ [ton CO2e], [ton CO2e/Km]│
└─────────────────┘   └─────────────────────┘   └─────────────────────────┘
```

Figure 3. Environmental impact assessment of the AB2P technologies.

Table 3. Number of interventions per case study over the 60 years.

	South EU—IT	Central EU—D	North EU—UK
Inlay Wearing Course	9	2	2
Inlay Wearing Course + Base Course	2	1	2
Rehabilitation	1	1	1
TOTAL	12	4	5

Table 4. Calculated total tonnes CO_2e footprints (and percentage of variation with respect to the Baseline) over 60 years for the all case studies with aspect.

Case study	South Europe: Italy		Central Europe: Germany		North Europe: England	
Baseline	2361	–	953	–	649	–
SMA IT-RA30add	2356	−0.20%	741	−22.30%	573	−11.70%
SMA IT-RA60add	2295	−2.80%	614	−35.50%	555	−14.50%
SMA IT-RA90add	2236	−5.30%	492	−48.40%	539	−17.00%
SMA D-RA30	2595	9.90%	822	−13.70%	656	1.10%
SMA D-RA60	2410	2.10%	670	−29.70%	598	−7.90%
SMA D-RA60add	2512	6.40%	697	−26.80%	628	−3.20%

sources, instead of using the default value. Furthermore, this tool does not carry out a complete LCA analysis but is limited to the carbon footprint, and does not take into account the Use phase. Despite its complexity and some limitations, asPECT comes with a very well explained manual so that it was possible to understand, replicate all the calculations and double check all the outputs. Results of the analysis performed by using asPECT are shown in Table 4.

4.4 ECORCE M

ECO-comparator applied to Road Construction and Maintenance is a full process, customizable road pavement LCA tool based on a database populated with data coming from researches conducted in France. The main inputs concern pavement volumetric data, transport distances and modes, and mixtures recipes. All the other data included in the database, are average data obtained from current manufacturing and maintenance operations in France and the tool does not allow modifying them. On the one hand, this makes the tool really easy to use for non-expert users, even without a manual. On the other hand, not having a quick access to the database's references makes it more difficult to fully understand results obtained; for the same reason, it is not possible to change CO_2e values of the mixture components or to add some specific element such as fibers, adhesive enhancers, emulsifiers, thickeners, fluxing agents, etc. Therefore, ECORCE M is very much user-friendly and it possibly needs the addition of a universal/open features allowing higher level of customization to expert users. However, despite these limitations, the tool provides benefits such as including the analysis of the earthworks and soil treatments, it can be used for comparison between maintenance operations and above all it carries out a full process LCA analysis, not restricted only to CF. In order to have a comparison with the other tools, Table 4 shows only the results CO_2e emissions for all case studies (Baselines and AB2P mixtures).

Table 5. Calculated total tonnes CO$_2$e footprints (and percentage of variation with respect to the Baseline) over 60 years for the all case studies with ECORCE M.

Case study	South Europe: Italy		Central Europe: Germany		North Europe: England	
Baseline	1492	–	739	–	502	–
SMA IT-RA30add	1400	–6.1%	552	–25.4%	418	–16.6%
SMA IT-RA60add	1310	–12.2%	448	–39.4%	391	–22.0%
SMA IT-RA90add	1250	–16.2%	356	–51.8%	375	–25.3%
SMA D-RA30	1638	9.8%	632	–14.5%	501	–0.1%
SMA D-RA60	1509	1.2%	525	–29.0%	462	–8.0%
SMA D-RA60add	1509	1.2%	525	–29.0%	462	–8.0%

From Table 4 and Table 5 it can be noticed that, even though absolute values obtained from ECORCE M are lower than asPECT's ones, the trend line and ratios between scenarios are very close to each other, especially for the German and English case studies. ECORCE M shows always lower absolute values of CO$_2$e. However, it has to be highlighted that the authors were not able to input additives and fibers present in the mixes, so that results from the two software can't be fully comparable. In conclusion, ECORCE M doesn't allow great level of customization, but it is very much user-friendly, provides similar results to asPECT, in terms of CF, and it is the only freely available tool (over those analyzed) that allows performing a full process LCA of road pavement in both scenarios: new construction and maintenance of existing assets.

4.5 CARBON ROAD MAP

CARBON ROAD MAP allows estimation of the CF of maintenance operations of existing road pavement at European level. Its database is filled with information acquired from West-European Countries (Netherlands, Denmark and United Kingdom); however the software allows using an "expert mode" with which it is possible to customize the database through Excel. Its strength lies in the few amount of data required that makes it a user-friendly tool, especially for inputs concerning maintenance strategies. Furthermore, as explained before, this tool is the only one that includes the possibility of considering the entire life cycle analysis period (e.g. 60 years), including maintenance strategies and traffic change. The outputs, also in form of graphs, allow comparing the total amount of CO$_2$e obtained by considering different maintenance strategies in different countries. Despite the concept and architecture of the tool are remarkable, unfortunately the tool is not recommendable because the copy received from the authors of the CEREAL project, is not free from bugs and this makes the software not stable and therefore not recommendable. Furthermore, the tool shows also many limitations and drawbacks:

- The User manual is not exhaustive,
- Users can't change the specific CO$_2$e inventory values,
- It is not possible to define precise length of the road sections (1 km and multiples).
- Users should access the database to enter CO$_2$e values of customized pavement components and these that would need to be calculated separately (for instance using asPECT).

For these reasons, in order to carry out the analysis, as performed with the other two tools, authors needed to first modify the database by entering in the expert mode and introducing the kgCO$_2$e/t values of each asphalt mixture as calculated with asPECT. The software then requires details of project definition (case study details and analysis period), construction data (pavement structure, design life and traffic data), maintenance strategies and road dimensions. Authors created one file for each alternative and run the software to obtain the summary of the total CO$_2$e emissions, also grouped per type of maintenance intervention. Results obtained were not satisfactory as shown in Table 6. In fact, it can be noticed that the total tonnes of CO$_2$e are of one order of magnitude bigger and also the proportions of CO$_2$e values between operations are really different from those obtained from asPECT and ECORCE M; in particular Use of equipment and Production materials look over-propor-

Table 6. Calculated total tonnes CO_2e footprints (and percentage of variation with respect to the Baseline) over 60 years for the all case studies with Carbon Road Map.

Case study	South Europe: Italy		Central Europe: Germany		North Europe: England	
Baseline	1206654	–	191764	–	404942	–
SMA IT-RA30add	1206643	–0.001%	191621	–0.074%	404856	–0.021%
SMA IT-RA60add	1206561	–0.008%	191534	–0.120%	404836	–0.026%
SMA IT-RA90add	1206481	–0.014%	191453	–0.162%	404817	–0.031%
SMA D-RA30	1206968	0.026%	191677	–0.045%	404950	0.002%
SMA D-RA60	1206716	0.005%	191571	–0.100%	404884	–0.014%
SMA D-RA60add	1206854	0.017%	191590	–0.091%	404919	–0.006%

Table 7. Key information of LCA/CF tools analyzed.

	asPECT	ECORCE M	Carbon Road Map
Full LCA	Only CF	Yes	Yes
Materials and processes LCI	Tool library (asPECT manual)	Tool library (ECORCE M manual)	Tool library (www.cereal.dk)
Customization	High	Medium	High
User-friendly	Medium	High	High
Entire life-cycle period within the tool	No	No	Yes
Reliability	High	High	Low

tioned relative to the Transport. As a result, CARBON ROAD MAP needs to be "labeled" as well structured and very promising tool, but under development. Its use is still not recommended due to several software bugs and above all, results are not comparable with those obtained with the other tools and with other researches in literature.

5 COMPARISON OF THE TOOLS: SUMMARY

Freely-available road pavement specific LCA/CF tools built in Europe in the last 4–5 years were presented. Each tool has its own benefits and limitations, however results obtained for specific AB2P case studies were very different (Tables 4,5,6). It can be quickly noticed that, results coming from Carbon Road Map are over-estimated compared to the other tools and this leads to the suspicion that there are some mistake in the tool database. On the other hand, results obtained from asPECT and ECORCE M are comparable and they lead to similar considerations and recommendations. The only issue of using ECORCE M in the analyzed case studies is restricted to the limitation of specifying the incorporation of fibers and additives within asphalt mixtures. Nevertheless, this can't justify the difference in results considering that from the results obtained with aspect, the emissions due to fibers account for less than 1% of the total. Table 7 provides a summary of all features including pro and cons of the tools compared in the present study. This information can be useful to select the tool that best fit interest of practitioners or decision makers while helping further developments of such tools within European framework. LCI for material and processes include the embedded value and is available within library of each software.

6 CONCLUSIONS

Based on this investigation the following recommendations can be drawn out for practitioners:

- For its flexibility, asPECT is the best tool for performing cradle-to-grave carbon footprinting analyses of road pavement components in EU. If maintenance strategies and end-to-life phase are needed, then system boundaries can be widened as explained in Figure 3.

- ECORCE M is the most user-friendly tool and allows performing a full LCA, which takes into account several environmental impact indicators. Also in this case, maintenance and end-of-life can be added by widening the system boundaries.
- In terms of system boundaries CARBON ROADMAP is the tool better structured to perform decision-making based on environmental impact within pavement asset management. However, despite the tool is well structured, it is not reliable yet and would need further development.

Overall, this paper provides a comprehensive comparison between three free-available road pavement LCA/CF tools. This analysis was able to highlight advantages and disadvantages of the considered tools and it also highlighted the issue of benchmarking these LCA/CF studies at the countries level inside Europe. As a result, the authors believe that any other attempt to build continental-wide LCA tools should first of all pay attention on the definition of the tool architecture so to include appropriate system boundaries. Furthermore, enhancing cooperation between the involved countries could allow decreasing issues related to geographical benchmarking by the definition of common databases and impact assessment methodologies. Such step would be fundamental to encourage a greater deployment of LCA within the road engineering and asset management industries.

ACKNOWLEDGMENTS

The research presented in this paper was carried out as part of the CEDR Transnational Road research Programme Call 2012. The funding for the research is provided by the national road administrations of Denmark, Finland, Germany, Ireland, Netherlands and Norway.

REFERENCES

ALLBACK2PAVE 2016 WP5 deliverables D5.1; D5.2; D5.3 – http://allback2pave.fehrl.org
ANAS. Highway Maintainance group reposnsible, Carlo Piraino Palermo, (July 2015).
BASt. Section S3 – Asphalt Pavements—Federal Highway Research Institute (BASt) responsible, Oliver Ripke (July 2015).
EARN D5. Wayman M LDCS. CEDR; 2014.
ECORCE M. Jullien A, Dauvergne M. IFSTTAR; 2014.
Huang Y, Spray A, Parry T. Sensitivity analysis of methodological choices in road pavement LCA. International Journal of Life Cycle Assessment. 2013;18:93–101.
Laurent A, Olsen S, Hauschild M. Limitations of Carbon Footprinting as Indicator of Environmental Sustainability. Environmental Science and Technology. 2012;46:4100–4108.
Re-Road.: End of Life Strategies of Asphalt Pavements. [Internet]. 2012 Available from: http://re-road.fehrl.org/.
Santero NJ. Pavements and the Environment: A Life-Cycle Assessment Approach, dissertation, 2009. Berkeley, CA: University of California; 2009.
Spray A. Global Warming Potential Assessment. Methodology and Road Pavements. PhD Thesis. University of Nottingham; 2014.
Spriensma, R., Gurp, C. v., & Larsen, M.R. (2014). CEREAL: Final Report and User Guide. The Netherlands: ERA-NET ROAD.
Trupia L., Parry T., Neve L. Lo Presti D. Rolling resistance contribution to a road pavement life cycle carbon footprint analysis. The International Journal of Life Cycle Assessment http://link.springer.com/article/10.1007/s11367-016-1203-9 - October 2016.
Wayman M, Schiavi-Mellor I, Cordell B. Protocol for the calculation of whole life cycle greenhouse gas emissions generated by asphalt. Berks, UK: Transport Research Laboratory; 2014.

Route level analysis of road pavement surface condition and truck fleet fuel consumption

Federico Perrotta, Laura Trupia, Tony Parry & Luis C. Neves
Nottingham Transportation Engineering Centre, Faculty of Engineering, University of Nottingham, Nottingham, UK

ABSTRACT: Experimental studies have estimated the impact of road surface conditions on vehicle fuel consumption to be up to 5% (Beuving et al., 2004). Similar results have been published by Zaabar and Chatti (2010). However, this was established testing a limited number of vehicles under carefully controlled conditions including, for example, steady speed or coast down and no gradient, amongst others. This paper describes a new "Big Data" approach to validate these estimates at truck fleet and route level, for a motorway in the UK. Modern trucks are fitted with many sensors, used to inform truck fleet managers about vehicle operation including fuel consumption. The same measurements together with data regarding pavement conditions can be used to assess the impact of road surface conditions on fuel economy. They are field data collected for thousands of trucks every day, year on year, across the entire network in the UK. This paper describes the data analysis developed and the initial results on the impact of road surface condition on fuel consumption for journeys of 157 trucks over 42.6 km of motorway, over a time period of one year. Validation of the relationship between road pavement surface condition and vehicle fuel consumption will increase confidence in results of LCA analyses including the use phase.

1 INTRODUCTION

In many countries, environmental questions have become an important part of the decision-making process for design and maintenance of highways (Beuving et al., 2004). Therefore, fuel efficiency and limiting Greenhouse Gas (GHG) emissions of road pavements has become a central focus of many projects and studies all over the world. Life-Cycle Assessment (LCA) aims at evaluating the impacts associated with all stages of a product's life. This method has been used to estimate the long-term impact of pavements on the environment (Santero et al., 2011), but different methodologies can lead to different conclusions (Trupia et al., 2016). This can be because different studies consider different phases of the life of a pavement in their analyses (Santero and Horvath, 2009, Trupia et al., 2016), because of inadequate information available, or because of different models for estimating the effect of road pavement conditions on vehicle fuel consumption (Zaabar and Chatti, 2010).

In the past many pavement LCA studies omitted the Pavement Vehicle Interaction (PVI) and its effects on vehicle fuel economy. PVI represents the impact of the interactions between pavements and vehicles during the use phase of a road. Although the energy losses due to the PVI can be mainly tracked to the tyre properties, the characteristics of the pavement surface, in terms of roughness and macrotexture, can also significantly affect the rolling resistance and therefore the vehicle fuel consumption (Sandberg et al., 2011).

This area of study is particularly interesting for pavement engineering and road agencies, because of the opportunity to reduce the fuel consumption associated with the road surface condition through conventional maintenance strategies. Pavement condition improvements, based on the reduction of rolling resistance, by controlling pavement roughness and texture depth, can be made rapidly using available technology, and have the potential of generating relevant energy and cost savings, and reductions in GHG emissions. By contrast, approaches involving improvements in vehicle technology or traffic reductions can be more complicated and require longer implementation periods.

Calculating the impact of pavement surface properties on the rolling resistance and then on vehicle fuel consumption is complex, although some studies have been performed over the last years (Wang et al., 2012a; Wang et al., 2012b; Hammarström et al., 2012) to analyze this component and its impact in a pavement LCA. These studies, as well as showing the relevance of the impact of pavement surface condition on a road LCA, have developed and implemented some models correlating pavement surface properties to vehicle fuel consumption.

A recent study (Trupia et al., 2016 (in press)) has analyzed the implications of using different rolling resistance models calibrated in different geographical locations for a UK case study. They concluded that some methodological choices and site-specific elements can play a significant role in the development of these models, producing rolling resistance and fuel consumption models that are not suitable for all geographic locations. In addition, the LCA studies are sensitive to the chosen model and can generate significantly different findings, reducing confidence in their use for LCA studies. For UK roads, there are not yet any general rolling resistance and fuel consumption models, able to predict the relationship between pavement surface properties and fuel economy, based on local conditions. Further research is needed in this area before introducing this component in the pavement LCA framework with confidence.

Recent studies assessed that road surface conditions account for approximately 5% of the total fuel consumption of road vehicles (Beuving et al., 2004 and Zaabar and Chatti, 2010 In England, the 2% reduction in fuel consumption assumed by Zaabar and Chatti (2010), would mean a saving of up to £1 billion (considering the current cost of fuel) which corresponds to a quarter of the funding spent in maintenance of local highways (House of Commons, 2011).

Zaabar and Chatti (2010) calibrated their model for US conditions, using a limited number of vehicles tested under carefully controlled conditions (e.g. steady speed) along selected road segments (with selected geometry). This can reduce the range of validity of the study first to the US and second to specific vehicle models and road conditions.

Other studies on the topic (Hammarström et al., 2012) used coast-down measurements (in order to exclude the impact of road gradient on the fuel consumption measurements) or cruise control (no change in direction and vehicle speed) performing the tests only in good weather conditions (e.g. wind speed <4m/s). This controls the variables in an experimental method to improve the repeatability of results but does not reflect what happens at route level under real driving conditions.

Nowadays, truck fleet managers analyse fleet performance to reduce vehicle operating costs by: 1) training drivers and 2) maintaining vehicles. Data are collected by sensors that are installed on trucks as standard (SAE International, 2002), and measure the vehicle fuel usage, the vehicle speed, its direction and position, the engine performance, among many other parameters. This data is collected constantly during vehicle use. Road agencies monitor road surface condition for decision making about pavement maintenance. Data are usually collected on an annual basis, including measurements of the pavement surface and structural strength, etc.

Using these data, it may be possible to validate the results of experimental studies for specific routes and vehicle types during operation. This is important because for instance, the truck fleet in the US is different of England. Different payloads are allowed, different tyres are used, different engines are installed, and different speed limits are set. As assessed by many different authors in the past from Sandberg (1990) to Zaabar and Chatti (2010), the impact that these variables can have on fuel economy is much larger than that of pavement surface conditions. Using truck fleet data may allow fuel efficiency models to be calibrated for different geographical areas and fleet composition.

In this paper we report the results of an initial study to test the feasibility of using truck fleet data for a motorway in England, to establish the impact of pavement surface texture depth and roughness on truck fuel efficiency. The results are compared to those of previous studies and discussed in terms of LCA and its role in decision making for road maintenance strategies.

2 DATA

The truck data is recorded every time an 'event' is triggered at any brake, stop, anomaly, or routinely each 2 minutes (120 s) or 2 miles (~3,219 m). From hundreds of gigabytes of data that each single vehicle's performance database contains, this study considers:

- the vehicle profile, containing its main characteristics,
- the tracker ID reference for the system of sensors installed,
- the geographical position of the truck, (5 m GPS precision),
- the distance travelled by the vehicle since the previous event (m),
- the time spent by the vehicle to travel to the current position from the previous event (s),
- the total fuel consumed until the current event is recorded (0.001 litres precision, rounded to 0.1 litres for the purpose of reducing database size),
- the air temperature (0.1°C precision),
- the current gear,
- the current engine torque percentage,
- the engine revolutions (revs/mins).

The Highway Asset Performance Management System (HAPMS) is the database owned and used by Highways England to monitor the condition of the strategic road network in England. The database contains historical information and condition measurements, including:

- a road identifier code,
- a direction code,
- the year of construction,
- the latest date of significant maintenance,
- the construction materials,
- roughness measurements,
- texture measurement,
- skid resistance measurement,
- deflection measurement (not considered in this study).

In this initial study only information about road surface condition is taken from HAPMS. This includes measurements of roughness (Longitudinal Profile Variance, LPV, at 3 and 10 metres in mm^2), texture (Sensor-Measured Texture Depth (SMTD) in mm), skid resistance (SCRIM Coefficient), and road gradient (0.1% resolution). These are the most common parameters used in England for road condition monitoring. Previous studies typically used IRI (International Roughness Index) as a measurement of roughness and MPD (Mean Profile Depth) as a measurement of texture. Benbow et al. (2006) and Viner et al. (2006) established that these roughness and texture parameters are closely related.

3 METHOD

In this initial study, data from trucks driven on M18 (near Doncaster, England) in 2015 where considered. The M18 motorway has been chosen because of its wide range of pavement surfaces, including asphalt and concrete. To the initial 910,591 records available the following filters were applied:

- 3-axle tractor and 3-axel trailer articulated trucks with Euro 5 or euro 6 engine,
- measurements recorded at the default time or distance (i.e. no other driving event (e.g. harsh braking or cornering) triggered the record),
- speed is steady and set to an average of 85 km/h (±2.5 km/h),
- no gear change (gear 12, the most commonly used gear at 85 km/h was selected).

These filters were applied in this initial study to reduce the effect of other variables and isolate the effect of pavement surface condition on truck fleet fuel economy. Data from 157 articulated trucks driving along the M18 in 2015 remained to be considered in this study. In all, 3677 records are available.

Of these, 1707 data points are for articulated trucks equipped with 12,419cc, Euro 6 engine and 1970 data points are available for the same type of vehicles but with euro 5 engine.

Based on the literature review the payload and the gradient are two of the most influential variables on fuel consumption (Sandberg, 1990, Beuving et al., 2004). However, no data about the payload is currently available for this study. Therefore, the generated engine torque (as percentage of the maximum) is used instead.

Engine torque percentage, the road gradient, LPV at 3 and 10 metres wavelength and SMTD texture measurements are considered. Performing a backward analysis based on the Aikake Information Criterion (AIC (Aikake, 1973)), a predictive model for fuel consumption (l/100 km) has been generated. Among all the possible models, the one that shows the lowest AIC coefficient is considered. Finally the two subsets of data for the two engine types, their distribution, and the generated models are compared.

4 RESULTS

The two datasets (Euro 5 and Euro 6 engine) show very similar mean fuel consumption and standard deviation (see Table 1). However, the fuel consumption data is multi-modal, which probably reflects the resolution of the recorded values. A Kolmogorov-Smirnov two-sample test excludes the hypothesis that the two datasets come from the same population. In fact, different shapes characterize the two distributions of data.

Separate backward AIC analyses have been performed on the two datasets. It is possible to see that in both cases the generated models include only the Profile Variance (LPV) at 10 metres and the texture (SMTD) in the model. Therefore, of those included, they may be identified as the most impactful road condition measurements on fuel consumption. This is a confirmation of previous studies (e.g. Sandberg, 1990, Beuving et al., 2004, Zaabar and Chatti, 2010).

For the Euro 5 dataset the following equation has been obtained:

$$FC = 19.18 + 0.066T\% + 6.85g\% + 1.91LPV10 + 2.77t \qquad (1)$$

where, FC = predicted fuel consumption [l/100 km]; $T\%$ = engine torque percentage [%]; $g\%$ = road gradient [%]; $LPV10$ = Longitudinal Profile Variance at 10 m wavelength [mm^2]; t = texture depth [mm].

In this case, the correlation coefficient (r) between the predicted and the measured fuel consumption is 0.54 (Figure 1).

For the Euro 6 dataset the following model has been generated:

$$FC = 17.17 + 0.067T\% + 7.57g\% + 1.40LPV10 + 3.10t \qquad (2)$$

The correlation coefficient (r) between the predicted and the measured fuel consumption is 0.66 (Figure 2).

Table 1. Average and standard deviation for fuel consumption data of Euro 5 and Euro 6 articulated trucks driven on M18 at 85 km/h using gear 12.

Vehicles	Avg fuel consumption (l/100 km)	Standard deviation (σ)
Euro 5	28.93	9.58
Euro 6	28.66	9.47

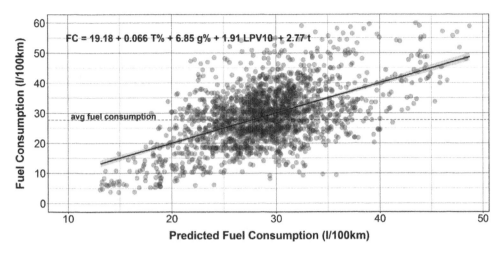

Figure 1. Comparison between predicted value and fuel consumption measurements for articulated trucks, equipped with 12419cc engines Euro 5 driven at 85 km/h using gear 12 only.

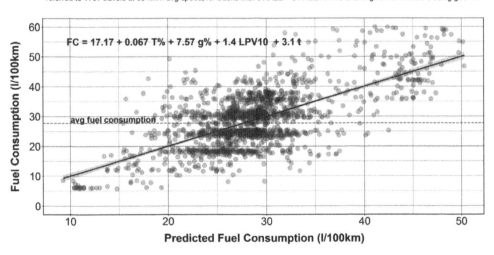

Figure 2. Comparison between predicted value and fuel consumption measurements for articulated trucks, equipped with 12419cc engines Euro 6 driven at 85 km/h using gear 12 only.

5 DISCUSSION

The two generated models (equations (1) and (2)) show similar results. The model for the Euro 5 trucks shows that 4.5% of the fuel consumption depends on the level of roughness along the M18. On the other hand 4.1% of fuel consumption depends on roughness along the M18 for the considered Euro 6 trucks. These first results agree with those found in previous studies (Beuving et al., 2004, Zaabar and Chatti, 2010) and this gives us confidence in the using a Big Data approach.

While these two initial models have identified the same pavement surface condition measurements as the most significant from those tested, the correlation coefficients between the measured and predicted fuel consumption show the models are far from explaining the variability in the data. This will in part be due to the influence of unmeasured factors, including for instance, wind speed and direction and other meteorological factors. It is also probably due to the lack of a direct measure of payload and the use of engine torque as a surrogate.

Further work will attempt to address this factor by estimating the payload using a physical-mechanical model. It may be possible to validate such a model for a limited number of trucks using fleet manager's information about delivery schedules or with limited trials of trucks with measured payloads. It may also be possible to record more frequent and accurate fuel consumption data for a limited number of trucks. It will also be necessary to test non-linear models. It is hoped that these approaches will provide further confidence in the use of this approach to estimating PVI fuel consumption.

The initial results obtained in this paper demonstrate the feasibility of using the Big Data approach to make a fleet and route level analysis to estimate fuel consumption due to PVI. The further work that is planned is intended to generate estimates that will be suitable to be introduced into pavement LCA studies, including the use phase, for UK case studies. A similar approach should be possible in other countries. Assessing this impact at a more general level using a Big Data approach will represent an important step in the development of consistent and accurate PVI fuel consumption models. Further research, to extend this approach to all types of vehicles is necessary to improve the confidence in introducing the PVI component into road pavement LCA studies.

ACKNOWLEDGMENTS

The authors would like to thank all the partners involved in the project for their help and support in data collection, data analysis and in the interpretation of these first results. These are Alex Tam of Highways England for giving us permission to use data from HAPMS, Emma Benbow, David Peeling and Helen Viner from TRL Ltd for their help in the data analysis and interpretation of results, and Mohammad Mesgarpour and Ian Dickinson from Microlise Ltd for allowing us to use an anonymized part of their truck telematics database and for their support in this initial part of the research.

 This project has received funding from the European Union Horizon 2020 research and innovation programme under the Marie Skłodowska-Curie grant agreement No. 642453 and it is part of the Training in Reducing Uncertainty in Structural Safety project (TRUSS Innovative Training Network, www.trussitn.eu).

REFERENCES

Akaike H, 1973. Information theory and an extension of the maximum likelihood principle. In Second International Symposium on Information Theory, ed. B. N. Petrov and F. Csaki, 267–281. Budapest: Akailseoniai–Kiudo.
Benbow E, Nesnas K, and Wright A, 2006. Shape (surface form) on Local Roads. Report PPR131. TRL Ltd, UK.
Beuving E, De Jonghe T, Goos D, Lindhal T, and Stawiarski A, 2004. Environmental Impacts and Fuel Efficiency of Road Pavements. Industry Report. Eurobitume & EAPA Brussels.
Chatti K and Zaabar I, 2012. Estimating the Effects of Pavement Condition on Vehicle Operating Costs, National Cooperative Highway Research Program, Report nr 720. Washington, DC.
Du Plessis HW, Visser AT, and Curtayne PC 1990. Fuel Consumption of Vehicles as Affected by Road-Surface Characteristics. ASTM STP 1031, pp. 480–496.
Haider M, Conter M, and Glaeser KP 2011. Discussion paper what are rolling resistance and other influencing parameters on energy consumption in road transport, Models for Rolling Resistance in Road Infrastructure Asset Management Systems (MIRIAM), AIT, Austria.

Hammarström U, Eriksson J, Karlsson R, Yahya MR, 2012. Rolling resistance model, fuel consumption model and the traffic energy saving potential from changed road surface conditions. VTI rapport 748A, Swedish National Road and Transport Research Institute (VTI), Linköping, Sweden.

House of Commons, 2011. House of Commons Committee of Public Accounts Departmental Business Planning.

Laganier R and Lucas J 1990. The Influence of Pavement Evenness and Macrotexture on Fuel Consumption. ASTM STP 1031 pp. 454–459, USA 1990. SAE International, 2002. Surface Vehicle Recommended Practice, 4970, pp. 724–776.

SAE International 2002. Vehicle Application Layer—J1939-71 — Surface Vehicle Recommended Practice Rev. Aug. 2002.

Sandberg, USI. 1990. Road Macro- and Megatexture Influence on Fuel Consumption. ASTM STP 1031, pp. 460–479.

Sandberg U, Bergiers A, Ejsmont JA, Goubert L, Karlsson R, Zöller M 2011. Road surface influence on tyre/road rolling resistance. Deliverable 4 in Sub-phase 1 of project MIRIAM. Download from http://www.miriam-co2.net/Publications/MIRIAM_SP1_Road-Surf-Infl_Report%20111231.pdf. Accessed 18 July 2016.

Santero NJ, Harvey J, and Horvath A, 2011. Environmental policy for long-life pavements. Transportation Research Part D: Transport and Environment, 16(2), pp. 129–136.

Santero NJ, and Horvath A, 2009. Global warming potential of pavements. Environ. Res. Lett. 4 (2009) 034011, Environmental Research Letters, IOP Publishing, doi:10.1088/1748-9326/4/3/034011. Accessed 08 December 2015.

Trupia L, Parry T, Neves LC, and Lo Presti D. 2016 (in press). "Rolling Resistance Contribution To A Road Pavement Life Cycle Carbon Footprint Analysis", The International Journal of Life Cycle Assessment Ms. No. JLCA-D-16-00118R1.

Viner H, Abbott P, Dunford A, Dhillon N, Parsley L, and Read C, 2006. Surface Texture Measurements on Local Roads. Report PPR148. TRL Ltd, UK.

Wang T, Lee I-S, Harvey J, Kendall A, Lee EB, Kim C 2012a. UCPRC Life Cycle Assessment Methodology and Initial Case Studies for Energy Consumption and GHG Emissions for Pavement Preservation Treatments with Different Rolling Resistance. UCPRC-RR-2012-02. University of California Pavement Research Center, Davis and Berkeley.

Wang T, Lee I-S, Kendall A, Harvey J, Lee E-B, Kim C 2012b. Life cycle energy consumption and GHG emission from pavement rehabilitation with different rolling resistance. Journal of Cleaner Production 33:86–96.

Zaabar I and Chatti K, 2010. Calibration of HDM-4 models for estimating the effect of pavement roughness on fuel consumption for U. S. conditions. Transportation Research Record, (2155), pp. 105–116.

Zaniewski JP, 1989. Effect of Pavement Surface Type on Fuel Consumption. Portland Cement Association—Research & Development Information. Special report. FHWA.

Pavement Life-Cycle Assessment – Al-Qadi, Ozer & Harvey (Eds)
© 2017 Taylor & Francis Group, London, ISBN 978-1-138-06605-2

Impact of PCC pavement structural response on rolling resistance and vehicle fuel economy

D. Balzarini, I. Zaabar & K. Chatti
Michigan State University, East Lansing, Michigan, USA

ABSTRACT: Reduction in vehicle fuel consumption is one of the main benefits considered in technical and economic evaluations of road improvements considering its significance. Surface roughness, texture, and structural response are the main pavement characteristics influencing rolling resistance. This project investigates the increase in vehicle energy consumption caused by the structural response of a cement concrete pavement. The rotation and deformation of the slabs on a viscoelastic subgrade, which is represented as a damped Winkler foundation, cause the vehicle to consume additional energy to overcome the slope formed by the local deflection basin. The structural rolling resistance is calculated on sections with different mechanical characteristics at different speeds, temperature, loading conditions and subsequently converted into fuel consumption excess. The maximum deflection-induced energy consumption is about 0.1% of the total consumption for articulated trucks. Note that the effects of curling and Load Transfer Efficiency (LTE) below 100% were not considered in this study.

1 INTRODUCTION

The theme of sustainable development is of primary importance today. The chance to reduce fuel consumption required by vehicles for their operation and thus to lower energy prices has its effect on consumers, businesses and environment.

To reduce the environmental impact of mobility on the road, one of the objectives is to reduce the energy consumption of vehicles. For this purpose, the EU and the US have set binding emission targets for new fleets of cars and vans, leading all major automobile manufacturers to compete with massive investment in technological development and innovation in the energy sector to reduce emissions throughout the life cycle of the vehicle. However, the search for higher performance engines with lower fuel consumption is not the only way forward. Many factors, beyond the thermodynamic efficiency of the engine, have an impact on the fuel consumption of a vehicle. The main ones are the air resistance and the rolling resistance.

Rolling resistance is a force acting in the direction opposite to that of motion, during the rolling of the tire on the road pavement. It includes mechanical energy losses due to aerodynamic drag associated with rolling, friction between the tire and road and between the tire and rim, and energy losses taking place within the structure of the tire.

The characteristics of the tire that affect the rolling resistance the most are the tire material, shape, width and the inflation pressure. It has been proven that the rolling resistance at the pavement—wheel interface is also significantly affected by the surface of the pavement.

A report published by Eapa/Eurobitume Task Group (2004) states that different textures of road surfaces influence fuel consumption by up to 10%. The NCHRP 720 study (Chatti & Zaabar 2012) proved from field data that different surface characteristics provide a major contribution to the rolling resistance. Since 2012 more studies have been conducted on the Structural Rolling Resistance (SRR), which can be estimated using two methods:

– Considering the stress-strain history, as the energy dissipated in the hysteresis loop of the viscoelastic material, in a finite volume of pavement (Coleri et al. 2016, Pouget et al. 2012, Shakiba et al. 2016).

- Considering the deflection basin, in terms of the energy required for a rolling wheel to move uphill, facing a positive slope caused by the delayed deformation of the viscoelastic pavement. (Louhghalam et al. 2013, Chupin et al. 2013).

All of these studies assumed the asphalt pavement to be a homogeneous continuous viscoelastic medium, which is a valid hypothesis for asphalt pavements but can't be applied to rigid pavements, due to the discontinuities caused by the joints. The study carried out in this paper investigates the effect of the SRR on rigid pavements, namely the increase of vehicle energy consumption induced by the pavement structural response due to the deformation of subgrade materials and the rotation of the concrete slabs under passing vehicles. This study is part of a research that aims to quantify rolling resistance due only to the structure of the pavement. In parallel with the study shown in the paper, different models are being developed to evaluate the SRR on flexible pavement. The findings of these studies will finally be compared and checked versus field measurements to establish if different types of pavement (rigid or flexible) or different pavement structures could lead to a change in rolling resistance and therefore in fuel consumption. The approach used to estimate the fuel consumption excess of a vehicle caused by the SSR is in three folds:

1. Compute the concrete pavement response due to a moving vehicle of three types of vehicles (car, SUV, truck) at different speeds and positions on the slab using a finite element solution (DYNASLAB).
2. Calculate the energy dissipated in the pavement, which is equal to the energy needed by the vehicle to overcome the additional traction forces caused by the pavement's deformation.
3. Estimate the fuel consumption excess due to such energy dissipation.

2 CONCRETE PAVEMENT RESPONSE

The calculation of the response of the pavement is performed using the 2D finite element software DYNASLAB (Chatti 1992). The program can analyze pavements with one or two layers resting on a damped frequency-dependent Winkler foundation, modeled by uniformly distributed springs and dashpots. The concrete slab is modeled by rectangular medium-thick plate elements. Each node contains three degrees of freedom: a vertical translation in the s-direction and two rotations about the x and y axes, respectively. The program can also analyze multiple slabs with variable load-transfer mechanisms across cracks and joints: a bar element to represent dowel bars or a vertical Kelvin-Voigt element (spring and dashpot connected in parallel) to represent the aggregate interlock. The moving load is simulated using finite-element shape functions at successive time-dependent positions of the vehicle.

In this paper, three different concrete pavements sections located in I-5 and US-50 near Sacramento were used (Figure 1). Their mechanical characteristics were backcalculated from falling weight deflectometer tests, which have been conducted during daytime and nighttime, so that the effects of daily temperature change could be accounted for.

2.1 Effects of the joints

Any concrete pavement requires joints. Through the joints, the bending and shear stresses are transferred between slabs. When a slab is loaded, the adjacent slabs also deflect, and the Load Transfer Efficiency (LTE) is defined as

$$LTE(\%) = \frac{\Delta_{i+1}}{\Delta_i} \cdot 100 \tag{1}$$

where Δ_i is the deflection on the edge of the loaded slab and Δ_{i+1} is the deflection on the edge of the adjacent slab. In this study the joints are modeled by a Kelvin-Voigt element, consisting of a spring and dashpot connected in parallel. DYNASLAB does not allow entering the LTE value as an input, therefore a sensitivity analysis has been conducted to establish the

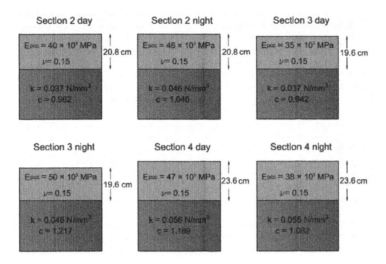

Figure 1. Pavement sections.

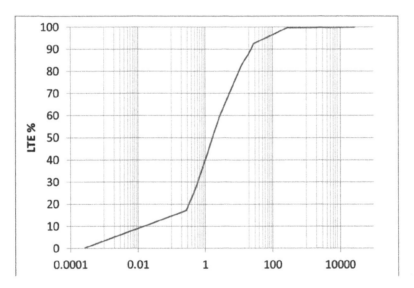

Figure 2. Relation between LTE and spring stiffness.

relation between the k, c values of the spring and dashpot and the load transfer efficiency. FWD tests have been simulated in DYNASLAB using different values of k and c, and the LTE associated with each of those values was calculated. The sensitivity analysis showed that LTE is highly sensitive to the stiffness but not sensitive to the damping coefficient (Fig. 2).

The joints have a major impact on the structural behavior of the pavement. While asphalt pavements can be assumed to be represented as a continuous medium, PCC pavement cannot be considered as a semi-infinite in the longitudinal or in the transversal direction. Joints have a significant impact on the response and even more on the energy dissipation, since relative rotation between two consecutive slabs is allowed even when the deflection is continuous along the pavement (LTE = 100%). From the point of view of a load moving along a semi-infinite slab, the deflection basin would be the same for the entire duration of travel; the maximum deflection would be constant and located under the load. Instead, considering a finite length slab, the deflection basin would depend on the position on the slab. At any different location of the load on the slab the rotation parameter changes, and so does the

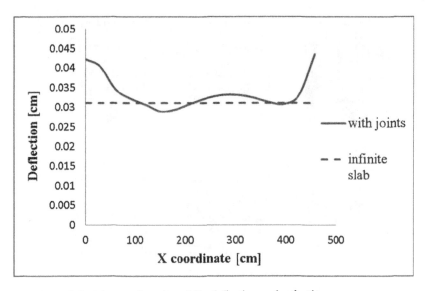

Figure 3. Effects of the joints on the value of the deflection under the tire.

deflection basin. Figure 3 shows the deflection of both jointed and infinite PCC pavement under the front axle of a truck pavement without joints. When the load is closer to the edges (x = 0, x = 450) the rotation of the slab is maximum, and thus the deflection is higher.

3 CALCULATION OF THE ENERGY LOSS OF THE VEHICLE

It has been shown that, with the assumptions of a quasi-static regime, and non-dissipative vehicle tires, the power dissipation of a wheel due to the structural rolling resistance can be evaluated as (Chupin et al. 2013):

$$P_{RR}^{str} = \int_S p \frac{dw(x,y,z,t)}{dt} dS \qquad (2)$$

where p is the pressure applied on the pavement, S is the area of the tire print and $w(x,y,z,t)$ is the deflection of the pavement.

As well as the deflection basin, also the average slope under the tire depends on the position of the wheels on the slab. In Figure 4a the slope as seen by the wheels of the front of a trailed tandem axle moving at 100 km/h is shown. The slope is maximum at the moment the tire print is entirely on the slab (x = 30), while it is minimum as the wheel is leaving the slab. The rear axle surmounts the slab when the front axle is in x = 152, and it can be noted that the slope does not change significantly, however, when the LTE is lower than 100%, the slope has a sudden increase. Figure 4b shows that the slope as seen by the rear axle is negative for most of the time. It is due to the fact that the maximum deflection of the slab is generally located between the two axles. There is no gain of energy, since the effects of the two axles cannot be decoupled.

To take into account the dependency of the slope on time and the position on the slab, the slab (of length L) is divided into m intervals of length. The energy dissipated on the slab is calculated as

$$W_{RR} = \sum_{i=1}^{n} p_i S_i \sum_{j=1}^{m} \left\langle \frac{dw(x_j,t_j)}{dx} \right\rangle \cdot \Delta x \qquad (3)$$

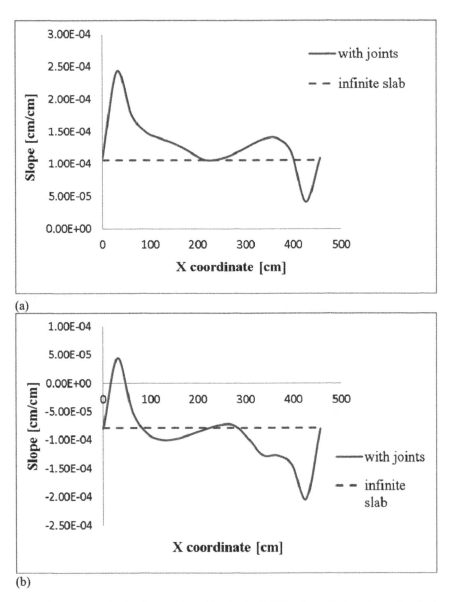

Figure 4. Slope as seen by the front (4a) and back wheels (4b) of a trailed tandem axle of a loaded truck at 100 km/h, central loading.

The total energy dissipated by the vehicle per mile can be calculated as

$$W_{diss}[MJ/km] = W_{RR} \cdot \frac{1000}{L} \tag{4}$$

where L is the length of the slab.

4 ESTIMATION OF THE FUEL CONSUMPTION EXCESS

The fuel consumption excess caused by the dissipation of energy can be evaluated as following:

$$Fuel_{RR} = \frac{W_{diss}}{\xi_b} \quad (5)$$

The factor ξ_b is the effective calorific value of the combustible, and is a function of the engine technology. According to Baglione (2007), the maximum efficiency of Gasoline engines is about 25–30% while it is about 40% for Diesel engines; those percentages represent the energy released by the Gasoline and Diesel engines that will be available to move the vehicle. Since the calorific value of Diesel is about 40 MJ/L and the one of Gasoline is about 34 MJ/L, the value used for ξ_b is 16 MJ/L for a Diesel engine and 10.5 MJ/L for a Gasoline engine.

5 RESULTS

Typical slab length (4.50 m, 15 ft.) and width (3.6 m, 12 ft.) are considered in the simulation. Three types of vehicles are simulated moving along the sections (medium car, SUV and the trailed tandem axle of a loaded truck) at two different speeds: 50 km/h and 100 km/h. For the truck, only one tandem axle has been considered; the other axles do not affect the response significantly since different axles are never located on the same slab at the same time. The characteristics of the loading for each type of vehicle used are shown in Table 1. For each vehicle two simulations are made with a different position of the wheels on the slab, as shown in Figure 5.

Table 1. Characteristics of the vehicles and their fuel consumption.

	Vehicle characteristics				Fuel consumption			
					F_c [ml/km]		Max Fuel$_{excess}$ [%]	
Vehicle	Number of axles	Number of tires	Load per Axle [kN]	Load per Tire [kN]	50 km/h	100 km/h	50 km/h	100 km/h
Car	2	4	7.15	3.58	70.0	95.6	0.002	0.002
SUV	2	4	12.25	6.13	78.7	120.9	0.004	0.006
Truck	1	4	151.42	37.85	273.4	551.7	0.072	0.081

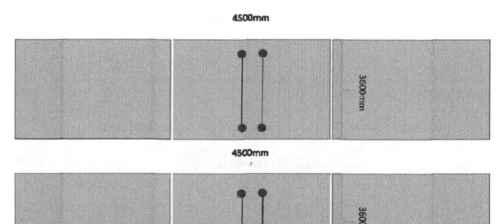

Figure 5. Offset and edge loading.

The fuel consumption excess is calculated as a percentage of the consumption ($Fuel_C$) of the vehicle calculated using the HDM4 Model (Chatti & Zaabar 2012):

$$Fuel_{excess} = \frac{Fuel_{RR}}{Fuel_C} \cdot 100 = \frac{W_{diss}}{\xi_b \cdot Fuel_C} \cdot 100 \qquad (6)$$

The model provides an estimation of the fuel consumption of a vehicle at different speeds based on the power required to overcome the traction forces P_{tr} (which include rolling resistance), the power required for engine accessories P_{accs} (e.g. fan belt, alternator etc.) and the power required to overcome internal engine friction P_{eng}. The F_c value used in the paper was obtained using standard parameters for the vehicles to account for P_{accs} and P_{eng}; no grade, curvature or acceleration were considered and the road surface is in good condition (IRI = 1 m/km, 0.5 mm texture).

The estimations of the fuel consumption of the vehicles are shown in Figures 6–8. The results are shown for the sections with and without joints. Although it is not realistic to consider concrete pavement without joints (except for continuously reinforced concrete pavements), the comparison points out how the joints increase the energy dissipation due to the structural rolling resistance.

The fuel consumption is directly proportional to the speed. Although the deflection is higher at lower speeds, the slope seen by the wheels increases with the velocity. For an increase of 100% of the velocity of the vehicle, the fuel consumption excess increases of 83–129%, where the higher increase occurs for heavier loads.

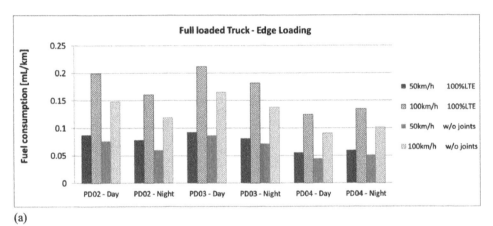

(a)

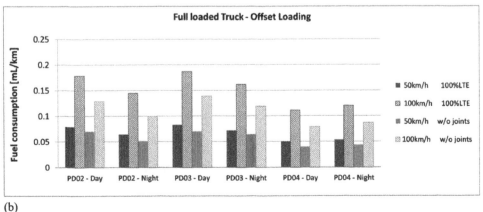

(b)

Figure 6. Comparison between fuel consumption excess due to structural response of a jointed concrete pavement with 100%LTE and an infinite slab to a moving tandem axle at different speeds.

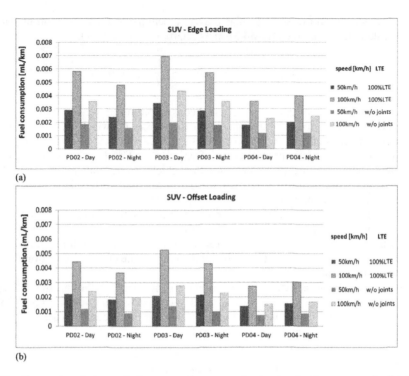

Figure 7. Comparison between fuel consumption excess due to structural response of a jointed concrete pavement with 100%LTE and an infinite slab to a SUV at different speeds.

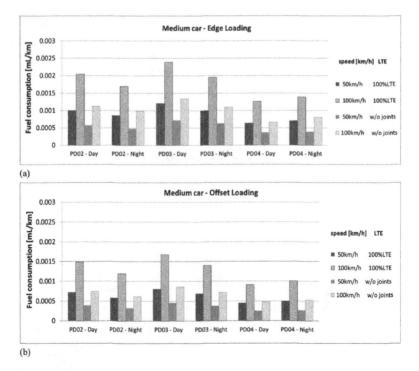

Figure 8. Comparison between fuel consumption excess due to structural response of a jointed concrete pavement with 100%LTE and an infinite slab to a medium car at different speeds.

The position of the wheels on the slab also affects the fuel consumption excess: for the pavement with joints the difference between edge and offset loading is 9–12% for a truck, 23–25% for the SUV and 27–33% for the medium car.

The temperature does not seem to be affecting the structural rolling resistance significantly. The difference in fuel consumption between day and night is small and does not show a specific trend, for sections 2 and 3 the fuel consumption is higher during daytime, while for section 4 it is higher during nighttime.

The section that shows the least structural rolling resistance is section 4, which has a thicker concrete slab and higher values of the Winkler foundation parameters k, c when compared to the other sections.

In general the results show that the fuel consumption excess due to the structural rolling resistance on a rigid PCC pavement does not exceed 0.22 mL/km per axle for a truck, which, when compared to the consumption of the entire vehicle, corresponds to a $Fuel_{excess} \cong 0.08\%$ (see Table 1), which is a very small quantity.

6 CONCLUSIONS

This paper presented a methodology to calculate the deflection-induced fuel consumption excess of trucks, SUVs, and cars travelling on concrete pavements. Using the finite element model DYNASLAB, different types of vehicles moving along sections with different mechanical properties are simulated at different speeds and load positions on the slab. The response of the pavement was used to calculate the dissipation of energy and fuel consumption caused by the structural rolling resistance.

The results show that the excess of fuel consumption of a vehicle travelling on concrete pavements due to the SRR is a very small quantity, less than 0.1% of the total fuel consumption of the truck. While this excess fuel consumption due to the structural rolling resistance is very small, the paper showed that:

- By increasing the speed, the fuel consumption excess increases.
- The position of the vehicle on the slab has an effect on the results. The fuel consumption increases as one of the wheels is closer to the edge of the slab. This increment is amplified as the velocity of the vehicle increases.
- The effects of temperature do not follow a specific trend. The differences in fuel consumption related to temperature changes are very small if compared to the effects of the other factors. Note that the effects of temperature on curling were not considered in this analysis.
- The structural rolling resistance is lower for sections with thicker concrete slabs.

REFERENCES

Baglione, M. 2007. Development of System Analysis Methodologies and Tools for Modeling and Optimizing Vehicle System Efficiency.

Brundtland, G. et al. 1987. Our common future. *Brundtland report*.

Chatti, K. & Zaabar, I. 2012. Estimating the Effects of Pavement Conditions on Vehicle Operation Costs. *(NCHRP) Report 720*.

Chatti, K., Lysmer, J. and Monismith, C. L. 1994. Dynamic Finite-Element Analysis of Jointed Concrete Pavements. *Transportation Research Record, No. 1449*; pp. 79–90.

Chatti, K. 1992. Dynamic Analysis of Jointed Concrete Pavements Subjected to Moving Transient Loads. *Ph.D. dissertation, University of California-Berkley*.

Chupin, O., Chabot, A., Piau, J.M. 2013. Evaluation of the Structure-induced Rolling Resistance (SRR) for pavements including viscoelastic material layers.

Coleri, E., Harvey, J. T., Zaabar, I., Louhghalam A., and Chatti, K. 2016. Model Development, Field Section Characterization, and Model Comparison for Excess Vehicle Fuel Use Attributed to Pavement Structural Response. *Transportation Research Record: Journal of the Transportation Research Board, No. 2589, Transportation Research Board, Washington, D.C., 2016*, pp. 40–50. DOI: 10.3141/2589-05.

Eapa/Eurobitume Task Group Fuel Efficiency. 2004. Environmental Impactes and Fuel Efficiency of Road Pavements.

LaClair, T. J. 2006. The Pneumatic Tire.
Louhghalam A., Akbarian, M., Ulm F.-J. 2013. Scaling Relations Of Dissipation-Induced Pavement-Vehicle Interactions. *trrjournalonline.trb.org/doi/abs/10.3141/2457*
Lysmer, J. 1965. Vertical Motion of Rigid Footings. *Ph.D. Dissertation, University of Michigan.*
Pouget, S., Sauzéat, C., Benedetto, H., and Olard, F. 2012. Viscous Energy Dissipation in Asphalt Pavement Structures and Implication for Vehicle Fuel Consumption. *J. Mater. Civ. Eng., 10.1061/(ASCE)MT.1943-5533.0000414, 568–576.*
Shakiba M, Ozer H, Ziyadi M, Al-Qadi IL. 2016. Mechanics based model for predicting structure-induced rolling resistance (SRR) of the tire-pavement system. *Mechanics of Time-Dependent Materials.* U.S. Greenhouse Gas Inventory Report: 1990–2013.
Zaabar, I. & Chatti, K.. 2010. Calibration of HDM4 Models for Estimating the Effect of Pavement Roughness on Fuel Consumption for U.S Conditions. *Transportation Research Record 2155, Journal of the Transportation Research Board,* pp. 105–116.

Cool pavement LCA tool: Inputs and recommendations for integration

H. Li, J.T. Harvey & A. Saboori
University of California Pavement Research Center, Davis, CA, USA

ABSTRACT: To advance the adoption of strategies to reduce the greenhouse gas and air pollutant emissions and urban heat island effects of pavement systems within California, a collaborative research project was conducted between the University of California Pavement Research Center (UCPRC), Lawrence Berkeley National Laboratory (LBNL), and University of Southern California (USC) to develop a tool for comparing environmental impacts of alternative decisions at the local government level in California. This paper details results of the pavement management survey, albedo of different pavement treatment materials, dynamic modeling of albedo of public pavement for different local governments in California, and Life Cycle Assessment (LCA) models of such materials and common pavement surface treatments to capture their environmental impacts. This information is intended for use in the urban heat island LCA tool and for inputs into climate modeling in that tool's use stage.

1 INTRODUCTION

The construction, use, and maintenance of California's roadways and parking lots are responsible for substantial energy and resource consumption and emissions of Greenhouse Gases (GHGs) and other air pollutants. In addition, pavements—which cover about one-third of a typical U.S. city (Akbari et al. 2009) can have a strong influence on local temperatures and air quality.

Research has identified opportunities to reduce the environmental impacts of pavements. Previous research indicated that "cool" pavements with high solar reflectance can reduce ambient temperatures, slow the temperature-dependent formation of smog, decrease air conditioning and peak electricity demand, and induce negative radiative forcing that cools the atmosphere (Akbari et al. 2009, Rosenfeld et al. 1998). Moreover, cooler asphalt pavements may be less prone to rutting and cracking, and under certain conditions may also have lower rolling resistance due to viscoelastic energy dissipation under heavy truck loading (Lenke et al. 1986, Pouget et al. 2012).

Recognizing the potential for cool pavements to reduce greenhouse gas emissions and improve heat islands and air quality, California local governments are beginning to adopt cool pavement strategies in their Climate Action Plans. Chula Vista, Vallejo, and Santa Rosa are a few of the local governments that have already identified cool pavements as an important strategy both for mitigating and adapting to climate change. Despite this interest, the greenhouse gas, local climate, and air quality impacts of cool pavements remain largely unquantified.

As California re-engineers its local government practices to reduce GHG emissions and air pollution and to adapt to climate change, decision-making requires a strong understanding of the life-cycle environmental impacts of conventional and cool pavements. Evaluating the environmental impacts of pavement in California and estimating the potential impact of GHG reduction strategies present an opportunity to reduce greenhouse gas emissions, reduce ambient temperatures, improve air quality, and protect public health. This will help the California Air Resources Board (CARB) meet its short- and long-term greenhouse gas emissions reduction targets; help regions and the state meet air pollution standards; and help local governments adapt to increasing temperatures.

University of California Pavement Research Center (UCPRC) in a collaboration with Lawrence Berkeley National Laboratory (LBNL) and University of Southern California (USC) conducted a study on benefits and environmental impacts of cool pavements in urban areas in California. The project, funded by California Air Resources Board and Caltrans, was aiming at developing a tool to compare alternative pavement management strategies for reducing urban heat island. With an analysis period of 50 years, the scope of the tool consisted of pavement material production, transportation to the site, pavement construction activities, the changes in urban temperature due to cool pavement strategies implemented, and the resulting changes in building energy consumption throughout the analysis period. In addition to the total primary energy demand GHG emissions, air quality impacts were investigated by comparing smog and particulate matter formation under each scenario. The research project seeks to progress the adoption of strategies to reduce the greenhouse gas and air pollutant emissions and urban heat island effects of pavement systems within California. The following tasks will be completed to achieve this objective:

1. Review the existing literature for cool pavements and pavement LCA, and convene an expert panel that will inform the goal and scope of the LCA analysis.
2. Develop a scenario-modeling tool to analyze, for a wide range of pavement characteristics, the GHG emissions inventories and the air quality, Urban Heat Island (UHI), and building energy use impacts of pavement albedo over a wide range of California city characteristics.
3. Create a pavement strategy guidance tool for local government officials based on the scenario results that can be used to estimate the potential impact of cool pavement adoption.
4. Create clear guidelines for the continual maintenance of the modeling and guidance tools.

This paper covers part of Tasks 2 and 3.

2 PAVEMENT MANAGEMENT PRACTICE

A pavement management survey was conducted with several local California governments to obtain general information about the pavement treatment practices in current use. The main questions included in the pavement management survey for different local governments concerned the following:

1. The size of the pavement network managed by the local government (any units, lane-miles, square feet, centerline miles, etc.).
2. The portion of the network that in a typical year gets any kind of treatment. For example, "treat 7.5 lane-miles per year, or treat 5 percent of the network per year."
3. The approximate breakdown of the treatments used, for example: slurry seal, 70 percent or 7 lane-miles.

Table 1 summarizes the results of the pavement management survey and Table 2 shows a summary of the pavement treatment surface materials, the recommended thickness or the user specifies the thickness, and approximate ranges of the expected time between replacements.

Table 1. Summary of pavement treatment practice currently used by local governments in California.

City	Public Pavement Network Lane-Miles (Centerline Miles)[1]	Portion of Network Treated Every Year	Portion of Each Treatment Used in Total Network Treated					
			Slurry Seal	Sand Seal	Chip Seal	Cape Seal	Asphalt Overlay	Reconstruction
City of Bakersfield	(1,264)	20%	–	75%	–	–	13%	12%
City of Berkeley	453 (216)	7.4%	31%	–	–	–	41%	28%
City of Chula Vista	(461)	3.9%	28.3%	–	46.4%	0.5%	21.8%	3%
City of Fresno[2]	(1,548)	1.3%	–	–	–	–	100%	–
City of Los Angeles	28,000	7.4%	60.7%	–	–	–	35.4%	3.9%
City of Richmond	576	5.2%	47.1%	–	0.7%	0.5%	45.9%	5.9%
City of Sacramento	3,065	4.3%	82.4%	–	–	–	17.6%	–
City of San Jose	4,264	5%	80%	–	–	–	20%	–
Average	–	6.8%	41.2%	9.4%	5.9%	0.1%	36.8%	6.6%

[1]Use multiplier 2.2 to convert centerline miles to lane-miles. The lane width is assumed 12 ft.
[2]Forty (40) centerline miles asphalt overlay up to 2009, then 20 centerline miles asphalt overlay since 2009.

Table 2. Summary of pavement treatments.

Treatment Type	Range of Treatment Life[1]	Thickness (mm) or Application Rate (Asphalt, Aggregate)
Conventional Asphalt Concrete Overlay	2–12 years (1–2 inch) Varies with traffic and design (> 2 inches)	User gives thickness
Rubberized Asphalt Concrete Overlay	2–12 years (1–2 inch) Varies with traffic and design (> 2 inches)	User gives thickness
Asphalt Concrete or Overlay with Reflective Coating	2–12 years (1–2 inch) Varies with traffic and design (> 2 inches)	User gives thickness
Chip Seal	1–10 years	9 mm stone
Slurry Seal	1–10 years	6 mm
Cape Seal	2–15 years	Chip plus slurry
Fog Seal	1–5 years	–
Sand Seal	1–6 years	–
Portland Cement Concrete Whitetopping	Varies with traffic and design 10–20 years (3–5 inches) Varies with traffic and design (> 5 inches)	User gives thickness User gives thickness
User Defined Material	User input	User input

[1]Adapted from Treatment Selection for Flexible Pavements. www.pavementpreservation.org/library/getfile.php?journal_id=941 for local streets, parking lots, etc., not for highways.

3 ALBEDO DATA FOR DIFFERENT PAVEMENT TREATMENTS

There are two ASTM standard test methods for determining the solar reflectance of a surface: ASTM C1549 (Standard Test Method for Determination of Solar Reflectance near Ambient Temperature Using a Portable Solar Reflectometer) (ASTM 2009) and ASTM E1918 (Standard Test Method for Measuring Solar Reflectance of Horizontal and Low-Sloped Surfaces in the Field) (ASTM 2006). A modified method was developed by UCPRC that is in accordance with ASTM E1918. This modified method essentially follows the standard method except for two differences: it uses a dual-pyranometer instead of a single pyranometer and it uses a Data Acquisition System (DAS) composed of a datalogger powered by a battery and connected to a computer to record data automatically. These modifications provide a way to monitor the solar reflectivity of a surface over long time periods.

Albedo data were collected from three sources: LBNL, the Federal Highway Administration (FHWA), and UCPRC. The LBNL Heat Island Group has compiled a pavement albedo database that includes sets of measurements from laboratory samples of various cool pavement treatments taken using spectrophotometer, from field samples taken using the pyranometer test method (ASTM E1918), and compiled from various sources such as field testing and literature.

An on-going FHWA project, entitled "Quantifying Pavement Albedo" (Solicitation Number: DTFH61-12-R-000050), is measuring the albedo of different pavement materials. Some initial albedo data were provided by the project contractor, Iowa State University, and included asphalt and concrete materials with different ages measured using the pyranometer test method (ASTM E1918).

An on-going study on cool pavements being conducted at UCPRC is devoted to investigating the thermal behavior and cooling effect of different pavement types (including asphalt, concrete, and block paver) and different designs (conventional impermeable and novel permeable designs), to using the field measurement data to validate the heat-transfer modeling, to employing the validated model to simulate the thermal behavior and cooling effect of different pavements in various contexts (climates and surroundings), and to examining the effect of cool pavements on human thermal comfort (Li et al. 2013). Nine test sections were the primary sections for albedo measurements at UCPRC. These nine test sections include three different pavement surfacing materials, namely interlocking concrete pavers (surfacing Type A), open-graded asphalt concrete (surfacing Type B), and portland cement concrete (surfacing Type C). More details on the materials can be found in reference (Li et al. 2013). Along with these nine sections, several extra pavement sections with conventional impermeable asphalt and concrete surfacing were also included in the study for the field measurement of albedo. For comparison, albedo has also been measured on other land cover materials, including gravel, soil and grass. Some of these materials were of different ages when solar reflectivity measurements were conducted on them. In May 2014, more field albedo measurements were performed around Davis, California, and these included slurry seal, fog seal, cape seal, chip seal, and more PCC and AC materials.

The steady-state (the final stable albedo value remained after a certain time of weathering and trafficking) albedo of the different pavement materials summarized across all data sources are shown in Table 3.

The main findings on albedo include:

- The most commonly used pavement treatments currently have relatively low steady-state albedo, ranging from 0.05 to 0.15 with average of 0.1 for asphalt concrete, and from 0.1 to 0.24 with average of 0.15 for chip seal. Albedos for slurry seals were measured in the City of Davis and ranged from 0.07 to 0.1, with an average of 0.08. Albedos for cape seals measured in the City of Davis ranged from 0.05 to 0.15, with an average of 0.06. Albedos for fog seals measured in the City of Davis ranged from 0.04 to 0.07, with an average of 0.06. Data were not available for sand seals. For experimental coatings, albedos ranged from 0.2 to 0.3 with average of 0.25 for asphalt concrete with reflective coating, and from 0.15 to 0.35 with average of 0.25 for concrete with reflective coating.
- Although the initial albedo of treatments with reflective coatings can be very high (e.g., up to 0.7 or higher), albedo will decrease very quickly and significantly to a low value due to weathering and tracking with current technology.

Table 3. Summary of steady-state albedo of different pavement treatment materials with different data sources.

Material Type	Albedo (LBNL) Range	Avg.	Albedo (FHWA) Range	Avg.	Albedo (UCPRC) Range	Avg.	Albedo (Typical) Range	Avg.
Asphalt Concrete or Overlay	0.1–0.15	0.12	0.05–0.15	0.1	0.06–0.15	0.1	0.05–0.15	0.1
Asphalt Concrete or Overlay with Reflective Coating	0.2–0.3	0.25	–	–	–	–	0.2–0.3	0.2
Cape Seal	–	–	–	–	0.05–0.15	0.06	0.05–0.15	0.06
Chip Seal	0.1–0.2	0.15	–	–	0.14–0.24	0.18	0.1–0.24	0.15
Conventional Interlocking Concrete Pavement	–	–	–	–	0.25–0.3	0.26	0.25–0.3	0.26
Fog Seal	–	–	–	–	0.04–0.07	0.06	0.04–0.07	0.06
Permeable Asphalt Pavement	–	–	–	–	0.08–0.12	0.1	0.08–0.12	0.1
Permeable Concrete Pavement	–	–	–	–	0.18–0.28	0.25	0.18–0.28	0.25
Permeable Interlocking Concrete Pavement	–	–	–	–	0.25–0.3	0.26	0.25–0.3	0.26
Portland Cement Concrete	0.15–0.25	0.2	0.2–0.3	0.25	0.18–0.38	0.25	0.15–0.35	0.25
Sand Seal	–	–	–	–	–	–	0.07–0.1	0.08
Slurry Seal	–	–	–	–	0.07–0.1	0.08	0.07–0.1	0.08

4 DYNAMIC ALBEDO CHANGE MODELING FOR URBAN PAVEMENT

Urban area land cover is a term used to include buildings, pavements, vegetation, and bare soil, etc. Urban area land cover is illustrated in Figure 1. Urban area pavements include both public and private ones. Public pavements are usually managed by a government agency, and some portion of them receives regular treatment to improve pavement performance. A number of treatments are used on these public pavements each year, and many of these treatments have different albedos. Because of this, the total pavement albedo and consequently the urban albedo change over time. This dynamic process can be modeled with a dynamic albedo model, as shown in Equations (1) through (4). This report will focus on the pavement albedo using Equations (3) and (4). Figure 2 shows an example dynamic albedo of public pavement with the assumptions that 10 percent of the network is treated annually and 50 percent of the slurry seal and asphalt overlay are replaced with reflective materials.

$$\alpha_t = (1 - R_{p,t}) \alpha_{np,t} + R_{p,t} \alpha_{p,t} \qquad (1)$$

$$\alpha_{p,t} = (1 - R_{pp,t}) \alpha_{npp,t} + R_{pp,t} \alpha_{pp,t} \qquad (2)$$

$$\alpha_{pp,t} = (1 - R_t) \alpha_{t-1} + R_t \alpha'_t \qquad (3)$$

$$\alpha'_t = \sum n_{i,t} \alpha_{i,t} \qquad (4)$$

where:
$R_{np,t}$ is the total non-pavement area portion in urban land surface in year t;
$R_{p,t}$ is the total pavement network area portion in urban land surface in year t;
$R_{pp,t}$ is the total *publicly managed* pavement network portion in urban land surface in year t;
R_t is the portion of pavement network for treatment in year t;
$n_{i,t}$ is the portion of treated pavement network which use treatment i in year t;
α_t is the albedo of total urban area in year t;
$\alpha_{p,t}$ is the albedo of total pavement area in year t;
$\alpha_{np,t}$ is the albedo of total non-pavement area in year t;
$\alpha_{npp,t}$ is the albedo of non-public pavement area in year t;
$\alpha_{pp,t}$ is the albedo of public pavement area in year t;
$\alpha_{i,t}$ is the albedo of pavement treatment i in year t;
α_t is the average albedo of pavement network in year t;

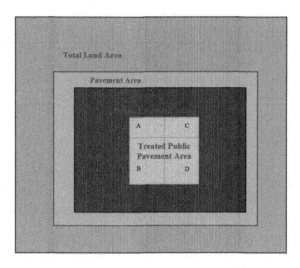

Figure 1. Urban land cover at year t.

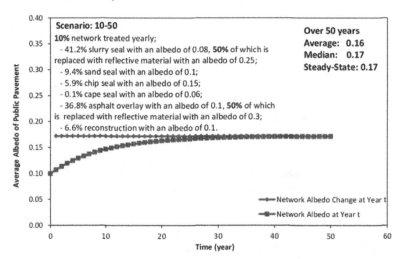

Figure 2. Example dynamic albedo of public pavement with assumptions that 10 percent of the network is treated annually and 50 percent of the slurry seal and asphalt overlay are replaced with reflective materials.

α'_t is the average albedo of pavement treated in year t;
i is the pavement treatment type (A, B, C, D, ...); and
t is the year (1, 2, ..., 50).

5 LIFE CYCLE INVENTORY FOR PAVEMENT TREATMENTS

LCA models were either developed or taken from GaBi software for the construction materials, energy sources, transport modes, and surface treatments. Table 4 and Table 5 show the list of the items included in the database. These models were calibrated to better represent local practice in California in terms of electricity grid mix, plant fuel mixes, and construction processes. Model for materials are cradle to gate. Surface treatment models include material production, transportation to the site (assumed to be 50 miles, the average hauling distance in California), and construction activities. Extensive literature survey was conducted to identify common reflective coatings currently in practice around the world and to determine the components of each. This was used to develop the Life Cycle Inventories (LCI) of the reflective coatings (Li & Saboori 2014). Mix design and construction process for each of the surface treatments were taken from Caltrans' Maintenance Technical Advisory Guide (MTAG) (Caltrans 2007) or through field investigations and inquiries from local experts.

Following California's implementation of the Renewables Portfolio Standard (RPS), it was decided that the LCA models in this study should also be developed taking into account the state's projected grid mix for the year 2020. The new grid mix was developed based on projections in reports from the California Public Utility Commissions (CA PRS website). All the models of materials and treatments were updated based on the new electricity grid mix, resulting in two separate datasets, one based on 2012 grid mix and one based on 2020.

Table 4. Indicators and flows to be used in the final tool.

Impact Category/Inventory	Abbreviation	Unit
Global Warming Potential	GWP	kg of CO_2e
Photochemical Ozone Creation (Smog) Potential	POCP	kg of O_3e
Particulate Matter, less than 2.5 micrometers in diameter	PM2.5	kg
Primary Energy Demand from Renewable & Non-Renewable Resources (net calorific value excluding feedstock energy)	PED (total)	MJ
Feedstock Energy	FE	MJ

Table 5. List of materials, energy sources, transport modes, and surface treatments for which LCA models were developed.

Materials and Energy Sources			Surface Treatments
	Aggregate—Crushed		Cape Seal
	Aggregate—Natural		Chip Seal
	Bitumen		Fog Seal
	Bitumen Emulsion		Conventional Asphalt Concrete (mill and fill)
	Crumb Rubber Modifier (CRM)		Conventional Asphalt Concrete (overlay)
	Dowel & Tie Bar		Conventional Interlocking Concrete Pavement
	Energy Sources	Diesel Burned in Equipment	Permeable Asphalt Concrete
		Electricity	Permeable Portland Cement Concrete
		Natural Gas Combusted in Indust. Equip.	Portland Cement Concrete
	Limestone		Portland Cement Concrete w. SCM
	Quicklime		Reflective Coating—BPA
	Paraffin (Wax)		Reflective Coating—Polyester Styrene
	Portland Cement	Regular	Reflective Coating—Polyurethane
		Slag Cement (19% Slag)	Reflective Coating—Styrene Acrylate
		Slag Cement (50% Slag)	Rubberized Asphalt Concrete (mill and fill)
	Portland Cement Admixtures	Accelerator	Rubberized Asphalt Concrete (overlay)
		Air Entrainer	Sand Seal
		Plasticizer	Slurry Seal
		Retarder	Thin White Topping
		Superplasticizer	Barge Transport (Transport)
		Waterproofing	Heavy Truck (24 tonne) (Transport)
	Styrene Butadiene Rubber (SBR)		Ocean Freighter (Transport)

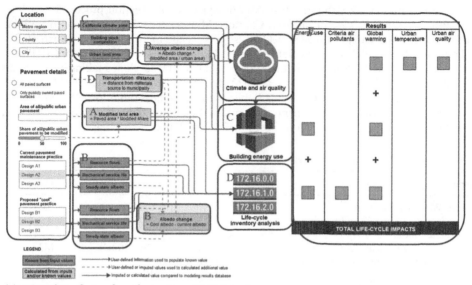

(a) User interface of cool pavement LCA tool

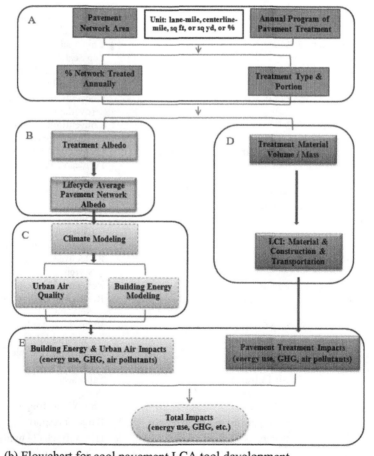

(b) Flowchart for cool pavement LCA tool development

Figure 3. Cool pavement LCA tool development.

Furthermore, the whole database developed under this study and previous LCA studies at UCPRC were verified by a 3rd party committee. The details of all the assumptions and final results are available in the documentation of the UCPRC LCI database (Saboori et al., in prep.).

6 INTEGRATION OF RESULTS INTO COOL PAVEMENT LCA TOOL

Figure 3 shows the proposed integration of the pavement management practices along with the life-cycle inventories and impact calculations developed in this project. Figure 3(a) shows the integration of all the elements in the user interface. Figure 3(b) shows the recommended flow of calculations for using the results of this study in calculating the life-cycle flows and impacts of interest for the complete life-cycle—including the materials production, construction, and use stages, where the use stage includes building energy use.

7 SUMMARY AND RECOMMENDATIONS

This study includes these key findings:

- Based on responses from eight of the California local governments surveyed, most local governments treat a small portion of the public street pavement network every year, ranging from 1.3 percent to 20 percent with an average of 6.8 percent. The survey consisted of responses from cities as opposed to counties.
- The main treatments used by these local governments include slurry seal, chip seal, cape seal (chip seal plus microsurfacing or slurry seal), asphalt overlay, sand seal, and reconstruction (including Asphalt Concrete [AC], Rubberized Asphalt Concrete [RAC], Full-Depth Reclamation [FDR], and Cold In-place Recycling [CIR]).
- Slurry seal is the major treatment used by most local governments, ranging from 28 percent to 82 percent of the pavement miles treated and an average of 41 percent. Asphalt overlay is another major treatment that most local governments use, ranging from 13 percent to 100 percent of mileage treated with an average of 36.8 percent. Chip seals makes up on average 5.9 percent of the mileage treated, ranging from 0.7 percent to 46.4 percent, while cape seal is used on less than 1 percent on average. Sand seal was used by one city for the majority of their program, and they are the city that treats 20% of their network each year, which is much larger than all other local governments surveyed. On average, reconstruction is used for 6.6 percent of the total treatments, ranging from 3 percent to 28 percent.
- Most pavement treatments currently used have relatively low steady-state albedo, with a range from 0.05 to 0.15 and an average of 0.1 for asphalt concrete, and from 0.1 to 0.24 with an average of 0.15 for chip seal. All reconstruction treatments have an asphalt overlay surface and should have similar albedos as AC. Albedos for cape seals and slurry seals were measured in the City of Davis and ranged from 0.06 to 0.15, with an average of 0.12. Data were unavailable for sand seals. Experimental treatments include asphalt concrete with reflective coating, which had albedo ranging from 0.2 to 0.3 with an average of 0.25, and concrete with reflective coating, which ranged from 0.15 to 0.35 with an average of 0.25.
- Although the initial albedo of treatments with reflective coating can be very high (e.g. up to 0.7 even higher), the albedo will decrease very quickly and significantly down to a low value due to weathering and tracking with current technology. This information represents currently available technologies.
- Due to the small portion of pavement network treated every year with treatments of relatively low steady-state albedo, the final steady-state albedo increase of the pavement network in the 50 years is relatively low, ranging from 0.03 to 0.14. The 50-year average increase of the pavement network albedo is even lower, ranging from 0.02 to 0.12.

ACKNOWLEDGEMENTS

The research presented in this paper was requested and sponsored by the California Department of Transportation (Caltrans) and the California Air Resources Board (CARB). Caltrans and CARB sponsorship of that work is gratefully acknowledged. Special thanks are for Nick Burmas, Joe Holland, and Deepak Maskey for their support for and contributions to this research project. The contents of this paper reflect the views of the authors and do not necessarily reflect the official views or policies of the State of California, or the Federal Highway Administration.

REFERENCES

Akbari, H., Menon S., & Rosenfeld A. 2009. Global Cooling: Increasing World-Wide Urban Albedos to Offset CO_2. *Climatic Change*, Vol. 94, No. 3, pp. 275–286.
ASTM, 2006. ASTM E1918-06 Standard Test Method for Measuring Solar Reflectance of Horizontal and Low-Sloped Surfaces in the Field. American Society for Testing and Materials.
ASTM, 2009. ASTM C1549-09 Standard Test Method for Determination of Solar Reflectance near Ambient Temperature Using a Portable Solar Reflectometer. American Society for Testing and Materials.
CA RPS webpage available at: http://www.cpuc.ca.gov/PUC/energy/Renewables/
Caltrans Maintenance Technical Advisory Guide (MTAG) 2007. California Department of Transportation, Sacramento, CA.
Lenke, L. & Graul, R. 1986. "Development of runway rubber removal specifications using friction measurement and surface texture for control." The Tire Pavement Interface, ASTM STP 929, MG Pottinger ans TJ Yagger, Eds, American Society of Testing and Materials: 72–88.
Li, H., Harvey, J.T. & Kendall, A. 2013. Field Measurement of Albedo for Different Land Cover Materials and Effects on Thermal Performance. *Building and Environment*, Vol. 59, No. 2013, pp. 536–546.
Li, H., Saboori, A., & Cao, X. 2014. Reflective Coatings for Cool Pavements: Information Synthesis and Preliminary Case Study for Life Cycle Assessment. *2014 Pavement LCA Symposium*, University of California Davis, Davis, CA.
Pouget, S., C. Sauzéat, H. Benedetto, and F. Olard., Viscous Energy Dissipation in Asphalt Pavement Structures and Implication for Vehicle Fuel Consumption. Journal of Materials in Civil Engineering, Vol. 24, No. 5, 2012, pp. 568–576.
Rosenfeld, A.H., Akbari H., Romm J.J., & Pomerantz, M. 1998. Cool Communities: Strategies for Heat Island Mitigation and Smog Reduction. *Energy and Buildings*, Vol. 28, No. 1, pp. 51–62.
Saboori, A. in prep. Documentation of the UCPRC Life Cycle Inventory (LCI) Used in CARB/Caltrans LBNL Heat Island Project and other Caltrans' LCA Studies. University of California Pavement Research Center, Davis, CA.

Rolling resistance and traffic delay impact on a road pavement life cycle carbon footprint analysis

Laura Trupia, Tony Parry, Luis C. Neves & Davide Lo Presti
Nottingham Transportation Engineering Centre (NTEC), Faculty of Engineering, University of Nottingham, Nottingham, UK

ABSTRACT: The application of Life Cycle Assessment (LCA) to road pavements has been continuously evolving and improving over the last years, however there are several limitations and uncertainties in the introduction of some components in the framework, such as road pavement rolling resistance—in terms of pavement surface properties—and traffic delay during maintenance activities. This paper analyses the influence of methodological assumptions and the model used to estimate the increased emissions for traffic delay and road pavement rolling resistance on the results of an LCA. The Greenhouse Gases (GHG) emissions related to these two phases of a pavement LCA will be calculated for a UK case study, using different models, and a sensitivity test is performed on some specific input variables. The results show that the models used and the input variables significantly affect the LCA results, both for the rolling resistance and the traffic delay.

1 INTRODUCTION

The rapid increase in pavement life-cycle studies in the literature shows the growing interest in improving the sustainability of this critical infrastructure system (Santero et al. 2011b).

While LCA represents a commonly accepted standard method (International Organization for Standardization (ISO) 2006), there are no widely accepted standards for pavement LCA (Harvey et al. 2016). The first pavement LCA studies were limited to the extraction and production of pavement materials. New studies have included other phases, such as traffic delay, vehicle-pavement interaction, and pavement albedo showing promising reduction opportunities (Santero et al. 2011a). The lack of standardized procedures makes it difficult to perform comparable assessments, thus creating a synergistic set of literature that continuously builds upon itself rather than producing conclusions that are independent of the approach taken. In addition, the current knowledge gaps related to some phases makes the implementation of LCA principles complex and characterized by uncertainty. The impact of the traffic delay related to the work zone during the construction and the maintenance phase and the rolling resistance due to the pavement surface properties during the use phase are two components that, despite their omission from many previous LCA studies, can have a significant impact on the results.

Traffic delay results from lane or road closures at construction and maintenance work zones due to queueing or detours, around the construction site. In order to estimate the impact of this component, a two-step method, including a traffic model and an emission model, is usually used. In previous studies, two approaches have been used:

- A more sophisticated approach, using a microsimulation model to describe the work zone, by defining the average queue length and the instantaneous speed of individual vehicles (Galatioto et al. 2015; Huang et al. 2014). Usually, these software tools include one or more emission models able to define the impact from the work zone.
- A simplified approach, based on the Demand-Capacity (D-C) model, defined in the Highway Capacity Manual (HCM) (Transportation Research Board 2010) that describes the

work zone average queue and speed. An emission model is used, based on the output provided by the traffic models. Actually, the HCM methods are not really suitable to assess construction activities or queues increasing over time (congested network), unless specific modifications are performed by the analyst. The Federal Highway Administration (FHWA), based on the Demand-Capacity (D-C) model, defined in the HCM, has developed a computational approach to analyse the user cost of work zone traffic delay, in Life Cycle Cost Analysis (LCCA) (Walls III and Smith 1998).

The advantage of using a simplified approach is the ease of implementation, requiring limited data input, such as hourly traffic volume, capacity and Traffic Management (TM) layout. However, the accuracy of the results could be compromised especially when the TM scheme is particularly complex (Wang et al. 2014b) or the area of impact is extensive and requires the modelling of a wider network.

By contrast, an approach based on microsimulation modelling is more flexible and accurate, producing disaggregated traffic data and it can readily include the wider network. However, these models are usually incorporated in commercial software that increase the cost of the analysis and require detailed traffic data, which can limit the size of the network model. The model used to calculate the emissions related to this component can affect the results, especially for high traffic volume roads.

The road pavement rolling resistance is the energy loss due the Pavement-Vehicle Interaction (PVI) and it is affected by the tire properties and by the pavement surface condition. Roughness and macrotexture, usually represented by International Roughness Index (IRI) and Mean Profile Depth (MPD), are the pavement surface properties affecting the rolling resistance. These parameters change over the life of a pavement and their variation may be different for each lane, depending on the traffic volume and type, the surfacing type and the regional climate.

In order to estimate the impact of the pavement surface properties on vehicle fuel consumption, several models have been developed (Chatti and Zaabar 2012; Hammarström et al. 2012; Wang et al. 2014a). However, there is still uncertainty concerning the lack of validated models used to analyse the vehicle emissions and the influence of specific variables and assumptions on the results. Indeed, the literature related to the influence of road surface properties on vehicle rolling resistance and emissions shows different results, possibly because road surface components are a relatively small part of the rolling resistance and of the total driving resistance, it is difficult to isolate the road surface effects from other effects and quantify their contribution and different methods of measuring rolling resistance can lead to different results.

In the UK, this component is not generally included in the pavement LCA framework and there are no general pavement deterioration models to predict the deterioration rate of IRI and MPD. Some empirical deterioration models have been developed for specific maintenance treatments and geographical areas (Lu et al. 2009; Tseng 2012). However, in these models both the IRI and MPD values tend to increase over time, so they are not applicable to a UK case study where the MPD will generally decrease with time. The model used to investigate this impact and the input deterioration model can influence the results of a study, making the conclusions unreliable.

Recent studies have included these two components (Santero and Horvath 2009; Santos et al. 2015; Trupia et al. 2016; Xu et al. 2015) and during the last Pavement LCA Symposium in 2014 (Harvey and Jullien 2014), papers related to the emissions due to the work zone traffic delay (Huang et al. 2014; Wang et al. 2014b) and the impact of the PVI rolling resistance during the use phase (Akbarian et al. 2014; Ciavola and Mukherjee 2014) were presented. Although efforts have been made to fill the research gaps related to these components, there is still a level of uncertainty concerning the methodological assumptions, the chosen methods to analyse the vehicle emissions and the parameters that can affect the results.

The aim of this paper is to explore the influence of the model used and the assumptions made to estimate the increased emissions due to work zone traffic delay and the PVI rolling resistance phases on pavement LCA results. The GHG emissions related to these two phases will be estimated for a UK case study, using different models from the literature, and a sensitivity test is performed on specific input variables (extent of the area of impact of the work zone traffic delay and surface condition deterioration rate for the PVI rolling resistance).

2 METHODOLOGY

The CO_2 emissions due to the traffic delay during a maintenance event and due to the influence of the pavement surface deterioration rate on the PVI rolling resistance will be estimated, using different models including a sensitivity test on specific input variables. This will allow a comparison of the models available in the literature to decide if the knowledge related to these components is sufficient to implement them in a standard pavement LCA framework in the UK.

The tailpipe GHG emissions are made up of over 99.8% CO_2 emissions, so in this study, only this component will be taken into account. The case study analyzed in this paper is a 4 km section of the dual carriageway A1 (M) motorway located in the North East of England, UK. The Annual Average Daily Flow in 2009 was 45,862 motor vehicles and 5,640 HGVs (heavy goods vehicles) making this segment a medium–high trafficked road. The original construction included a chipped hot rolled asphalt surface course. In 2009, a 40 mm overlay of thin surfacing was applied to a 4 km section of both carriageways and this is the event modelled in this study. The impact of these two construction events, in terms of CO_2e (carbon dioxide equivalent) emissions, amounts to 4031 t of CO_2e for the original construction and 223 t of CO_2e for the thin surfacing treatment.

2.1 *Work zone traffic delay*

Jean Lefebvre (UK) Technical Centre provided several potential TM solutions for the 2009 maintenance work. The one assumed in this paper involves a carriageway closure and contraflow on the other carriageway and requires 24 hours to install the 40 mm Thin Surface Course per 1 km, three days to deploy the TM and three days to remove it (Figure 1), resulting in 17 days work to resurface both carriageways. During the closure of one carriageway, both northbound and southbound traffic experiences a lane reduction, so the traffic delay in both directions was considered.

2.1.1 *Comparison of the traffic models*
The CO_2 emissions due to traffic delay were estimated using two different approaches: a simple approach using the D-C model and a vehicle emission model, Emission Factor Toolkit (EFT) (UK Department for Environment Food & Rural Affairs 2014); and a more

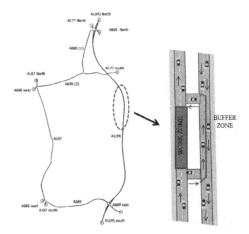

Figure 1. Work zone location and layout (northbound carriageway closure).

sophisticated approach using the microsimulation software AIMSUN (Advanced Interactive Microscopic Simulator for Urban and Non-Urban Networks) (Transport Simulation System (TSS) 2015) that includes a vehicle emission model. A detailed description of these models can be found in the references, but the key elements are described here. In both methods, the calculation of the impact of the work zone is based on the difference between the emissions during normal conditions (no work zone) and operational condition (maintenance event).

The first approach, the LCCA procedure developed by the FHWA, consists of several calculations:

1. Project future year traffic demand;
2. Work zone directional hourly demand;
3. Roadway capacity in normal condition and during maintenance;
4. Compare roadway capacity with hourly traffic demand and identify the work zone components (i.e. upstream traffic, queuing zone, slowing down zone) and the number of vehicles affected in each component.
5. Use the output from the traffic models in EFT to calculate the CO_2 emissions.

The emission factors for the CO_2 are those published by the UK Department for Transport (Boulter et al. 2009).

In this procedure, the calculation of the speed of the queue on which the fuel consumption and the CO_2 emissions depend—is based on the use of a graph (Forced-Flow Average Speed versus Volume to Capacity ratio) for level of Service F (congested condition) contained in the earlier versions of the HCM. According to the HCM, the curve in the graph is unstable and the values represent estimations. This fact generates a high level of uncertainty in the results, since the change in fuel consumption related to the variation of speed is much more significant at low speeds (0–25 mph) found in work zone queues.

The microsimulation approach, with AIMSUN, uses a graphical interface and is able to model the road network geometry and the behavior of individual vehicles on the network. An interesting feature of this software is the possibility to choose a different route selection for some vehicles. In order to compare the two approaches a base case scenario was identified where the network analysed was confined to the linear segment of the A1 (M), including just the 4 km work zone (see Figure 2 'mini network'.).

2.1.2 *Sensitivity test on network modelling boundary*

The traffic modelling requires the identification of the extent of the network impacted by the work zone. For a comprehensive understanding, the modelling should cover the whole network affected. During a maintenance event, the behavior of the vehicles is affected by the congestion occurring in the work zone and they could take alternative routes, thus affecting other roads. Or, in the worst case scenario, the congestion could extend to an area not

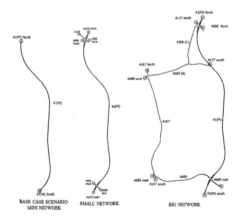

Figure 2. Network extension scenarios.

included in the modelling boundary. The microsimulation approach is more flexible, allowing the area of analysis to be extended, taking into account the interaction of elements, such as traffic lights, roundabouts, other junctions, etc.

In order to assess the impact of the network boundary, three different scenarios were considered: the 'mini network' that represents the base case, a 'small network' including two roundabouts at the A1 (M) junctions and joining traffic streams and a 'big network' that includes possible diversions that vehicles could take in case of congestion (Figure 2).

2.2 Rolling resistance

In order to estimate the effect of the pavement surface condition on vehicle fuel consumption, the CO_2 emissions were calculated with two different rolling resistance models; the UCPRC model developed at the University of California Pavement Research Center (UCPRC, Davis) (Wang et al. 2014a) and the model developed by the Swedish National Road and Transport Research Institute (VTI), within the European Commission project Miriam (Models for rolling resistance In Road Infrastructure Asset Management systems) (Hammarström et al. 2012). A similar analysis comparing these two models has been performed on another UK case study with a lower traffic volume (Trupia et al. 2016). The general features of the two models will be described here. Further details can be found in the references cited.

In the UCPRC model, the CO_2 emissions for a specific vehicle type can be calculated directly, based on the analysed pavement segment's MPD and IRI values by using equation (1) and multiplying by the vehicle mileage travelled.

$$T_{co_2} = a_1 \times MPD + a_2 \times IRI + Intercept \qquad (1)$$

where T_{CO_2} is the tailpipe CO_2 emission factor, the terms a_1, a_2 and *Intercept* are the coefficients derived from a regression analysis and are different for each combination of the categorical variables (pavement, road and road-access type, vehicle type), *IRI* is the road roughness (m/km) and *MPD* is the macrotexture (mm).

The model was developed using two software calibrated for US conditions or based on empirical US data; the Highway Development and Management Model—version 4 (HDM-4) (PIARC 2002) (an empirical—mechanistic model to perform cost analysis for the maintenance and rehabilitation of roads) used to calculate the rolling resistance; and MOVES (Motor Vehicle Emission Simulator) (EPA's Office of Transportation and Air Quality (OTAQ) 2014), the US EPA highway vehicle emission model based on national data, used to model the vehicle emissions as a function of the rolling resistance. This model is included in the Pavement LCA Framework proposed for the USA by the FHWA (Harvey et al. 2016) as an appropriate modeling approach to calculate the impact of roughness and texture depth during the use phase.

The second approach is based on the VTI model to estimate the vehicle fuel consumption and emission factors proposed by International Carbon Bank & Exchange (ICBE 2010) to convert the vehicle fuel consumption into CO_2 emissions. The model, based mainly on empirical data from coastdown measurements from other projects and drum measurements, includes a general rolling resistance model (equation (2)) to estimate the contribution of the rolling resistance to the total driving resistance and a fuel consumption model (equation (3)) to calculate the vehicle fuel consumption (Hammarström et al. 2012).

$$F_r = m_1 \times g \times (0.00912 + 0.0000210 \times IRI \times V + 0.00172 \times MPD) \qquad (2)$$

where m_1 is the vehicle mass (kg) and v is the vehicle speed (m/s).

$$F_{CS} = 0.286 \times \left(\begin{pmatrix} 1.209 + 0.000481 \times IRI \times V + 0.394 \times MPD + 0.000667 \times V^2 \\ +0.0000807 \times ADC \times V^2 - 0.00611 \times RF + 0.000297 \times RF^2 \end{pmatrix}^{1.163} \right) \times V^{0.056} \qquad (3)$$

where ADC is the average degree of curvature (rad/km) and RF is the road gradient (m/km).

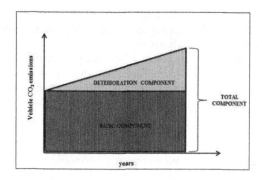

Figure 3. Total CO_2 emissions, divided into basic component (dark grey area) and deterioration component (light grey area) (from (Trupia et al. 2016)).

Table 1. Pavement deterioration rate, in terms of IRI and MPD, during the analysis period.

Scenario	MPD mm	IRI m/km
Average deterioration	1.8–0.8	1.0–2.3
Worst deterioration	1.5	1.0–5.0
No deterioration	1.8	1.0

This second model may better represent UK conditions, since the texture depth deterioration, vehicle types and emission factors are considered more similar to the UK conditions than the USA values. However, the Swedish fleet, both for cars and HGVs, is different from the EU average.

The models allow the total CO_2 emissions related to the IRI and MPD to be calculated (Figure 3), here defined as the "total component" (total area).

It represents the sum of the "basic component" (dark grey area) representing the emissions if the IRI and MPD do not change over time—no deterioration, and the "deterioration component" (light grey area) representing the emissions due to the deterioration of the pavement condition during the analysis period, in terms of IRI and MPD. To better understand the behaviour of the two models, all the components will be assessed in this study.

Since in the UK there are no published models to predict pavement condition deterioration rate, the time progression of IRI and MPD on the assessed road segments over the analysis period (20 years) is generated based on literature data for other maintenance strategies (Aavik et al. 2013; Jacobs 1982; Wang et al. 2014a).

A sensitivity test is performed on different scenarios of deterioration of IRI and MPD for the two case studies (see Table 1), to take into account the uncertainty related to these parameters and the range of potential impact. The average deterioration values and the IRI values in the worst deterioration scenario include an initial and final condition value and a linear change with time is assumed. In the average deterioration scenario, the MPD decreases over time; this is typical in the UK were high MPD values are specified for new surfacings to assist in provision of high-speed wet skidding resistance. The MPD in the worst deterioration scenario and the MPD and IRI for the no deterioration scenario are considered constant.

3 RESULTS

3.1 *Work zone traffic delay*

Table 2 shows the extra CO_2 emissions due to the work zone, obtained by running the two models for the 'mini-network'.

Table 2. CO_2 emissions due to traffic delay during the work zone.

	Emission of CO_2 (ton)	
	Microsimulation	D-C model
TM (17 days)	48.58	329.67
Per day (average)	2.86	19.39

Table 3. Sensitivity test on network boundary.

	Emission of CO_2 (ton)		
	Mini	Small	Big
TM (17 days)	48.58	53.85	41.88
TM (1 day)	2.86	3.17	2.46

The D-C model generates CO_2 emissions higher than the microsimulation model. Clearly, this difference will have a big impact on the results when the TM involves many days of work.

As explained above, the calculation of the emissions with the D-C model is strongly affected by the speed of the queue, since a small change in this value in congested conditions (between 0 and 25 mph) can generate significantly different results, in terms of fuel consumption and CO_2 emissions. However, the curve used in the LCCA procedure to calculate this value does not allow a precise evaluation of the queue speed. In order to use this model in a confident way, this procedure should be updated by using more accurate methods of evaluating queue speed.

Other factors explaining the results obtained are the different emissions factors used in the two models, the different approaches used to estimate the queue speed (in Aimsun, it is based on other factors, such as the car-following models) and the fact that in the HCM, several parameters (such as capacity and average speed) are based on empirical data on USA roads.

Table 3 shows the results obtained performing a sensitivity test on the network boundary, using the microsimulation, in order to investigate if and how the area of impact of the work zone can affect the results and how microsimulation software can be helpful in this process.

The results obtained are sensitive to the definition of the area of impact of the work zone. The Mini network is composed of a linear segment in the A1 (M) that includes the work-zone area but does not consider any potential diversions for the vehicles. The Small network takes into account the traffic generated by the two roundabout junctions to the North and the South of A1(M) and the associated traffic streams, but it does not allow any change in route choices. This network, compared to the Mini network, estimates larger emissions, because it considers also the emissions produced at the roundabouts due the extension of the congestion from the work zone. By contrast, in the Big network, the extra emissions estimated are smaller than in the Mini network, because the vehicles have the possibility to change their route during congestion to reach the same destination point.

3.2 *Rolling resistance*

Table 4 and Figure 4 show the results for PVI rolling resistance obtained for the different pavement surface condition deterioration rates, using the VTI and the UCPRC models.

The values obtained for the basic components show that, even when not considering the deterioration rate over time, the two models produce different results. The reason for this difference may be the different approaches used to develop the models. In the VTI model the rolling resistance parameters were estimate using coastdown and drum measurements for 90 different types of tyre to better correspond to a representative vehicle. In the UCPRC model, in order to develop the equation function the authors have modelled a series of IRI and MPD values for combinations of specific variables using MOVES. The estimated

Table 4. CO_2 emissions due to pavement surface condition (average deterioration).

A1

Model	Emission of CO_2 (ton)		
	Basic	Deterioration	Total
VTI	109344	−4205	105139
UCPRC	18058	4586	22645

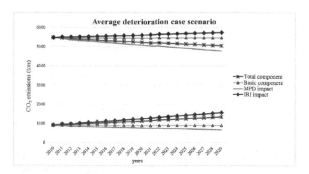

Figure 4. Impact of IRI and MPD in the VTI (above) and UCPRC (below) models.

emission factors depend on different variables, including the tyre rolling resistance represented by a default coefficient. This default value was obtained through dynamometer tests on a smooth surface and therefore, it only takes into account the influence of the tyre on the rolling resistance, neglecting the effect of the pavement properties. In order to calculate the emissions under different IRI and MPD conditions, the default rolling resistance coefficient was updated in the MOVES database by using the formula adopted in the HDM-4 software that includes the effect of the pavement properties on the rolling resistance.

The difference in the basic component for the two models also affects the calculated values of the total component, which are also significantly different. The deterioration component for the VTI model is negative and this means that the evolution of the pavement surface condition generates an overall reduction in the total CO_2 emissions. This result is due to two factors; in this case study the MPD decreases over time and the VTI model assigns to the MPD term a greater impact on the rolling resistance and on the emission estimate than for IRI (even at high speed, which increases the impact of the IRI). In the UCPRC model, instead, the decrease of the MPD term over the years is offset by the increment of the IRI term that has a larger impact in this model. So the different weight that the two models give to the IRI and MPD terms can affect the results, especially for a case study where the MPD evolution is negative.

The potential impact of the pavement surface condition on the results is confirmed by the sensitivity test (see Table 5).

The results are very sensitive to the pavement deterioration rate, in particular:

- The two models generate the lowest emissions under different deterioration scenarios, average for the VTI model and no deterioration in the UCPRC. In the VTI model the deterioration component decreases over time, producing in the average scenario an overall reduction of the emissions.
- The highest emissions for the two models occur for the worst case scenario for both cases. Under this scenario, the VTI model does not generate a negative term for the deterioration component, since the IRI effect is larger than the MPD effect.

Table 5. Sensitivity Test Results.

Scenario	Emission of CO_2 (ton)					
	VTI			UCPRC		
	B*	D**	T***	B*	D**	T***
Average	109344	−4205	105139	18058	4586	22645
Worst	109344	4716	114059	18058	19634	37693
No	109344	0	109344	18058	0	18058

*Basic component, **Deterioration component, ***Total component.

These results confirm the findings for the previous case study with a lower level of traffic (Trupia et al. 2016). The results are significantly sensitive both to the pavement deterioration rate assumed and the rolling resistance/fuel consumption model used. The different validation of the two models, together with the different approaches used, can be considered the main reasons for this significant difference in the results, indeed, the models were calibrated for different countries with different input data, in terms of weather, vehicles, and roads.

4 CONCLUSIONS

The methodological assumptions and the models chosen for a pavement LCA, in terms of traffic delay due to a maintenance work zone and PVI rolling resistance model and pavement condition deterioration, significantly affect the results.

For the traffic delay, the type of traffic model used can affect the results. In addition, the extent of the road network modelled is an important factor in the analysis of the traffic delay component. Further research is necessary in this area to understand if it is possible to standardize this element and the type of traffic model required for a specific LCA study.

The results related to the comparison of the rolling resistance models confirm the findings of a previous case study with a lower level of traffic. The results are sensitive both to the model used to estimate the PVI rolling resistance CO_2 emissions, and to the surface deterioration rate chosen. Site specific elements and methodological choice affect the development of the rolling resistance and fuel consumption models, meaning they are not suitable for all geographical locations. In the UK, pavement deterioration models and rolling resistance models need to be developed in order to introduce this component into the pavement LCA framework.

ACKNOWLEDGMENT

This study was funded from an Engineering and Physical Sciences Research Council (EPSRC) Doctoral Training Grant (DTG)—Faculty of Engineering, University of Nottingham. The authors would like to thank staff at Highways England and John Lefebvre (UK) for providing information for the case study.

REFERENCES

Aavik A, Kaal T, Jentson M (2013) Use of Pavement Surface Texture Characteristics Measurement Results in Estonia. Paper presented at the XXVIII International Baltic Road Conference, Vilnius, Lithuania 26–28 August.

Akbarian M, Louhghalam A, Ulm FJ (2014) Mapping The Excess-Fuel Consumption Due To Pavement-Vehicle Interaction: A Case Study of Virginia's Interstate System. Paper presented at the International Symposium on Pavement Life Cycle Assessment, Davis, California, USA.

Boulter PG, Barlow TJ, McCrae IS, Latham S (2009) Emission factors 2009: Final summary report PPR361. Transport Research Laboratory.

Chatti K, Zaabar I (2012) Estimating the effects of pavement condition on vehicle operating costs. NCHRP. Transportation Research Board.

Ciavola B, Mukherjee A (2014) Assessing the Role of Pavement Roughness in Estimating Use-Phase Emissions. Paper presented at the International Symposium on Pavement Life Cycle Assessment, Davis, California, USA.

EPA's Office of Transportation and Air Quality (OTAQ) (2014) User Guide for MOVES 2014.

Galatioto F, Huang Y, Parry T, Bird R, Bell M (2015) Traffic modelling in system boundary expansion of road pavement life cycle assessment. Transportation Research Part D: Transport and Environment 36:65–75.

Hammarström U, Eriksson J, Karlsson R, Yahya M-R (2012) Rolling resistance model, fuel consumption model and the traffic energy saving potential from changed road surface conditions VTI Rapport 748 A.

Harvey J, Jullien A Papers from the International Symposium on Pavement Life Cycle Assessment 2014. In: International Symposium on Pavement Life Cycle Assessment, Davis, California, USA.

Harvey JT, Mejijer J, Ozer H, Al-Qadi IL, Saboori A, Kendall A (2016) Pavement Life-Cycle Assessment Framework FHWA-HIF-16-014. FHWA.

Huang Y, Galatioto F, Parry T (2014) Road pavement maintenance life cycle assessment—a UK case study. Paper presented at the International Symposium on Pavement Life Cycle Assessment, Davis, California, USA.

ICBE (2010) tCO_2 in Gaseous Volume and Quantity of Fuel Type. http://www.icbe.com/carbondatabase/volumeconverter.asp.

International Organization for Standardization (ISO) (2006) BS EN ISO 14040:2006 Environmental managment—Life cycle assessment—Principles and framework. British Standards Institution.

Jacobs F (1982) M40 High Wycombe By-pass: Results of a Bituminous Surface-texture Experiment. Transport and Road Research Laboratory, UK.

Lu Q, Kohler ER, Harvey JT, Ongel A (2009) Investigation of Noise and Durability Performance Trends for Asphaltic Pavement Surface Types: Three-Year Results UCPRC-RR-2009-01.

PIARC (2002) Overview of HDM-4. The Highway Development and Management Series Collection.

Santero N, Masanet E, Horvath A (2011a) Life-cycle assessment of pavements Part II: Filling the research gaps. Resources, Conservation and Recycling 55:810–818.

Santero N, Masanet E, Horvath A (2011b) Life-cycle assessment of pavements. Part I: Critical review. Resources, Conservation and Recycling 55:801–809.

Santero NJ, Horvath A (2009) Global warming potential of pavements. Environmental Research Letters 4:034011.

Santos J, Ferreira A, Flintsch G (2015) A life cycle assessment model for pavement management: methodology and computational framework. International Journal of Pavement Engineering 16:268–286.

Transportation Research Board (2010) Manual-HCM, Highway Capacity. National Research.

Transport simulation system (TSS) (2015) Aimsun traffic simulation software. http://www.aimsun.com/.

Trupia L, Parry T, Neves L, Lo Presti D (2016 (in press)) Rolling Resistance Contribution to a Road Pavement Life Cycle Carbon Footprint Analysis The International Journal of Life Cycle Assessment doi:10.1007/s11367-016-1203-9

Tseng E (2012) The construction of pavement performance models for the California Department of Transportation new pavement management system. Department of Civil and Environmental Engineering—University of California Davis.

UK Department for Environment Food & Rural Affairs (2014) Emissions Factors Toolkit User Guide.

Walls III J, Smith MR (1998) Life-Cycle Cost Analysis in Pavement Design-Interim Technical Bulletin. Federal Highway Administration (FHWA).

Wang T, Harvey J, Kendall A (2014a) Reducing greenhouse gas emissions through strategic management of highway pavement roughness Environmental Research Letters 9:034007.

Wang T, Kim C, Harvey J (2014b) Energy consumption and Greenhouse Gas Emission from highway work zone traffic in pavement life cycle assessment. Paper presented at the International Symposium on Pavement Life Cycle Assessment, Davis, California, USA.

Xu X, Gregory J, Kirchain R (2015) Role of the Use Phase and Pavement-Vehicle Interaction in Comparative Pavement Life Cycle Assessment. Paper presented at the Transportation Research Board 94th Annual Meeting, Washington DC, United States.

LCA case study for O'Hare International Airport taxiway A&B rehabilitation

John Kulikowski
U.S. Air Force, Tyndall AFB, FL, USA

ABSTRACT: The ability to measure and quantify the environmental impacts of such projects is in higher demand. Life Cycle Assessment (LCA) studies/tools are developed for highway infrastructure and pavements but a limited number of studies/tools have been developed for airports and few for their pavement facilities. This paper utilizes an airfield pavement LCA tool called LCA-AIR 1.0 to quantify the sustainability strategy for O'Hare International Airport's rehabilitation of Taxiway A and B. LCA-AIR was used to evaluate three rehabilitation options consisting of rubbilization of existing Portland Cement Concrete (PCC) with mill/Asphalt Concrete (AC) inlay, Precast Concrete Panel (PCP) replacement and full-depth reconstruction. The results showed the strategy with the least impact for Global Warming Potential (GWP) was as follows: PCP replacement, rubblization with a mill/AC inlay, and full-depth reconstruction. The ranking of least impact for energy for the three strategies was rubblization with a mill/AC inlay, PCP replacement, and finally, full-depth reconstruction. The GWP potential for rubblization was 2,395 kg CO_2/yd^2 (4.310×10^{-10} kg CO_2/lb-mile), for reconstruction was 2,409 kg CO_2/yd^2 (4.729×10^{-10} kg CO_2/lb-mile), and for PCP was 2,413 kg CO_2/yd^2 (4.736×10^{-10} kg CO_2/lb-mile). The energy consumed for rubblization was 0.1861 TJ/yd^2 (3.576×10^{-8} TJ/lb-mile), for PCP was 0.1863 TJ/yd^2 (3.657×10^{-8} TJ/lb-mile) and for reconstruction was 0.1864 TJ/yd^2 (3.658×10^{-8} TJ/lb-mile).

1 INTRODUCTION TO LCA

In the United States the transportation sector generates 27% of total US Greenhouse Gas emissions (GHG) of which, aircraft account for 8.2% of the total (Federal Highway Administration 2015b). The US has an estimated 13,112 airports (Federal Aviation Administration 2015a). Airports world-wide (U.S.) process 3.3 billion passengers (838 million), 3.6 trillion passenger-miles (963 billion) and 55 million short-tons of freight annually, driving approximately 3.0% of the worlds GDP (International Civil Aviation Organization 2014; Senguttuvan 2011). Significant quantities of environmental impacts are produced from the transportation sector. Quantifying these impacts requires a transparent method, more specific than rating systems. Many rating systems currently exist for facilities with a limited number for infrastructure. These rating systems help agencies and other organizations find sustainable and innovative solutions to minimize their impacts on the environment. Currently, airport pavements do not have such rating systems or methods to quantify environmental impacts. A Life Cycle Assessment (LCA) is a method to quantify these environmental impacts from processes and projects throughout its life (Santero N. J. 2009). LCA studies/tools have been completed and developed for highway pavements, such as PaLate (Horvath 2004), PE-2 (Mukherjee & Cass 2011), Athena Pavements LCA (Athena Sustainable Materials Institute 2013), and recently the Illinois Tollway tool, ICT-LCA (Al-Qadi et al. 2015) but airport pavement LCA studies and tools are needed. LCA-AIR opens the field for improvements and expansion. Using a LCA methodology, LCA-AIR analyzed three different rehabilitation strategies for Chicago's O'Hare International Airport (ORD) taxiways. LCA results can be used in conjunction with other tools available to airport owners, operators, engineers and contractors to make educated decisions regarding infrastructure investment and their influence on environmental impacts.

2 INTRODUCTION TO ORD TAXIWAYS

The pavement structure of ORD Taxiways A&B, surrounding the main terminal, is reaching the end of its performance life and is in need of rehabilitation. There are various distress types and severity levels in the pavement structure. In addition to loading and environmental factors, inadequate site, surface, and subsurface drainage may be contributing to the pavement structural condition and impact future rehabilitation choices. Based on as-built data provided by ORD engineers, the Taxiways (TWs) were constructed between 1986 and 1988 over cohesive soils with varying amounts of sand. The approximate length of each taxiway is 11,088 feet. Per as-built drawings, typical pavement section consists of 21 inches of PCC over six inches of a bituminous aggregate mixture, over six inches of an aggregate base over the subgrade. Coring data from 2007 and 2013 showed PCC surface ranges from 19-1/2 inches to 25.5 inches with an average of 21.96 inches. The asphalt base ranges from 0 inches to 9 inches thick with an average of 5.31 inches. The aggregate base layer ranges from 0 inches – 20 inches with an average of 5.51 inches thick (MACTEC, 2007; Chicago Airports Resources Enterprise, 2014).

After evaluating a wide suite of rehabilitation strategies (mill PCC and in-lay AC, mill PCC and in-lay bonded/unbonded PCC with macrofibers, reconstruction, selective slab replacement, mill PCC and provide geo-fabric or stress absorbing interlayer membrane interlayer with AC surface, partial/full depth repair) for ORD's Taxiways A&B, three strategies were selected as cases for further investigation: Precast Concrete Panel (PCP) replacement, rubblization of existing PCC with mill/AC inlay and full-depth reconstruction. PCP consists of removing the slabs and placing a precast slab in the void after preparing the base from damage done during removal to ensure compacted, uniform support. Rubblization consistes of using a guillatine and multi-head breakers to rubblize the existing PCC layer. This also facilitates more rapid milling prior to AC inlay. Milling will go to a depth of five inches to void entaglement in the welded-wire mesh. Reconstruction consists of demolition of all pavement and base layers down to the subgrade prior to reconstruction. These methods were selected as feasable alternatives based on impact to airlines, long term performance and elevation constraines from adjacent features. The material quantities and construction methods differ for each alternative and will have different associated environmental impacts. These rehabilitations occur at the 30-year point and will extend the taxiway's life to 50 years. The rubblization with mill/AC inlay has a potential to last 10 to 15 years, depending on the traffic level and thickness of the AC inlay, after which a milling of the old asphalt/new inlay will occur. For the assessment, this milling of the inlay occurs at year 40 to ensure the pavement reaches 50 years. The PCP or full-depth reconstruction will last between 20 to 40 years. For this assessment a 20-year extension was used. Current data on the remaining slabs that were rehabilitated especially in low traffic path zones still have good remaining life and are performing well. It is probable the remaining slabs will require significant work after this 50 year period. Therefore, extending the PCC repair analysis in conjunction with long term plans to 30 or 40 years was not performed. Doing so, would reduce the impacts per functional units compared to the impacs of one to two more mill and inlays to reach another 10 to 20 years. An extensive cost analysis was not part of this study. This paper focuses on the LCA for each case in the rehabilitation for 200, keel section slabs (125,000 ft^2) on the southern side of each TW. Each strategy rehabilitates the same area per TW but the depth and materials are dependent on the method. Figure 1 shows the pavement cross-sections for original construction and each rehabilitation strategy with the legend to the right. The red dashed line indicates the depth limit of each strategy.

Information on the development of LCA-AIR can be found in the 11th International Conference on Concrete Pavements (ICCP) from the paper titled, Development of LCA-AIR – An Airport Pavement Life Cycle Assessment Tool (Kulikowski, et al., 2015). The inputs for each phase are discussed. The results for Tool for the Reduction and Assessment of Chemical and other Environmental Impacts 2.1 (TRACI) indicators were calculated and presented with a focus on energy and GWP with respect to functional units of square yard and pound-mile due to the unique aspects of airfield pavements. Square yard was used for the Material Production (MP) and Construction, Maintenance and Rehabilitation (CMR) phases. The second unit, pound-mile, was used as the functional unit for the use (U) phase accounting for aircraft weight and travel distance as it is not a direct result of pavement conditions as in

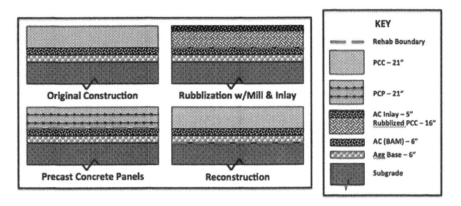

Figure 1. Rehabilitation strategy cross-sections.

roadway LCAs. The total impacts were divided by the area of the TWs and the total impacts for the use (U) phase were divided by the total pound-mile traveled. The MP impacts for initial construction and U phases are assumed to be the same for each case because initial construction is not different and the maintenance does not change the U phase components. Therefore, the case study will focus on the impact results from the CMR phase that includes the impact from all machinery used in initial construction, the maintenance and rehabilitation in addition to the new materials used in the repairs and rehabilitation.

2.1 *Geometry, pavement structure and mix design inputs for initial construction*

The geometry and pavement structure used for initial construction was based on the as-built drawings with the width and length of each TW being 75 feet and 11,088 feet. The concrete, asphalt and aggregate base densities were assumed to be 150, 145 and 140 pounds/ft^3, respectively. These slabs had doweled transverse joints and for the most part include welded wire mesh 6.5 inches from the surface. Individual slab weight is 165,000 pounds with 942 pounds from steel contribution (0.0571%). The original mix designs were not available, so the following mix designs in Table 1 were used for the analysis in the construction and maintenance materials. A prime coat was used on the crushed aggregate base and a tack coat was used between asphalt layers. It was assumed the six – seven inch asphalt layers were placed in two lifts. The distance from the concrete and asphalt plant to the construction site was five miles.

As discussed in the ICCP, the impacts for the CMR phase is a function of the equipment used, their productivity/fuel efficiency and the impacts associated with the repair materials. The productivity and fuel consumption is impacted by the operator's competency level, the material type and cycle times performing each activity. Based on experience, industry averages, interviews and literature the productivity and fuel consumption were calculated (Muench, et al., 2011; Ross, 2011; Pullen, Edwards, Rutland, & Tingle, 2014; CATERPILLAR, 2015; Craftco, 2015; Bockes, 2015; Shinners, 2015; Shinners, 2015; Wirtgen Group, 2015). When fuel consumption was not available, EPA NONROADS database was used based on the engine horsepower (Environmental Protection Agency, 2008).

The initial construction impacts of the TWs are the same for each case. The AC portions consisted of: the AC base under the PCC surface and the AC shoulders on each side of the TWs. The PCC portion was only used for the three, paving lane slabs on each TW. Equipment was selected for each task and then based on the geometry (volume and tonnage) and productivity/fuel efficiency the diesel combustion for each activity was calculated. The total fuel consumed during construction for the AC base (1,663,200 ft^2) and AC shoulders (1,108,800 ft^2) 11,899 gallons. The total fuel consumed during construction for the PCC 15,794 gallons. (A basis of square feet should not be used to compare the asphalt work vs. concrete work, as the volume is significantly different.)

Table 1. Mix designs for LCA case studies.

AC	% Weight of surface mix	% Weight of base mix
Virgin coarse crushed aggregate	71.325	74.178
Virgin fine crushed aggregate	20.922	20.922
Virgin natural sand	0	0
Coarse RAP (3.6% binder)	0	0
Fine RAP (6.4% binder)	0	0
Mineral Filler	2.853	0.571
Total binder	4.9	5
Distance from plant to site (miles)	5	5
PCC	JPCP (lbs/yd^3)	
Virgin coarse crushed aggregate	2,075	
Virgin fine crushed aggregate	1,056	
Portland cement	401	
Fly Ash	134	
Water	240	
Slag cement	0	
Distance from Plant to Site	5	

Maintenance and repair at airports occurs around the clock and very few activities are consistently scheduled to reoccur. A recurring maintenance/rehabilitation schedule was not available so schedules and quantities were established. For this study, the maintenance activities were aggregated over time and assumed to occur at specific years. Below is a list of the general activity intervals used for each case. The AC activities are on the shoulders with the exception of the reconstruction (base layer) and the rubbilization with mill/inlay.

PCC

– Restriping airfield markings—every ten years
– Joint and crack sealing—every eight years
– Full and partial depth repairs—every fifteen years
– Brooming—every other day

AC & AC Shoulders

– Restriping airfield markings—every ten years
– Crack sealing—every ten years
– Asphalt patching—every fifteen years
– Mill/inlay—every fifteen years
– Mill/inlay—10 years after the initial rubblization with mill/inlay section

2.2 *Use phase components*

As discussed in the ICCP paper, the U phase is broken into the following three categories: aircraft fuel consumption, snow removal operation (fuel consumption), and lighting (electricity usage for edge and centerline lights). Similarly, the U phase dominates the other phases because of the fuel consumed over thousands of flights per year for 50 years. This is especially case when including half of the time in flight to each airport. Fuel consumed in aircraft was modeled based on kerosene (heavy fuel, closest to jet fuel) combusted in industrial equipment because combustion of Jet A/Jet B or Jet Proppellant-8 was not defined in LCA data bases. However, the U phase has a small contribution when only the additional fuel consumed because of a change in roughness is calculated. Roadway LCAs attribute fuel consumption to the pavement facility for each vehicle because of constant pavement-vehicle interaction. However, this pavement-vehicle interaction is negligible for airport pavement facilities with the aircraft traversing the pavement surface approximately 20 minutes before taking off. Kulikowski, et al. discussed International Roughness Index (IRI) as applied in LCA-AIR (Kulikowski, et al., 2016).

Table 2. Use Phase aircraft information.

Aircraft type	Operation seight (75% MTOW) (lbs)	Yearly number of operations	Air flight duration (hrs)	Max weight (lbs)
B747-200B	624,750	1738	2	833,000
B777-200 ER	570,000	1738	2	760,000
A330-300 opt	390,525	1738	2	520,700
B737-700	115,875	62500	2	154,500

For each of the three cases the change in roughness of the pavement surface was assumed to be similar based on material deterioration. Greater resistance and increased fuel consumption for airplanes is experienced from aerodynamic drag (Goldhammer & Plendl 2014).

The aircraft selected for use and their associated weights/operations are seen Table 2. Based on the International Civil Aviation Organization (ICAO) the actual weight of 75% of the maximum Take-Off Weight (MTOW) was used. MTOWs were based on manufacture specifications. Four aircraft were selected to represent the eight aircraft categories (MTOW) trafficking TW A and B. The B737 represents groups one through four, the A330 represents groups five and six, the B777 represents group seven and the B747 represents group eight. The number of operations was based on the 2018-projected traffic. Due to the small quantity of operations in groups five through 8 the average operations were divided by three and attributed to each aircraft.

The other factors considered in the U phase were lighting and snow removal. There were 440 centerline lights and 887 edge lights used in the calculation of the electricity impact. The lights are used for night operations and during daytime inclement weather. Therefore, an average time of 12 hours per day was assumed for each case. Twenty snow occurrences were assumed for each case utilizing brooms and blows to clear every square foot of pavement as standard operating procedure. The efficiencies of equipment were considered and diesel combusted in industrial equipment was used for impact determination. Lighting and snow removal operation factors contributed minimally to the U phase impacts.

3 LIFE CYCLE ASSESSMENT FOR RUBBLIZATION WITH MILL/AC INLAY

This strategy consists of rubblizing the existing PCC pavement and providing an AC inlay. Due to the TWs connection to other features, changes in elevation can be detrimental to operations or require significant investment to remedy. The method for this rehabilitation consists of using a two-step breaking process which has proven successful at other airfields along with highways (Kulikowski, Development of A Pavement Life Cycle Assessment Tool for Airfield Rehabilitation Strategies, 2015). First, a guillotine style breaker will be dropped breaking the slab in the larger pieces (similar to the break and seat methods of highways). Immediately following this, a multi-head breaker will reduce the sizes varying from a few inches down to 12–15 inches at the bottom of the slab. Finally, milling machine removes the specified amount of rubblized PCC for paving operations (Figure 2).

The standard maintenance used was discussed previously. At 30 years, a large rehabilitation is performed on the TWs A&B. The following LCA analyzes the rubblization of 250,000 ft^2 (200 slabs per TW) and five inches of milling with a five-inch AC inlay. For ORD, milling deeper than five inches is not feasible because of the wire mesh at approximately six inches from the surface.

Appropriate equipment was selected for the CMR activities for PCC and AC, respectively. Figure 3 shows the fuel consumed for each activity. For many activities, it is a summation of the fuel consumption from multiple pieces of equipment. For example, the rubblization activity uses the fuel consumed from a guillotine breaker, multi-head breaker, z-grid roller and a steel drum roller. The highest fuel consumption activities during the CMR phase were brooming (112,721 gal), crack sealing (31,771 gal), restriping (16,746 gal) and land clearing (12,328 gal).

Converting the brooming machine to an electric source will eliminate the fuel consumed for brooming operations. There will be an adjusted impact for electrical consumption with use if it

Figure 2. Milling of rubblized PCC with an AC overlay (Hanson Engineering, 2015).

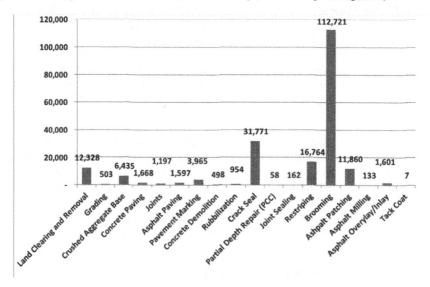

Figure 3. CMR fuel from rubblization.

is powered from a power plant. If it is by solar energy, the impacts will be reduced significantly. Frangible equipment could be utilized for solar generation as well as reducing green space maintenance and bird-air strike hazards. Adjusting the brooming schedule will also save fuel. Reducing the cleaning to once every five days will provide a 10% reduction in total fuel consumed. This must be balanced with increased risk of foreign object debris on the TWs. Crack sealing is a time and energy intensive activity. The tank must remain at a high temperature to ensure the sealant is still flowable. The sealant is placed by and individual with a wand walking every crack. The use of alternative, lower temperature sealant will reduce some of the impact.

Using asset optimization principles contracts can be structured to only include cracks of a certain size, which will reduce the impact as well. Similar asset management principles can be used to reduce the frequency or quantity of markings restriped. For land clearing, it was assumed the clearing for the entire depth of the pavement structure, in this case 33 inches. Stabilizing the subgrade may result in additional saving because of a reduction of aggregate subbase and total pavement thickness. The main rehabilitation activity of rubblization consumed 954 gallons of fuel and the additional AC inlays (apart from the shoulder inlays) consumed 553 gallons to place. The total fuel combusted was one component used to determine the environmental impacts for the CMR phase. The total fuel consumed for the CMR phase was 204,568 gallons.

The second component used to determine the total CMR impacts was the materials used in the maintenance and the rehabilitation. As expected, the rubblization with mill/inlay used the least amount of material because the 16 of the 21 inches of PCC remain in place. The only new material added for the rehabilitation is the prime/tack coats and the new five inches

Table 3. Total impacts per functional units—Rubblization with Mill/Inlay case.

Impact category	Unit	TW A&B		TW A&B	
		Total impact Per yd^2	Total impact Per lb-mile	Total impact Per yd^2 (ΔIRI Only)	Total impact Per lb-mile (ΔIRI Only)
Global warming	kg CO$_2$ eq	2.40E+03	4.31E-10	2.00E+02	3.93E-11
Primary energy consumption (renewable + non-renewable)	TJ	0.18612	3.58E-08	0.00518	1.02E-09

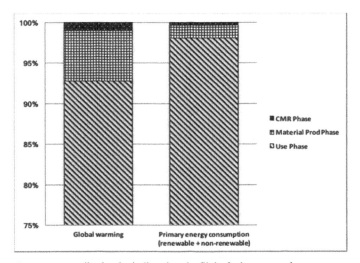

a) percent contribution including time-in-flight fuel consumption

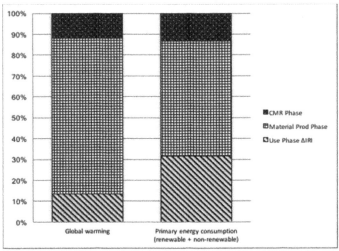

b) percent contribution including only aircraft fuel consumed from change in IRI

Figure 4. Phase impact contributions—Rubblization and Mill/Inlay Case.

of inlayed asphalt. Rubblization showed 24% and 30% less energy consumption compared to precast panels and reconstruction respectively. For GWP there was 43% lower production than precast panels and 37% lower production than reconstruction.

3.1 Rubbilization with Mill/AC Inlay impacts per phase and functional units

The total impact from each phase was summed and then divided by the functional units for the airport. Table 3 shows the total values for the impacts using both aircraft fuel consumed for one-half the time in flight as well as for only the aircraft fuel consumed because of an increase in International Roughness Index (IRI). Figure 4a and 4b show the percent contribution from each phase. As discussed previously, the fuel consumed by the aircraft over 40 years dominates the LCA. Although not in contact with the pavement there is still an environmental impact that doesn't disappear when the plane leaves the airport. When only using the fuel consumed by change in IRI (no aircraft fuel consumption while in air), MP phase accounts for about 56% of primary energy consumption, CMR phase accounts for 13% and the use phase is approximately 31%. This demonstrates a significant difference between highway and airfield pavement LCAs as noted previously by Kulikowski (Kulikowski, 2015).

4 LIFE CYCLE ASSESSMENT FOR PRECAST CONCRETE PANELS

Figure 1 shows the depth of repair (21 inches of PCC). The slab lift-out method was assumed for removal. Appropriate equipment was selected for the CMR activities for PCC and AC. Figure 5 shows the fuel consumed for each activity. The highest fuel consumption activities were brooming (112,721 gal), crack sealing (31,072 gal), restriping (16,222 gal) and land clearing (12,328 gal).

Brooming impacts are similar to the rubblization case study. Crack sealing is slightly less than the rubblization case because the PCP repair section will not require the same crack sealing as the AC inlay. The main rehabilitation activities were concrete pavement demolition, PCP placement, and diamond grinding. The concrete demolition consumed an additional 523 gallons of fuel. The PCP placement consumed 2,937 gallons and diamond grinding added 761 gallons of fuel. It was assumed the entire repair area was diamond ground to remove irregularities. Close coordination with the fabricator and installer can ensure tight tolerances are achieved and only spot grinding is necessary. The amount of patching is reduced on the rehabilitated section compared to rubblization because of the increased durability of concrete. The total fuel combusted was one component used to determine the environmental impacts for the CMR phase. The total fuel consumed for the CMR phase was 206,078 gallons, which is 2,052 gallons more than rubblization. The rubblization method is an efficient operation for demolition (leaves material in place) and paving is also fast providing for key reductions.

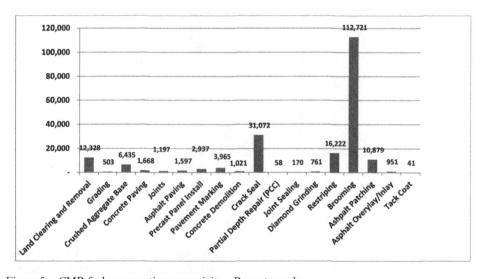

Figure 5. CMR fuel consumption per activity—Precast panels.

The second component used to determine the total CMR impacts was the materials used in the maintenance and the rehabilitation process. The precast strategy replaced the second most amount of material following reconstruction and subsequently will have higher impacts than rubblization as well as a longer service life. The new material includes some base leveling fine aggregate and new reinforcement in the PCC slabs. An 8% lower value in energy consumption and a 9% increase in GWP production was seen in the PCP method compared to reconstruc-

Table 4. Total impacts per functional units—Precast panel case.

Impact category	Unit	TW A&B Total impact Per yd²	TW A&B Total impact Per lb-mile	Total impact Per yd² (ΔIRI Only)	Total impact Per lb-mile (ΔIRI Only)
Global warming	kg CO₂ eq	2.41E+03	4.74E-10	2.18E+02	4.29E-11
Primary energy consumption (renewable + non-renewable)	TJ	0.1863	3.66E-08	0.0054	1.06E-09

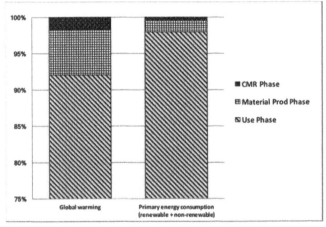

a) percent contribution per phase including time-in-flight fuel consumption

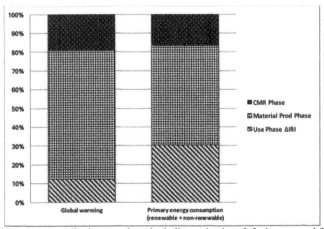

b) percent contribution per phase including only aircraft fuel consumed from change in IRI

Figure 6. Phase impact contributions—Precast panel case.

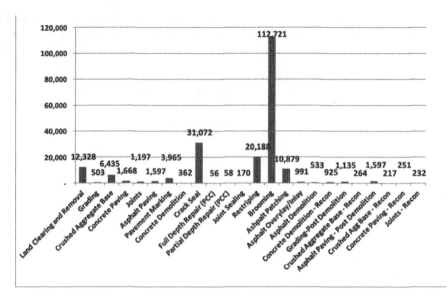

Figure 7. CMR fuel consumption per activity—Reconstruction.

tion. The lower energy value is attributed to a lower intensity installation process than reconstruction. This increase in GWP is attributed to the two mats of steel in the PCP panels.

4.1 Precast impacts per phase and functional units

The total impact from each phase was summed and then divided by the functional units for the airport. Table 4 shows the total values for the impacts using both aircraft fuel consumed for one-half the time in flight as well as for only the aircraft fuel consumed due to an increase in IRI. Figure 6a and 71b show the percent contribution from each phase. When only using the fuel consumed by change in IRI, MP phase accounts for about 53% of primary energy consumption, CMR phase accounts for approximately 17% and the U phase is approximately 30%.

5 LIFE CYCLE ASSESSMENT FOR RECONSTRUCTION

The reconstruction method assumes the PCC, AC base and base course layers are removed and rebuilt. The method for breaking the slabs was a hydraulic hammer on the end of an excavator. It is efficient, rapid and provides enough breakage to allow equipment to remove the debris. Appropriate equipment was selected for the CMR activities for PCC and AC, respectively. Figure 7 shows the fuel consumed for each activity. The highest fuel consumption activities were brooming (112,721 gal), crack sealing (31,072 gal), restriping (16,222 gal) and land clearing (12,328 gal).

Brooming impacts are similar to the rubblization case study. Crack sealing is slightly less than the rubblization case because the reconstruction (PCC) repair section (like PCP case) will not require the same crack sealing as the rubblization AC inlay. The most impactful rehabilitation activities were concrete/asphalt demolition, grading after demolition, and concrete/asphalt paving. These activities consumed 1,458, 1,135 and 1,848, respectively. The amount of patching is reduced compared to rubblization (same as PCP) because of the increased durability of concrete.

The total fuel combusted was one component used to determine the environmental impacts for the CMR phase. The total fuel consumed for the CMR phase was 205,201 gallons, which is 1,175 gallons more than rubblization and 877 gallons less than precast. The difference between precast and reconstruction is attributed to the speed of demolition and construction. Demolition and the placement of concrete requires less time than the placing of PCPs.

Table 5. Total impacts per functional units—Reconstruction case.

Impact category	Unit	TW A&B Total impact Per yd²	TW A&B Total impact Per lb-mile	TW A&B Total impact Per yd² (ΔIRI Only)	Total impact Per lb-mile (ΔIRI Only)
Global warming	kg CO_2 eq	2.41E+03	4.73E-10	2.15E+02	4.22E-11
Primary energy consumption (renewable + non-renewable)	TJ	0.1864	3.66E-08	0.0055	1.07E-09

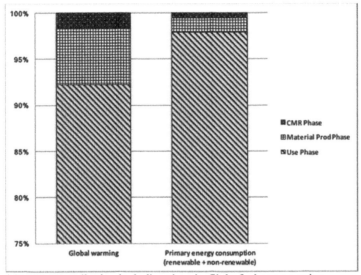

a) Percent contribution including time-in-flight fuel consumption

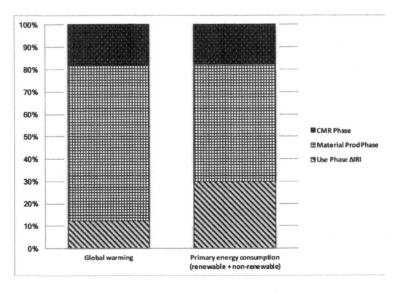

b) Percent contribution including only aircraft fuel consumed from change in IRI

Figure 8. Phase impact contributions—Reconstruction Case.

The second component used to determine the total CMR impacts was the materials used in the maintenance and the rehabilitation. As expected, reconstruction had the highest material impacts of all three cases, replacing the full pavement structure to the subgrade. A 30% higher value in energy consumption and a 37% higher value in GWP production were seen in the reconstruction method compared to rubblization. These increases are 6% higher than the PCP method.

5.1.1 *Reconstruction impacts per phase and functional units*

The total impact from each phase was summed and then divided by the functional units for the airport. Table 5 shows the total values for the impacts using both aircraft fuel consumed for one-half the time in flight as well as for only the aircraft fuel consumed due to an increase in IRI. Figure 8a and 8b show the percent contribution from each phase. When only using the fuel consumed by change in IRI, MP phase accounts for about 42% of primary energy consumption, CMR phase accounts for 28% and the use phase is approximately 30%.

6 CONCLUSION

LCA is one of many tools available to airport owners, operators, engineers and contractors to make educated decisions regarding infrastructure investment and their influence on environmental impacts. LCA-AIR was used to evaluate three rehabilitation options consisting of rubblization with mill/AC inlay, precast panel replacement and full-depth reconstruction for O'Hare International Airport TW A&B. The impacts were normalized to two functional units (square yard and pounds-mile traveled). Each case had the same values for the material production phase (initial construction) and use phase. The construction, maintenance and rehabilitation phase showed differences stemming from the quantity of materials, the construction methods and equipment used in the proposed rehabilitation methods. The rehabilitations at year 30 extended the pavement life to 50 years. The rubblization with mill/AC inlay required a second AC mill/inlay at year 40 to reach 50 years. The precast panels and reconstruction performance life was an additional 20 years from the time of rehabilitation. The results showed the rehabilitation with the least impact for GWP was as follows: precast concrete panel replacement, rubblization with a mill/AC inlay, and full-depth reconstruction. The ranking of least impact for energy for the three rehabilitation strategies were rubblization with a mill/AC inlay, precast concrete panel replacement, and finally, full-depth reconstruction. The GWP potential for rubblization was 2,395 kg CO_2/yd^2 (4.310 × 10^{-10} kg CO_2/lb-mile), for reconstruction was 2,409 kg CO_2/yd^2 (4.729 × 10^{-10} kg CO_2/lb-mile), and for PCP was 2,413 kg CO_2/yd^2 (4.736 × 10^{-10} kg CO_2/lb-mile). The energy consumed for rubblization was 0.1861 TJ/yd^2 (3.576 × 10^{-8} TJ/lb-mile), for PCP was 0.1863 TJ/yd^2 (3.657 × 10^{-8} TJ/lb-mile) and for reconstruction was 0.1864 TJ/yd^2 (3.658 × 10^{-8} TJ/lb-mile).

REFERENCES

Al-Qadi, I.Y., Kang, S., Ozer, H., Ferrebee, Roesler, J.R., Salinas, A., et al. (2015). Development of Present and Baseline Scenarios to Assess Sustainability Improvements of Illinois Tollway Pavements Using a Life Cycle Assessment Approach. *Transportation Research Record: Journal of the Transportation Research Board*.
Bockes, R. (2015, April 27). Owner/CEO—HEM Paving. (J. Kulikowski, Interviewer)
CATERPILLAR. (2015). *CATERPILLAR PERFORMANCE HANDBOOK* (Vol. 45). Peoria, Illinois, United States of America: CATERPILLAR.
Chicago Airports Resources Enterprise. (2014). *Taxiways A and B Subgrade Evaluation O'Hare International Airport (ORD) Chicago, Illinois*. Chicago, IL.
Craftco. (2015). *Craftco Equipment*. Retrieved April 2015, from http://www.crafco.com
Environmental Protection Agency. (2008). NONROAD Emissions Model.
Goldhammer, M.I., & Plendl, B.R. (2014). Surface Coatings and Drag Reduction. *Aero* (49), pp. 14–20.
Hanson Engineering. (2015, August 06). Milling and Paving Operations on Rubblized Concrete.

Horvath, A. (2004). A Life-Cycle Analysis Model and Decision-Support Tool for Selecting Recycled Versus Virgin Materials for Highway Applications. Research Report, University of California at Berkeley, Berkeley, CA.

International Civil Aviation Organization. (2014). Annual Report of the Council. Montréal, Quebec, Canada.

Kulikowski, J. (2015, December). Development of A Pavement Life Cycle Assessment Tool for Airfield Rehabilitation Strategies. 196. Urbana-Champaign, IL, United States of America: Univeristy of Illinois at Urbana-Champaign.

Kulikowski, J., Mohammed, S., Sladek, M., & Roesler, J. (2016, August). Development of LCA-AIR —An Airport Pavement Life Cycle Assessment Tool. *11th International Conference on Concrete Pavements* (p. 196). San Antonio, TX: International Society for Concrete Pavements.

MACTEC. (2007). *Taxiway 'B' Pavement Condition Report Chicago O'Hare International Airport*. Beltsville, MD.

Muench, S., Anderson, J., Weiland, C., Horvath, A., Pacca, S., Masanet, E., & Canapa, R. (2011, January 24). PaLATE v2.2 for Greenroads as Modified by the University of Washington (UW) Software and User Guide. Washington.

Mukherjee, A., & Cass, D. (2011, May 1). PE-2: Project Emmission Estimator.

Pullen, A.B., Edwards, L., Rutland, C.A., & Tingle, J.S. (2014). *Field Evaluation of Ultra-High Pressure Water Systems for Runway Rubber Removal*. US Army Core of Engineers, Engineer Research and Development Center—Geotechnical and Structures Laboratory. Vicksburg: US Army Core of Engineers.

Ross, M. (2011, April 26–28). *Penhall Company*. Retrieved April 22, 2015, from National Concrete Pavement Technology Center: http://www.cptechcenter.org/ncc/TTCC-NCC-documents/F2008-F2011/Equipment-Ross.pdf

Santero, N.J., Masanet, E., & Horvath, A. (2011). Life-Cycle Assessment Of Pavements Part II: Filling The Research Gaps. *Resources, Conservation and Recycling* (55), 810–818.

Senguttuvan, P. (2011). Global Trends in Air Transport: Traffic, Market Access & Challenges. *World Route Development Strategy Summit*. Berlin: International Civil Aviation Organization.

Shinners, M. (2015, Apr 12). Guillotine and Multi-Head Breaking at Airfields. (J. Kulikowski, Interviewer).

Shinners, M. (2015, April 10). Breaker Machine Efficiency. (J. Kulikowski, Interviewer).

Wang, T., Lee, I.-S., Harvey, J.T., Kendall, A., Lee, E.-B., & Kim, C. (2012). *UCPRC Life Cycle Assessment Methodology and Initial Case Studies on Energy Consumption and GHG Emissions for Pavement Preservation Treatments with Different Rolling Resistance*. University of California, Davis. Davis, CA: Institue of Transporation Studies.

Wirtgen Group. (2015, April). *Wirtgen Products*. Retrieved April 2015, from http://wirtgenamerica.com/us/products/wirtgen/product-entry-wirtgen.html

Yu, B., & Lu, Q. (2012). Life Cycle Assessment of Pavement: Methodology and Case Study. *Transportation Research Part D, 17*, 380–388.

Exploring alternative methods of environmental analysis

A.F. Braham
University of Arkansas, Fayetteville, AR, USA

ABSTRACT: While the majority of businesses focus on economic repercussions of business decisions, almost all Civil Engineering infrastructure projects have an impact on the environment as well. The most traditional method of analyzing environmental impacts of projects is through a Life Cycle Analysis, which tracks emissions such as Carbon Dioxide (CO_2) or Greenhouse Gasses through the production, construction, use, and end of life of projects. However, there are several other tools that can be used, including Ecological Footprint and Planet Boundary. These two tools are introduced with discussion on how to incorporate them into pavement design.

The concept of Ecological Footprint, or EF, originated in the early 1990s. The concept is based on nature's capital, and the fact that certain needs are necessary for human life. These needs include healthy food, energy for mobility and heat, fresh air, clean water, fiber for paper, and clothing and shelter. The goal of the EF was to develop a scientifically sound calculation and that could relate to clear policy objectives. In addition, it needed a clear interpretation, to be understandable to non-scientists, and to cover the functioning of a system as a whole. Finally, the metrics had to be based on parameters that are stable over long periods of time so that minor or local fluctuations would not compromise quantifications.

The concept of Planet Boundary was first proposed in 2009 and is defined as a "safe operating space" for humanity. According to this theory, if human activities stay within the safe space, the earth is able to absorb the human activities with no long-term harm to the environment; however, if the human activities move outside of the safe space, the Planet Boundary theory states that long-term harm may occur to the environment. These spaces are associated with the earth's biophysical subsystems and processes.

1 SUSTAINABILITY BACKGROUND

1.1 *What is sustainability – the United Nations*

The term "sustainability" is currently very popular. Both the public and private sector realize the benefits of protecting the future while succeeding in the present. In the present, sustainability is most often defined as incorporating three pillars into design: economics, environmental, and social. One path of this definition of the three pillars was developed by the United Nations (UN) through a series of conferences and forums.

The first significant milestone for sustainability within the UN was the World Conservation Strategy, developed in 1980 (IUCN, 1980). This strategy revolved around three goals that focused on the concept of protecting the environment, with terms such as ecological processes, life support systems, genetic diversity, species, and ecosystems. Seven years later the UN released the Brundtland Commission Report, which is probably the most recognizable milestone in the UN's sustainability development (Brundtland, 1987). The primary theme of the Brundtland Commission reads that sustainability "meets the needs of the present without compromising the ability of future generations to meet their own needs." This theme is independent of protecting the environment, but the concept of the environment is still woven into the fabric of the theme. In 2002 the UN hosted a World Summit on Sustainable Development, which for the first time defined what are called the three pillars of sustainability:

economics, environment, and social (UN, 2002). While the UN has continued to explore the concept of sustainably, the 2002 summit provided the foundation of the three pillars, which are generally accepted as the standard definition of sustainably. It is through this definition that organizations much closer to the pavement community, such as the American Society of Civil Engineers (ASCE), have also embraced the concept of sustainability.

1.2 What is sustainably – the American Society of Civil Engineers (ASCE)

ASCE was founded in 1852, is the oldest engineering society in the United States, and has more than 150,000 members across 177 countries. ASCE defines sustainability as: "A set of environmental, economic and social conditions in which all of society has the capacity and opportunity to maintain and improve its quality of life indefinitely without degrading the quantity, quality or availability of natural, economic, and social resources." This definition clearly incorporates the three pillars of sustainability (economics, environment, social) as developed by the UN. Not only does ASCE have a formal definition of sustainably, but the ASCE Code of Ethics (ASCE, 2006) mentions sustainably on multiple occasions.

The ASCE Code of Ethics has four fundamental principles and seven fundamental canons. Sustainability is mentioned at the very beginning of the Code in the first principle: "using [engineer's] knowledge and skill for the enhancement of human welfare and the environment." This principle directly addresses two of the three pillars of sustainability, environment and social. In addition to the first principle, sustainability is mentioned in several of the seven canons. Canon 1 says "engineers shall… strive to comply with the principles of sustainable development." Canon 1 also states that engineers need to work for the advancement of safety, health, and well-being of their communities (social pillar) and the protection of the environment (environment pillar). Canon 3 continues the sustainability theme by asking engineers to endeavor to extend public knowledge of engineering and suitable development (social pillar). By incorporating sustainably concepts into both the principles and canons, ASCE enforces the commitment of the civil engineering community in understanding and incorporating sustainable practices.

1.3 Quantifying and qualifying sustainability

Using the concept of the three pillars of sustainability, economic, environment, and social, many quantifications and qualifications have been developed. The economic pillar is by far the most developed, with concepts such as Life Cycle Cost Analysis, present/future/annual worth, rate of return, and benefit/cost ratio. On the other end of the spectrum, the social pillar is the least developed. While tools are available, such as the Oxfam Doughnut, Human Development Index, and Social Impact Assessment, there are limited metrics that either quantify or qualify the social aspect of civil engineering projects (Braham and Moon, 2016).

The development of the environment pillar lies somewhere in between the economic and social pillars. There has been significant work performed on Life Cycle Analysis (LCA). For example, LCA has been utilized to compare flexible pavement to rigid pavement (Weiland & Muench, 2010), pavement life (Harvey et al., 2016), and has provided the foundation for a Pavement Life Cycle Assessment Workshop held at the University of California Davis in May, 2010. Other tools have been developed in roadways as well, such as the Greenroads rating system. Greenroads, founded as a company in summer 2010 by Jeralee Anderson and Steve Muench, has eleven categories of project requirements and thirty-seven voluntary requirements. Project requirements that revolve around environment concepts range from runoff flow control to ecological connectivity to environmental training (Anderson & Muench, 2013). Finally, tools have been utilized in order to better capture environmental influences of pavements through Environmental Impact Assessments (EIA) (Moretti et al., 2013). EIAs allow for a systemic analysis of the impact of pavement design, production, construction, use, and end of life on the environment.

While tools such as LCA, Greenroads, and EIA have been utilized for pavements, there are two tools that have been developed that have not been utilized for pavements. These tools are Ecological Footprint and Planet Boundary. This report will provide an overview of these tools, along with recommendations for how to potentially leverage these tools to better understand the environmental impact of pavements.

2 ECOLOGICAL FOOTPRINT

Ecological Footprint, or EF, was developed at the University of British Colombia in the early 1990s (Wackernagel, 1994). The concept is based on nature's capital, and the fact that certain needs are necessary for human life. These needs include healthy food, energy for mobility and heat, fresh air, clean water, fiber for paper, and clothing and shelter. The goal of the EF was twofold: develop a scientifically sound calculation and clearly relate to policy objectives. In addition, it needed to have clear interpretation, be understandable to non-scientists, and cover the functioning of a whole system. Finally, the metrics had to be based on parameters that are stable over long periods of time so that local or other minor fluctuations would not compromise quantifications.

EF is based on taking specific economy or activity's energy needs, and converting that energy and matter to land and water needs. In short, this is determined through a five step calculation. First, the consumption of either a city, region, state, or country is calculated and split into food, housing, transportation, consumer goods, and services. Second, land area of the analysis zone is appropriated into either cropland, grazing, forest, fishing ground, carbon footprint, or built-up land. Cropland is land available to produce food and fiber for human consumption, feed for livestock, oil crops, and rubber. Grazing is land that can raise livestock for meat, dairy, hide, and wool products. Forest provides the land for lumber, pulp, timber products, and wood for fuel, while fishing ground covers the primary production area required to support the fish and seafood caught. While forest is one category for providing wood products, the carbon footprint is the amount of forest land required to absorb CO_2 emissions. Finally, the last category is built-up land, which is the area of land covered by human infrastructure. Once the consumption and land use is identified, both resource and waste flow streams are calculated, which is the third step in the calculation. The fourth step is the construction of a consumption/land-use matrix. This matrix shows all categories of both consumption and land use and indicates where there is not enough land for certain consumptions as well as which land is excess land. The deficiencies give numbers greater than one while the excess give numbers less than one. The fifth and final step sums all of the numbers and provides an estimate of EF for a region. These five steps are summarized in Table 1.

When considering EF from a country level, it is interesting to note that the highest EF countries are from the Middle East according to a 2010 report published by the Global Footprint Network (Ewing et al., 2010). This report states that the United Arab Emirates (UAE) and Qatar were producing EFs greater than 10.0 global hectares per person. This number states that if every person in the world was living the standard of living of the average UAE citizen living on UAE's resources, we would need over ten earths to sustain life. The next grouping down consists of western, fully developed countries, which required approximately 5–8 earths to maintain their standard of living. The list continues down through second world, developing, and third world countries. According to the report, it is interesting to

Table 1. Five-step calculation for Ecological Footprint.

Step	Description of each step
One	Consumption of food, housing, transportation, consumer goods, and services determined
Two	Land area appropriated into cropland, grazing, forest, fishing ground, carbon footprint, or built-up land
Three	Resource and waste flow streams calculated
Four	Construction of a consumption/land-use matrix
Five	Sum all of the numbers, provide an estimate of EF for a region

note that the countries requiring less than one earth is quite diverse both geographically and socio-economically, from the Democratic Republic of the Congo (population 63 million) to Bangladesh (population 158 million) to Puerto Rico (population 4 million).

One study that has been performed is using impervious surfaces, which includes pavements, as a proxy measure for EF (Sutton et al., 2009). Since it is relatively easy to calculate constructed areas per person from satellite images, it is convenient to use this measurement to determine EF instead of more difficult measures such as fiber and fuel wood consumption, two traditional inputs into an EF analysis. By using aerial photographs from thirteen cities in the United States, impervious surfaces (such as rooftops, sidewalks, parking lots, and roadways) were identified and the ratio of impervious surfaces to pervious surfaces (such as lawns, parks, and golf courses) over 100 random points across the image were classified. An R^2 value of 0.78 was found between the impervious surfaces and the EF, providing decent correlation between percentage of impervious surfaces and EF. Sutton et al. (2012) continued work in this area by developing a monetary correlation between impervious surfaces and consumption of ecosystem services.

There are, of course, some drawbacks to the EF concepts. First, the physical consumption-land conversion factor weights do not necessarily correspond to social weights. The analysis focuses one hundred percent on the metrics at hand, but do not consider the social choices people have to make. Second, the EF does not distinguish between sustainable and unsustainable use of land, only that land is being consumed. Therefore, forest could be clear-cut or sustainably harvested, two processes to extract wood from nature, but the EF would treat these practices as the same. A third criticism is that in the EF model there are many options to compensate for CO_2 emission and CO_2 assimilation, such as by forest, chemosynthesis, and autotrophs. However, the EF model only compensates for CO_2 emission and assimilation by forest, neglecting the other options. A fourth criticism is that there is a significant correlation between population density and resource endowment. As populations move away from rural living to urban living, the EF will increase significantly, especially as the analysis zone shrinks. This artificially inflates the EF of urban areas while perhaps underestimates the true EF of rural areas. The fifth and final criticism discussed here is that EF is hard to use as a planning device. While it is noble to attempt to decrease the EF, there are few tangible concepts that agencies can focus on to begin the reduction, making it difficult to leverage. While no measure is truly perfect, these deficiencies have led to the development of other metrics, including the Planet Boundary.

3 PLANET BOUNDARY

Planet Boundary as a tool to evaluate environmental impact was first proposed in 2009 and is defined as a "safe operating space" for humanity (Rockström et al., 2009a). According to this theory, if human activities stay within the safe space, the earth is able to absorb the human activities with no long-term harm to the environment; however, if the human activities move outside of the safe space, the Planet Boundary theory states that long-term harm may occur to the environment. These spaces are associated with the earth's biophysical subsystems and processes.

A major premise of Planet Boundary theory is that the environment has been unusually stable for past 10,000 years, commonly referred to as the Holocene period. During this time, the earth's temperatures, freshwater availability, and biogeochemical flows have all stayed within a narrow, stable range (Rockström et al., 2009b). However, with the beginning of the Anthropocene period, which is believed to have started after Industrial Revolution in the 1800s, human influence may have begun to damage the system that keeps earth within Holocene state according to Planet Boundary theory. This system was divided into nine subsystems, eight of which have been quantified. The nine subsystems have some overlap with both concepts learned in the LCA section and the EF section, but they also venture into new areas. The nine Plant Boundary subsystems are summarized and quantified in Table 2 for both 2009 and 2015 (Steffen et al., 2015), with a brief discussion following (Note: the nitrogen and phosphorus cycle are combined into one process called biogeochemical flows).

Table 2. Planet Boundary subsystems.

Earth-system process subsystems	Proposed boundary	2009 Status	2015 Status
Climate change (ppm CO_2)	350	387	396.5
Rate of biodiversity loss (extinctions per millions species-years)	10	>100	100–1000
Biogeochemical flows: Nitrogen cycle (millions tons/yr)	62–82	121	150
Biogeochemical flows: Phosphorus cycle (millions tons/yr)	6.2–11.2	8.5–9.5	14–22
Stratospheric ozone depletion (Dobson unit, DU)	276	283	As low as 200
Ocean acidification (carbonate ion concentration)	>2.75	2.90	2.89
Global freshwater use (km^3/yr)	4,000	2,600	2,600
Change in land use (percentage)	75	11.7	62
Atmospheric aerosol loading (aerosol optical depth)	0.25–0.50	Not determined	0.30
Chemical pollution	Not determined	Not determined	Not determined

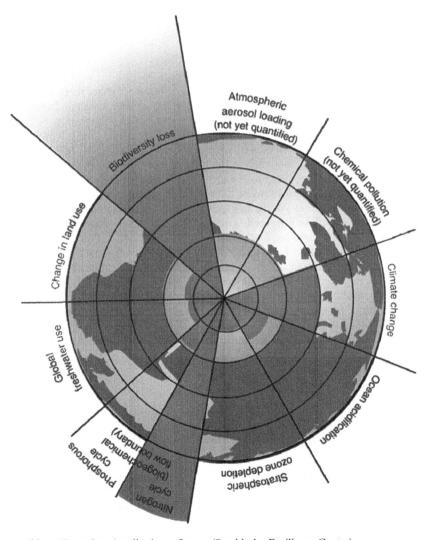

Figure 1. Planet Boundary (credit: Azote Images/Stockholm Resilience Centre).

In Table 2, climate change is quantified by measuring the atmospheric carbon dioxide concentration, with units of parts per million by volume. Another quantification discussed for climate change was the change in radiative forcing. This radiative forcing was listed at 1.5 in 2009 and is set at 1.0 for the boundary. Additionally, the rate of biodiversity loss was measured by the extinction rate, and is the number of species per million species per year lost. The nitrogen cycle is the amount of N_2 removed from the atmosphere for human use in millions of tons per year, while the phosphorus cycle is the quantity of P flowing into the oceans per year, in millions of tons. The stratospheric ozone depletion is the concentration of ozone, using the Dobson unit, while the global freshwater use is the consumption of freshwater by humans per year, in kilometers cubed. The change in land use is simply the percentage of global land cover converted to cropland from the natural state of the land. The atmospheric aerosol loading is the particulate concentration in the atmosphere on a regional basis, and the chemical pollution, which include emissions of everything from organic pollutants, to plastics, to heavily metals, and nuclear waste, has not yet been quantified. Figure 1 shows a visual image of the processes.

Like the EF, Planet Boundary theory has some pros and cons, but it is another tool that potentially can be used to quantify the influence of Civil Engineering infrastructure on the environment. At this point, no research has explicitly examined the influence of pavements on Planet Boundary, but it appears that the tool has promise in both pavements and other Civil Engineering applications.

4 CONCLUSIONS

The United Nations and the American Society of Civil Engineers have made sustainability a cornerstone of their organizations. These groups partially define sustainability as revolving around three pillars: economic, environment, and social. Unlike the variety of established tools available for the economic pillar of sustainability, there is only one well established tool to quantify the environmental aspects of sustainability, the Life Cycle Assessment (LCA). While other concepts are available, such as Ecological Footprint and Planet Boundary, these have not been implemented across a wide range of civil engineering applications. Ecological Footprint finds the relationship between human consumption, in the form of food, housing, transportation, consumer goods, and services and the land area available to produce these consumables, including cropland, grazing, forest, fishing ground, carbon footprint, and built-up land. While there have been no research project directly relating pavements to Ecological Footprints, relationships between impervious surfaces and Ecological Footprint have been developed and are relatively well correlated. Planet Boundary has established safe operating spaces in nine earth system processes (including concepts such as climate change, biodiversity loss, and ocean acidification). While there has been no direct research in establishing the impact of pavements on Planet Boundary, the concepts are conducive to exploring the application. Overall, there is promise in both Ecological Footprint and Planet Boundary, and these two tools may become more salient in the future of pavement research.

REFERENCES

Anderson, J. & Muench, S. 2013. Sustainability Trends Measured by the Greenroads Rating System. *Transportation Research Record: Journal of the Transportation Research Board*, No. 2357, Transportation Research Board of the National Academies, Washington, D.C., pp. 24–32.

ASCE. 2006. Code of Ethics. American Society of Civil Engineers, amended 2006.

Braham, A. & Moon, Z. 2016. Building a Framework for the Social Pillar of Sustainability in Transportation Engineering. Submitted to *Transportation Research Record: Journal of the Transportation Research Board*, July 2016.

Brundtland, G.H. 1987. Report of the World Commission on Environment and Development: Our Common Future. United Nations.

Ewing, B., Moore, D., Goldfinger, S., Oursler, A., Reed, A., & Wackernagel, M. 2010. The Ecological Footprint Atlas 2010, Oakland, CA, USA.

Harvey, J., Meijer, J., Ozer, H., Al-Qadi, I., Saboori, A., & Kendall, A. 2016. Pavement Life-Cycle Assessment Framework. *Federal Highway Administration*, Final report FHWA-HIF-16-014.

IUCN. 1980. World Conservation Strategy, Living Resource Conservation for Sustainable Development. International Union for Conservation of Nature and Natural Resources (IUCN), United Nations.

Moretti, L., Di Mascio, P., & D'Andrea, A. 2013. Environmental Impact Assessment of Road Asphalt Pavements. *Modern Applied Science*; Vol. 7, No. 11.

Rockström *et al.* 2009a. A safe operating space for humanity. *Nature*, Vol. 461, September.

Rockström *et al.* 2009b. Planet Boundaries: Exploring the Safe Operating Space for Humanity. *Ecology and Society*, Vol. 14, Issue 2.

Steffen *et al.* 2015. Planetary boundaries: Guiding human development on a changing planet. *Science*, Vol. 347, Issue 6223.

Sutton, P., Anderson, S., Elvidge, C., Tuttle, B. & Ghosh, T. 2009. Paving the planet: impervious surface as proxy measure of the human ecological footprint. *Progress in Physical Geography*; Vol. 33, pp. 510–527.

Sutton, P., Anderson, S., Tuttle, B. & Morse, L. 2012. The real wealth of nations: Mapping and monetizing the human ecological footprint. *Ecological Indicators*; Vol. 16, pp. 11–22.

UN. 2002. Report of the World Summit on Sustainable Development. United Nations, Johannesburg, South Africa.

Wackernagel, M. 1994. Ecological Footprint and Appropriated Carrying Capacity: A Tool for Planning Toward Sustainability. *Ph.D. Dissertation, The University of British Colombia*, Vancouver, Canada.

Weiland, C. & Muench, S. 2010. Life-Cycle Assessment of Reconstruction Options for Interstate Highway Pavement in Seattle, Washington. *Transportation Research Record: Journal of the Transportation Research Board*, No. 2170, Transportation Research Board of the National Academies, Washington, D.C., pp. 18–27.

An uncoupled pavement-urban canyon model for heat islands

S. Sen & J.R. Roesler
Department of Civil and Environmental Engineering, University of Illinois at Urbana-Champaign, Champaign, USA

ABSTRACT: Urban Heat Islands (UHIs) are a major environmental consequence of developing urban infrastructure, including pavements. The effect of a specific pavement on the urban environment depends on not just the pavement structure, but also the weather and urban form of the location. To develop a rational approach towards incorporating these variables in UHI analysis, a microscale, uncoupled pavement-urban canyon model was developed and applied on the warmest hour for 30-year representative weather data in Chicago. The UHI intensity was found to vary spatially not just with the aspect ratio of the urban canyon, but also its relative position, as well as the structure of the pavement. Furthermore, future weather scenarios such as warming trends, elevated the UHI intensity.

1 INTRODUCTION

According to the World Bank, the proportion of people living in urban areas around the world is over 50% and continues to rise (The World Bank, 2015), with the proportion in the US being over 80%. Previous studies (Kleerekoper, van Esch & Salcedo, 2012, Oke, 1988, Oke, 1982, Aflaki et al., 2016, Santamouris, 2015) have investigated the energetics that lead to the development of UHI and have described the UHI intensity (ΔT_{ur}), the difference between the urban air temperature and adjoining rural area, as ranging from less than 1°C to over 10°C. In particular, the role of pavements in developing UHI through absorption and storage of solar radiation, and methods to mitigate the effect through reflective pavements, has received great attention (Antonaia et al., 2016, Qin, 2015, Santamouris, 2013, Yang, Wang & Kaloush, 2015, Sen & Roesler, 2016).

Pavement-induced UHI has been studied through a microscale approach that uses Computational Fluid Dynamics (CFD) to model the movement of heat through the urban environment (Declet-Barreto et al., 2013, Herbert, Johnson & Arnfield, 1998, Saneinejad et al., 2012, Taleghani, Sailor & Ban-Weiss, 2016, Toparlar et al., 2015). In particular, a few studies have also investigated the role of the urban canyon and its form on temperature distribution (Uehara et al., 2000, Xie et al., 2006). However, these studies have two shortcomings: they model the pavement as a single layer, typically a thin layer on top of soil; and they select an arbitrary "warm" day of a year, or arbitrary values of surface temperatures, for the analysis. Previous studies (Sen & Roesler, 2014, Gui et al., 2007) have shown that the thermophysical properties of pavements have an impact on their surface temperature, which should also be captured in microscale UHI modeling.

Current UHI studies focusing on outdoor temperatures, including the impact of pavements, only take into account the present weather conditions without considering the impact of possible future climate change. In the closely-linked area of building energy modeling however, a few studies (Crawley, 2008, Chan, 2011) have looked at developing future weather scenarios, mostly by modifying existing weather data through additive or multiplicative transformations. In particular, some studies (Du, Underwood & Edge, 2012, Eames, Kershaw & Coley, 2010) have used global future climatic simulations to estimate outdoor air temperatures in the UK, without focusing on pavements, which in turn was used to simulate building energy performance.

At the 2014 Pavement LCA Symposium, the authors showed (Sen & Roesler, 2014) how a set of pavement structures with varying surface and sub-surface thermophysical properties, simulated under the same weather conditions, showed different surface temperature distributions. This study follows up on that work to demonstrate a rational method to select a "warm" day for modeling UHI, and extends a pavement thermal model to act as a boundary condition for an urban CFD model for various urban forms for a single "warmest" hour of the year. Furthermore, it uses an additive approach to extend the analysis for future weather scenarious.

2 MATERIALS AND METHODOLOGY

2.1 Scope

The aim of this study was to analyze the effect of pavement structure, surrounding urban form, and climate change on the average air temperature at 2 m in an urban canyon, where human activity takes place. Four pavement structures are analyzed representing various changes to the pavement surface and sub-surface layers, which are discussed in Section 2.2. These were analyzed in three urban canyon configurations representing different urban forms, as discussed in Section 2.3. The far field weather data was representative of average conditions over 30 years, and was modified to simulate possible variations in the future, as discussed in Section 2.4. Finally, all these variables were analyzed using a 1D pavement heat transfer model as an uncoupled input to a CFD solver, as discussed in Section 2.5.

2.2 Pavement cases

In this study, four pavement structures that were analyzed previously (Sen & Roesler, 2014) were considered. While details of the cases can be found in that paper, a brief summary is presented here:

a. Case P: This was a control case with a 100 mm thick concrete pavement of albedo 0.30 overlying a granular base of 150 mm, and a 300 mm granular subbase.
b. Case PL: This case was identical to the control case except that for the surface layer, lower-density concrete was used, which singifincatly lowered the density, thermal conductivity, and heat capacity of the layer.
c. Case PT: This case was identical to the control case except that the albedo of the surface was raised modestly to 0.35.
d. Case PC: In this case, the base layer of the control case was changed to a Cement Treated Base (CTB) to effectively change the thickness of the concrete surface layer from 100 mm to 250 mm.

2.3 Urban forms

The urban canyon, which is essentially a road surrounded by buildings on both sides, is the fundamental repeating unit of a typical urban form, which can be studied in either 2D or 3D. Several studies (Pearlmutter, Berliner & Shaviv, 2006, Kruger, Pearlmutter & Rasia, 2010, Kondo et al., 2001) have shown that the shape of the urban canyon as well as its relative configuration are important parameters to take into account for urban CFD modeling. The urban canyon configuration adopted for this study is shown in Figure 1.

The fluid domain, which will be modeled in Figure 1, is shown with the definition of each boundary indicated. The urban canyon consists of a road of width W surrounded on both sides by buildings of height H, for an Aspect Ratio (AR) of H:W. Three identical buildings were assembled for this model. The width of the road was fixed at $W = 10\ m$, so the only free variable was H. The domain height was set at 20H and the wake length at 4H in accordance with best practice guidelines (Franke et al., 2007).

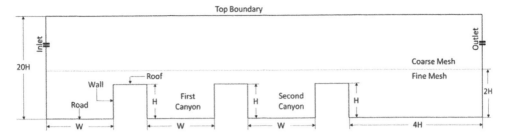

Figure 1. General urban canyon configuration with fluid domain.

In Figure 1, two canyons are marked as 'First Canyon' and 'Second Canyon'. The area to the left of the first canyon is used to allow the inlet wind profile to develop, and the area to the right of the second canyon is to allow for the development of a wake and to ensure that the outlet conditions do not affect the results. The top boundary is set at such a height as to ensure a 5% blockage ratio, as recommended. Three configurations were considered: $H:W = 0.5:1, 1:1,$ and $2:1$, by changing the height H progressively from 5 m to 20 m. The primary UHI metric of interest is the average air temperature at 2 m in the canyon, which represents the Canopy Layer UHI and the height impacting human health and activity.

2.4 *Far field weather conditions*

In the CFD model, far field weather conditions have to be set as boundary conditions, which represent the general mesoscale conditions in the urban areas without regard to microscale variations. As discussed previously, the current practice is to select these conditions from meterological data for an arbitrary "warm" day of the year. As an alternative, a method was used to make a more rational selection and apply it to the "warmest" hour of the year, which can later extended to an entire day. Specifically, temperature and wind speed profile need to be specified. In this study, the Typical Meterological Year Series 3 (TMY3) was adopted, which uses weather data from 1961–1990 to construct a weather file (Hall et al., 1978) and, as the name suggests, represents a typical year in terms of weather conditions.

As an illustration, the TMY3 data for Chicago, Illinois (O'Hare International Airport) was extracted from the National Renewable Energy Laboratory (NREL, 2005), and the warmest hour in the database (which corresponded to July 19, 3:00 PM) in terms of dry bulb temperature, was used as the uniform inlet temperature profile boundary condition. This temperature was $T_0 = 35°C$. The uniform inlet wind speed profile was set to a typical value of $U_0 = 2$ m/s normal to the inlet boundary. For determining ΔT_{ur}, TMY3 data for an adjoining rural area is needed. However, the TMY3 data is maintained only for major airports in the region. Therefore, data from DuPage Airport, which lies about 60 km west of Chicago and represents a suburban area, was used. The corresponding dry bulb temperature for DuPage Airport was 31°C for the warmest hour in Chicago (July 19, 3:00 PM). Thus, the mesoscale UHI intensity was 4°C (35°C – 31°C).

For this CFD models, three far field weather conditions were analyzed. The first used the TMY3 data with $T_0 = 35°C$ for Chicago and 31°C for DuPage, and this case was labeled 'TMY+0'. Then, the hourly dry bulb temperature was uniformly increased by 1°C and 2°C for 'TMY+1' and 'TMY+2' cases, for both Chicago and DuPage (thus, the mesocale UHI intenstity is kept constant at 4°C). This additive increase in air temperature, which follows from the approach suggested by (Crawley, 2008), is used to simulate possible future changes in climate. This is a simplistic approach and a much more intensive climatological study would be required to obtain better future weather files but suffcient for this study. For the TMY+1 and TMY+2 cases, as in the TMY+0 case, the warmest hour was selected for analysis, for which $T_0 = 36°C$ **and** $37°C$ for Chicago and $32°C$ and $33°C$ for DuPage, respectively. The inlet wind speed was kept at 2m/s for all cases.

2.5 Analysis

For each combination of pavement type, urban canyon form, and weather condition, two uncoupled analyses were conducted, as discussed next.

2.5.1 1D pavement thermal analysis

A 1D pavement thermal analysis was performed on each combination of pavement case and weather conditions using a 1D heat transfer model, ILLITHERM (Sen and Roesler, 2016), developed by the authors. In this case, the weather conditions corresponded to the weather files generated from the TMY3 data and adjusted accordingly for the three far field weather condition cases, TMY+0, TMY+1, and TMY+2. In addition to weather data, the model also requires thermophysical properties of each pavement type, which can be found in (Sen and Roesler, 2014). From ILLITHERM, the pavement surface temperaure corresponding to the warmest hour in the TMY3 series was extracted and used as a boundary condition for the CFD analysis, as described next. Because the two modeling steps were performed separately, this approach can be described as an uncoupled pavement-urban canyon model.

2.5.2 2D CFD analysis

The urban canyon geometry shown in Figure 1 was created using the commercial CFD software, ANSYS FLUENT (ANSYS, Inc., 2011). Before running a CFD simulation, a structured mesh was generated with elements of size 0.25 m below a height of 2H (marked as 'Fine Mesh' in Figure 1), and 5.0 m with an inflation factor of 10 above (marked as 'Coarse Mesh'). This mesh configuration balance the need for a fine resolution within the urban canyon with limitations to available computing capabilities, and was arrived at after a mesh convergence study. Mesh statistics revealed that the mesh was suitable for the simulation, and a cell size of 0.25 m was sufficiently fine for studying UHI at the Canopy Layer.

For the CFD analysis, the inlet boundary conditions were set as described in Section 2.4 and the road temperature was set as explained in Section 2.5.1. In order to isolate the effect of the pavement alone, the temperatures of the walls and roofs were set to be equal to the inlet condition T_0, although in reality, they too would vary like the road surface. A pressure condition was applied at the outlet with an operating pressure of 101,325 Pa (about 1 atm). Finally, the top boundary was set to a symmetry condition as per (Toparlar et al., 2015), although it was revealed in the mesh convergence study that this did not affect the average air temperature at 2 m in each canyon significantly.

The CFD analysis solved the complete Navier-Stokes Equations together with the Energy Equation as well as a turbulence closure model. A realizable $k-\epsilon$ turbulence model with standard wall functions was implemented for the solution. In order to prevent divergence, modest relaxation factors were used. As the Mach number anywhere in the domain was expected to be very small, a pressure-based steady state solver was used. Tolerances for scaled residuals were set at 10^{-3}, except for temperature, which was set at 10^{-6}, and the model was run to convergence.

3 RESULTS AND DISCUSSION

3.1 Pavement surface temperatures

The surface temperatures of each pavement case corresponding to the warmest hour in the TMY3 data were evaluated. A summary of the results is shown in Figure 2. Across the weather cases, TMY+0, TMY+1, and TMY+2, the far field air temperature was increased progressively by 1°C from the baseline temperature obtained from the TMY3 weather data series. The corresponding surface temperatures obtained from ILLITHERM, interestingly, also increased by approximated the same amount, although the actual magnitude varied with the pavement structure. The low density concrete, Case PL, showed the highest surface temperature, while the more reflective case (PT) showed the lowest, and cases P and PC showed similar temperatures. As expected, all the surface temperatures were higher than the far field air temperature.

3.2 *Average wind speed at 2 m*

The velocity field inside the canyon is coupled with the temperature field through advection, which affects the average temperature at 2 m, and hence the UHI. This is a coupled process, as the temperature field affects the thermal stratification and hence velocity field in the canyon as well (Uehara et al., 2000). The study by (Xie et al., 2006) demonstrated that the urban canyon significantly decreases the wind speed from the far field value because of the formation of dissipative vortices, thus entrapping more heat inside. For the uncoupled pavement-urban canyon model, the average wind speed at 2 m in each canyon for a given AR was found to be more or less the same irrespective of the type of pavement under consideration. These wind speeds, normalized to U_0, are shown in Figure 3.

Figure 3 shows there is no significant change in the wind speeds across the three weather cases, given the inlet wind speed U_0 was assumed to be fixed at 2 m/s. The wind speed varies with the Aspect Ratio (AR), increasing as AR goes from 0.5:1 to 1:1 and then decreasing to 2:1. As the AR increases, the turbulence increases, which pushes the air into the canyon.

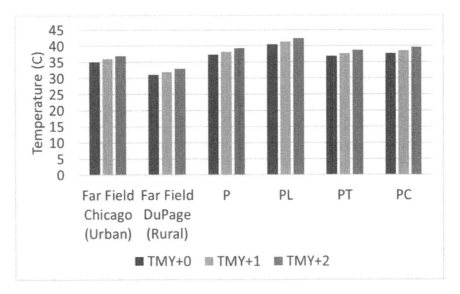

Figure 2. Pavement surface temperature in Chicago corresponding to the hour of maximum far field air temperature in Chicago and DuPage.

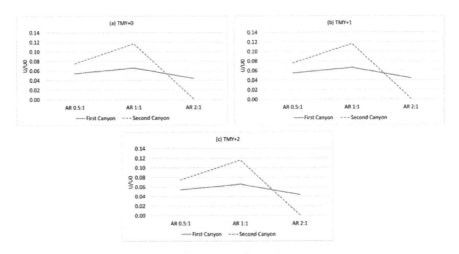

Figure 3. Normalized average wind speeds at 2 m for several aspect ratios (AR), H:W.

However as AR further increases to 2:1, the velocity dissipates before it can reach to 2 m above the canyon floor (i.e., the road).

The relative position of the canyon affects the average wind speed and temperature at 2 m (see Figure 4). This can be seen in Figure 4 (c) in more detail by considering the wind velocity contours for AR 2:1, Case P, for TMY+2. The building before the first canyon causes the velocity field to be displaced upwards, which eventually falls back downwards. Depending on the height of the first canyon, this velocity field may or may not fall back into the second canyon. For AR 0.5:1 and 1:1, it does fall back into the second canyon, whereas it does not for AR 2:1, which explains the curves obtained in Figure 3. Furthermore, the wind speed and temperature distributions shown Figure 4(b) respectively, which indicates that even within a canyon at 2 m, there can be variation from point to point.

3.3 *UHI intensity*

The average temperature at 2 m minus the far field air temperature in DuPage is a measure of the UHI intensity ΔT_{ur} in the corresponding canyon in Chicago. The uncoupled pavement-urban canyon model calculates the pavement temperature first, and then the temperature field in the canyon, from where the average temperature at 2 m can be extracted and ΔT_{ur} evaluated. Across all the cases, this is shown in Figure 5. For all the cases, for a given weather case and canyon form, the PL case shows the highest intensity, followed by cases PC, P, and PT. This correlates with the observations of surface temperature observed in Figure 2, indicating that in this uncoupled model, the air temperature strongly depends on the surface temperature. However, as can be seen in Figure 4(b), there is some variation in air temperature within the canyon even at 2 m, which is not captured when using an average value.

Next, consider the behavior of the intensity with AR. In Canyon 1, the intensity decreases with AR, whereas in Canyon 2, it increases. This behavior results because of the formation of vortices in each canyon, as demonstrated for a single canyon (Xie et al., 2006), which advects the temperature field differently in each canyon. These vortices can be complex for a series of canyons like the geometry considered here, and the resultant behavior is difficult to predict. Therefore, even though the average wind speed shows the same trends with AR for both canyons (first increasing and then decreasing in Figure 3), the temperature fields don't, as it depends on not just the magnitude but also the direction of the wind velocity.

The UHI intensity for a given weather case and AR is the highest for the Case PL and lowest fo the case PT, with the magnitude varying from 4.1°C to 4.6°C. The far field UHI

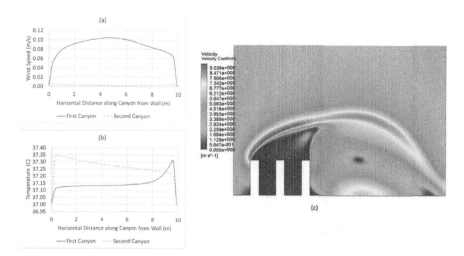

Figure 4. For pavement Case P and TMY+2 with AR 2:1, plots of (a) wind speed at 2 m (b) air temperatures at 2 m, and (c) contours of velocity magnitude.

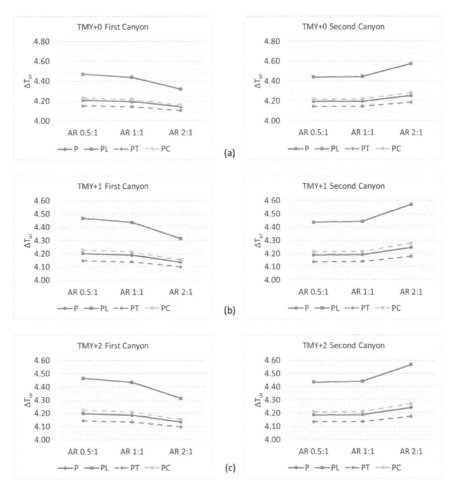

Figure 5. UHI intensity (in °C) at 2 m for (a) TMY+0 (b) TMY+1 and (c) TMY+2 cases in Chicago relative to DuPage. Plots on the left are for the first canyon and those on the right for the second canyon.

intensity, which does not consider the effect of AR and the pavement, would be 4°C, across the weather cases. Thus, pavement type (both surface and sub-surface properties) also effects the local microscale UHI intensity. Finally, the UHI intensity does not seem to vary significantly with the increase in far field temperature in the TMY+0, TMY+1, and TMY+2 cases. However, in these cases, both the urban and rural temperatures were increased by the same magnitude. If the urban temperature increased more than the rural, the UHI intensity would have also increased. The actual differences between urban and rural temperatures in future climatic scenarios is however, not very well-studied and requires more research.

4 CONCLUSION

The urban environment can be complex because of the assortment of building forms and interaction between the wind velocity and air temperature fields, as well as the pavement structures found in a city. Uncertainty in future climate adds to the complexity. To rationally analyze these factors, representative weather data for 30 years in Chicago, the TMY3 series, and predicted future weather scenarios, were input into an uncoupled pavement-urban canyon model for a microscale, CFD-based UHI analysis. To account for the surface temperature of road in UHI analysis, a pavement thermal analysis was performed using four different pavement types, which varied both the surface and sub-surface physical properties. The urban canyon CFD model idealized various urban forms by changing the Aspect Ratio (AR) of the canyon. The analysis was performed for combinations of these variables given the warmest hour of the year represented by the TMY3 data. The CFD-based UHI model estimated the impact of the pavement type, urban form, and weather data on the average wind speed and air temperature at 2 m in the canyon, which was then compared to a rural air temperature to determine the UHI intensity at that warmest hour.

For the TMY3 series at the warmest hour in Chicago, the difference in the 2 m temperature between Chicago and rural Dupage is 4°C. As the far field air temperature increased in Chicago, both the pavement surface temperature and average air temperature at 2 m in an urban canyon increased. If the far field air temperature increases equally in both rural and urban areas, as modeled here, the UHI intensity is expected to remain the same. If however, it increases more for urban areas, the UHI intensity would also increase.

Next, the UHI intensity varies with the type of pavement. The intensity was highest for the Case PL, which uses lower density concrete on the surface, and lowest for Case PT, which is a more reflective concrete. The difference between these cases was about 0.5°C on average, which is a significant difference for UHI intensity. The control pavement, Case P, and the Case PC, which had a CTB instead of a granular base, showed only a marginal difference between themselves for the hour analyzed.

Finally, the UHI intensity varied both with the aspect ratio of the urban canyon as well its position. In the First Canyon, the UHI intensity, ΔT_{ur}, was found to decrease with AR, whereas in the Second Canyon that was adjacent to the First, it was found to increase. The difference can be attributed to the development of complex vortices in the canyon and the resultant variation in wind velocity field in each. Prediction of this phenomenon can be quite complex without using microscale CFD.

The methodology proposed in this paper can be used for pavement engineers to work with communities, with unique urban forms and possible changes in future climate, to design better pavements and roadside features to mitigate UHI. In the future, the development of a coupled model to take into account convective and radiative heat transfer and shadowing is underway.

ACKNOWLEDGEMENTS

Funding for this study was provided by the US Department of Transportation (USDOT) through the University Transportation Center for Highway Pavement Preservation (UTCHPP) at Michigan State University with Contract Number DTR13-G-UTC44.

REFERENCES

Aflaki, A., Mirnezhad, M., Ghaffarianhoseini, A., Ghaffarianhoseini, A., Omrany, H., HuaWang, Z. and Akbari, H. 2016. Urban heat island mitigation strategies: A state-of-the-art review on Kuala Lumpur, Singapore and Hong Kong. *Cities*, In Press.
ANSYS, Inc. ANSYS FLUENT User Guide. Release 14.0. November, 2011.
Antonaia, A., Ascione, F., Castaldo, A., D'Angelo, A., De Masi, R.F., Ferrara, M., Vanoli, G.P. and Vitiello, G. 2016. Cool materials for reducing summer energy consumptions in Mediterranean climate: In-lab experiments and numerical analysis of a new coating based on acrylic paint. *Applied Thermal Engineering* 102: 91–107.
Chan, A.L.S. 2011. Developing future hourly weather files for studying the impact of climate change on building energy performance in Hong Kong. *Energy and Buildings* 43: 2860–2868.
Crawley, D.B. 2008. Estimating the impacts of climate change and urbanization on building performance, *Journal of Building Performance Simulation* 1(2): 91–115.
Declet-Barreto, J., Brazel, A.J., Martin, C.A., Chow, W.T. and Harlan, S.L. 2013. Creating the park cool island in an inner-city neighborhood: heat mitigation strategy for Phoenix, AZ. *Urban Ecosystems* 16(3): 617–635.

Du, H., Underwood, C.P. and Edge, J.S. 2012. Generating design reference years from the UKCP09 projections and their application to future air-conditioning loads. *Building Services Engineering Research and Technology* 33(1): 163–179.

Eames, M., Kershaw, T. and Coley, D. 2010. On the creation of future probabilistic design weather years from UKCP09. *Building Services Engineering Research and Technology* 32(2): 127–142.

Franke, J., Hellsten, A., Schlunzen, H. and Carissimo, B. 2007. Best Practice Guideline for the CFD Simulation of Flows in the Urban Environment. *COST Action 732: Quality Assurance and Improvement of Microscale Meteorological Models.*

Gui, J., Phelan, P.E., Kaloush, K.E. and Golden, J.S. 2007. Impact of Pavement Thermophysical Properties on Surface Temperatures. *Journal of Materials in Civil Engineering* 19(8): 683–690.

Hall, I.J., Prairie, R.R., Anderson, H.E. and Boes, E.C. 1978. Generation of a Typical Meteorological Year. *Proceedings of the Conference on Analysis for Solar Heating and Cooling*, San Diego, CA.

Herbert, J.M., Johnson, G.T. and Arnfield, A.J. 1998. Modelling the thermal climate in city canyons. *Environmental Modelling & Software* 13: 267–277.

Kleerekoper, L., van Esch, M. and Salcedo, T.B. 2012. How to make a city climate-proof, addressing the urban heat island effect. *Resources, Conservation and Recycling* 64: 30–38.

Kondo, A., Ueno, M., Kaga, A. and Yamaguchi, K. 2001. The influence of urban canopy configuration n urban albedo. *Boundary-Layer Meteorology* 100: 225–242.

Kruger, E., Pearlmutter, D. and Rasia, F. 2010. Evaluating the impact of canyon geometry and orientation on cooling loads in a high-mass building in a hot dry environment. *Applied Energy* 87: 2068–2078.

NREL 2005. National Solar Radiation Database, [Online] http://rredc.nrel.gov/solar/old_data/nsrdb/1991–2005/tmy3/ [1 August 2016].

Oke, T.R. 1982. The energetic basis of the urban heat island. *Quarterly Journal of the Royal Meteorological Society* 8(455): 1–24.

Oke, T.R. 1988. Street Design and Urban Canopy Layer Climate. *Energy and Buildings* 11: 103–113.

Pearlmutter, D., Berliner, P. and Shaviv, E. 2006. Physical modeling of pedestrian energy exchange within the urban canopy. *Building and Environment* 41: 783–795.

Qin, Y. 2015. A review on the development of cool pavements to mitigate urban heat island effect, *Renewable and Sustainable Energy Reviews* 52: 445–459.

Saneinejad, S., Moonen, P., Defraeye, T., Derome, D. and Carmeliet, J. 2012. Coupled CFD, radiation and porous media transport model for evaluating evaporative cooling in an urban environment. *Journal of Wind Engineering and Industrial Aerodynamics* 104–106: 455–463.

Santamouris, M. 2013. Using cool pavements as a mitigation strategy to fight urban heat island— A review of the actual developments. *Renewable and Sustainable Energy Reviews* 26: 224–240.

Santamouris, M. 2015. Analyzing the heat island magnitude and characteristics in one hundred Asian and Australian cities and regions. *Science of The Total Environment* 512–513: 582–598.

Sen, S. and Roesler, J.R. 2014. Assessment of pavement structure on urban heat island. *Papers from the International Symposium on Pavement Life Cycle Assessment*, Davis, CA, 191–200.

Sen, S. and Roesler, J.R. 2016. Contextual heat island assessment for pavement preservation. *International Journal of Pavement Engineering*, In Press.

Taleghani, M., Sailor, D. and Ban-Weiss, G.A. 2016. Micrometeorological simulations to predict the impacts of heat mitigation strategies on pedestrian thermal comfort in a Los Angeles neighborhood. *Environmental Research Letters* 11: 024003.

The World Bank 2015. Urban Population (% of Total) | Data, [Online] http://data.worldbank.org/indicator/SP.URB.TOTL.IN.ZS [29 August 2016].

Toparlar, Y., Blocken, B., Vos, P., van Heijst, G.F., Janssen, W.D., van Hooff, T., Montazeri, H. and Timmermans, H.J.P. 2015. CFD simulation and validation of urban microclimate: A case study for Bergpolder Zuid, Rotterdam. *Building and Environment* 83: 79–90.

Uehara, K., Murakami, S., Oikawa, S. and Wakamatsu, S. 2000. Wind tunnel experiments on how thermal stratification affects flow in and above urban street canyons. *Atmospheric Environment* 34: 1553–1562.

Xie, X., Liu, C., Leung, D.Y.C. and Leung, M.K.H. 2006. Characteristics of air exchange in a street canyon with ground heating. *Atmospheric Environment* 40: 6396–6409.

Yang, J., Wang, Z. and Kaloush, K.E. 2015. Environmental impacts of reflective materials: Is high albedo a 'silver bullet' for mitigating urban heat island? *Renewable and Sustainable Energy Reviews* 47: 830–843.

Pavement Life-Cycle Assessment – Al-Qadi, Ozer & Harvey (Eds)
© 2017 Taylor & Francis Group, London, ISBN 978-1-138-06605-2

The importance of incorporating uncertainty into pavement life cycle cost and environmental impact analyses

Jeremy Gregory
Massachusetts Institute of Technology, Cambridge, MA, USA

Arash Noshadravan
Texas A&M University, College Station, TX, USA

Omar Swei, Xin Xu & Randolph Kirchain
Massachusetts Institute of Technology, Cambridge, MA, USA

ABSTRACT: We present an approach for conducting probabilistic Life Cycle Cost Analyses (LCCA) and Life Cycle Assessments (LCA) and demonstrate its value with case study results. We define uncertainty quantities and methods for characterizing uncertainty for different types of parameters. The approach includes leveraging outputs from Pavement-ME to characterize uncertainty in pavement performance over time. Uncertainty in the input data and scenarios is used in a Monte Carlo analysis to quantify the uncertainty in life cycle costs and environmental impacts. The probabilistic results are then used to calculate several comparative metrics, including the statistical confidence that one alternative has a lower cost or environmental impact than another alternative, and to determine the parameters that contribute most to the variance of the results. The approach enables a wide analysis of the scenario space to determine which scenarios are most relevant to the comparison of alternatives, and iterative analyses that feature refined data selected in the influential parameter analysis. We demonstrate the value of the approach and the benefits of incorporating uncertainty into LCCAs and LCAs via results from cases in the literature.

1 INTRODUCTION

Uncertainty is pervasive in pavement life cycle cost and environmental impact analyses. Sources of uncertainty include inherent variation in data, the pedigree of data, the evolution of pavement performance and maintenance over time, the evolution of data over time, and analysis choices that require a value judgment. Incorporating uncertainty into these analyses is beneficial because it provides a more accurate representation of data and its evolution, enables quantitative risk assessments, and streamlines data collection by quantifying parameters that matter most to the results.

Literature exists on uncertainty in Life Cycle Assessment (LCA) (Gregory et al. 2016; Williams et al. 2009) and Life Cycle Cost Analysis (LCCA) (Ilg et al. 2016). However, there has been minimal application of these concepts to pavement LCA and LCCA. The US Federal Highway Administration (FHWA) describes an approach for conducting probabilistic pavement LCCAs (FHWA Pavement Division 1998), but there is limited guidance on how to characterize uncertainty in the inputs for such analyses. Similarly, FHWA guidance on pavement LCA (Harvey et al. 2016) describes sources of uncertainty and generic approaches for analyzing uncertainty from the LCA literature, but there is no detailed information on how to characterize uncertainty for input parameters or conduct a probabilistic pavement LCA.

We have developed approaches for conducting probabilistic LCCAs (Swei et al. 2013) and LCAs (Noshadravan et al. 2013; Gregory et al. 2016). They leverage outputs from Pavement-ME to characterize inputs, particularly the uncertainty in pavement performance over time. In addition, we have developed methods to estimate uncertainty in initial and future costs (Swei et al. 2016) and life cycle inventory data. Uncertainty in the input data is used in a Monte Carlo analysis to quantify the uncertainty in life cycle costs and environmental impacts. The probabilistic results are then used to calculate several comparative metrics, including the statistical confidence that one alternative has a lower cost or environmental impact than another alternative, and to determine the parameters that contribute most to the variance of the results. The approach enables a wide analysis of the scenario space to determine which scenarios are most relevant to the comparison of alternatives, and iterative analyses that feature refined data selected in the influential parameter analysis. In this paper we present a summary of both the probabilistic LCA and LCCA approaches, along with examples of case study results to demonstrate the benefits of incorporating uncertainty into LCCAs and LCAs. We begin with an overview of uncertainty concepts that are the foundation for any probabilistic LCA and LCCA, and then move to concepts that are specific to pavement analyses.

2 UNCERTAINTY QUANTITIES, CHARACTERIZATION, AND ANALYSIS

2.1 *Uncertainty quantities*

The LCA literature (see (Lloyd & Ries 2007) for a summary) has coalesced around three types of uncertainty for both Life Cycle Inventories (LCI) and Life Cycle Impact Assessment (LCIA) methods: parameter (uncertainty in input data), scenario (uncertainty in choices), and model (uncertainty in mathematical relationships) uncertainty. Differentiating these types of uncertainty can be challenging because of the overlap among them. All forms of uncertainty are expressed as uncertainty in a parameter value, even if there is an aggregate of multiple types of uncertainty.

We found guidance on uncertainty quantities from the work of Morgan and Henrion (1990), which defines quantities used in uncertainty analyses for risk and policy analysis. They define eight types of uncertainty quantities. The five that are of most relevance to LCA and LCCA are listed in Table 1. There will likely be a single decision variable and only a few outcome criteria for each analysis. However, there will almost certainly be numerous empirical, model domain, and value parameters. Some empirical quantities will be used directly in life cycle inventories, such as quantities of material inputs or emission outputs; these are *inventory parameters*. However, other empirical quantities are actually *model parameters*, such as pavement thickness or vehicle fuel efficiency, which are used to calculate inventory parameters.

Table 1. Summary of types of quantities in LCAs and LCCAs. Content adapted from (Morgan & Henrion 1990).

Quantity	Description	Example
Decision variable	Frame the decision—what is the best outcome?	Which product has lowest cost or environmental impact.
Outcome criterion	Metric for measuring performance.	Metric from a life cycle impact assessment method.
Empirical parameter	Measurable (in principle) with a *true* value.	Electricity consumption, particulate emissions, material cost.
Model domain parameter	Define scope of system with an *appropriate* value.	Temporal or geographic boundaries.
Value parameter	Represent aspects of the preferences of the analyst with an *appropriate* value.	Discount rate, allocation factor, model form.

2.2 *Uncertainty characterization*

Uncertainty characterization for quantities in LCAs and LCCAs depends on the type of quantity, but also the source of uncertainty. There are multiple sources of uncertainty; Morgan and Henrion (1990) define seven such sources. We build off of these and other work in the LCA community in defining types, sources, and methods for characterizing uncertainty in inventory parameters, listed in Table 2. Measurement uncertainty derives from variation and stochastic error in input data. This can be caused by geographic, temporal, and process variation or from inaccuracy in the measurement process itself. Inventory Quantities Application Uncertainty (IQAU) stems from the pedigree, or quality, of the inventory data amounts. The pedigree is related to the appropriateness of the input and output amounts in an inventory from a particular data source. Intermediate Flows Application Uncertainty (IFAU) also deals with data pedigree, but it refers to the appropriateness of other life cycle inventory data sets to represent intermediate flows (i.e., cumulative LCI data for upstream processes) in the inventory under consideration.

The methods listed in Table 2 for characterizing uncertainty are particularly relevant for empirical parameters. Indeed, it is possible to combine estimates of measurement uncertainty, IQAU, and IFAU into a single quantitative uncertainty characterization for a parameter (see section S2 of the supporting information of (Gregory et al. 2016) for details).

Empirical quantities may be represented by a weighted probability distribution (such as a normal or log-normal) because they have a true value. By contrast, model domain and value parameters should be defined as a range of continuous or discrete values with equal likelihood (i.e., an unweighted or uniform distribution) because one cannot state that one quantity is more likely than another.

There may be instances where it is not possible to quantify uncertainty in any type of parameter because there is no clear representative value and/or distribution. In these cases, a rough distribution should be defined, erring on the side of overestimating uncertainty. Such a situation may occur for parameters that deal with activities that will occur in the future (e.g., vehicle fuel efficiency or pavement degradation rate). If such a parameter is found to be influential, discrete values may be defined within a uniform distribution in order to enable analysis of specific scenarios.

2.3 *Uncertainty analysis*

The conventional uncertainty analysis approach is to conduct a probabilistic parameter uncertainty analysis for a few selected scenarios. While this approach may be effective for

Table 2. Sources of and methods for characterizing uncertainty in parameters.

Types of uncertainty	Sources of uncertainty	Methods for characterizing uncertainty
Measurement uncertainty	Variation and stochastic error: geographic, temporal, & process variation; measurement inaccuracy.	Use actual data on variation or use estimates from an expert panel (e.g., ecoinvent estimates).
Inventory quantities application uncertainty	Pedigree of inventory data amounts: appropriateness of input/output amounts from data source applied to inventory.	Pedigree matrix approach: qualitative characterization of data pedigree translated into quantitative distribution.
Intermediate flows application uncertainty	Appropriateness of intermediate flows: appropriateness of other LCIs applied to represent intermediate flows.	Extension of pedigree matrix approach.
Cutoff uncertainty	Incomplete or missing data: exclusion of input or output processes or substances.	Conduct hybrid LCA.
Human error uncertainty	Human errors: mistakes in data entry.	Unknown.

preliminary analyses, we propose a more robust approach that accounts for uncertainty in parameter, scenario, and model uncertainty through the simultaneous analysis of empirical, model domain, and value parameters across a wide range of the scenario space. Details of the approach can be found in (Gregory et al. 2016); a summary is presented here.

The methodology is outlined in Figure 1. It is intended for comparative analyses of two or more alternatives, which is typical for most life cycle cost and environmental impact assessments. The process is for a single set of decision variables and outcome criteria (e.g., impact assessment methods or net present value) and therefore must be repeated for different sets of decisions or criteria. It may be necessary to iterate the process several times before drawing final conclusions.

The methodology begins with an aggregated probabilistic scenario-aware analysis. This is a simultaneous analysis of uncertainty in empirical, model domain, and value parameters using a probabilistic analysis of the relative performance of the alternatives. The probabilistic analysis can be accomplished using any sampling-based method (such as a Monte Carlo or structured sampling). Care must be taken in the analysis to correlate parameters that are common between the two alternatives.

The first question asked in the approach is whether the alternatives can be resolved for all scenarios (i.e., we can statistically resolve the difference in the cost or impact of two alternatives). To do this we calculate the probability that one alternative has a lower impact than another across all of the simulations. This is accomplished using the comparison indicator, CI_L, defined as the ratio of the cost or environmental impact two products:

$$CI_L = \frac{Z_{L,B}}{Z_{L,A}} \quad (1)$$

where $Z_{L,A}$ and $Z_{L,B}$ are the cost or environmental impact for alternatives A and B, respectively, using the cost or environmental impact metric L. CI_L is calculated for a single simulation and therefore, can capture correlation in the parameters used in both alternatives in each simulation. We define β as the frequency that alternative B has a lower cost or impact than A across a set of simulations:

$$\beta = P(CI_L < 1) \quad (2)$$

We can state that B and A are resolvable if β or $(1-\beta)$ is greater than a threshold value, β_{crit}. This threshold, β_{crit}, is a decision parameter that controls the level of confidence in the decision and should be set by the analyst for a given context.

Returning to the first question in the approach (whether the alternatives can be resolved for all scenarios), if $\beta = 1$, then one alternative clearly has a lower cost or impact than the other and the analysis is complete. If they cannot, the next question is whether the alterna-

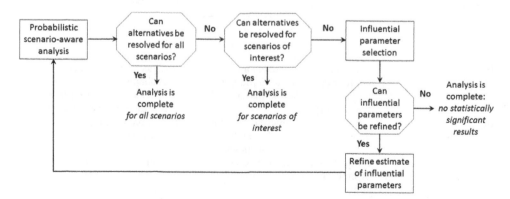

Figure 1. Methodology for evaluating uncertainty in comparative LCAs and LCCAs. Adapted from (Gregory et al. 2016).

tives can be resolved for scenarios of interest. It is unlikely that the two alternatives will be resolvable for all scenarios. By contrast, it is likely that some scenarios are of more interest to a particular set of decision makers (e.g., because they feel that a particular set of framing conditions are likely to be considered valid). If the alternatives can be resolved for the scenarios of interest, then the analysis is complete and the scenarios under which one alternative has a lower impact than another can be identified as statistically significant. This requires an approach to identify scenarios of interest, which can be done by manually analyzing sets of simulation results for given values of model domain and value parameters (i.e., sets of scenarios). This will likely be tedious for large sets of scenarios. Statistical software packages have Category and Regression Tree (CART) algorithms that may be used to analyze results across many scenarios. CART algorithms identify a succinct description of the statistically differentiable subpopulations within the scenario populations by recursively partitioning the space of input data and fitting a simple regression model within each partition.

If the alternatives cannot be resolved for the scenarios of interest, then the influential parameters for all scenarios need to be identified in order to determine the parameters that are worthy of further refinement because of their influence on the result. Influence can be assessed using different methods of sensitivity analysis.

Once influential parameters are identified, an assessment needs to be made as to whether resources are available to improve the fidelity of the analysis (third question in Figure 1). This would manifest in the refinement of uncertainty characterization for influential parameters (e.g., more data collection). If the influential parameters cannot be refined, then the analysis is complete and the outcome is that there are insufficient statistically significant results for the scenarios of interest. If they can be refined, then the entire process should be repeated using the refined uncertainty characterizations. This iterative process of influential parameter identification and refinement represents one of the major benefits of using uncertainty analysis in LCA and LCCA: resource-intensive data collection is only required for parameters influential parameters.

3 UNCERTAINTY CHARACTERIZATION AND ANALYSIS FOR PAVEMENT LCA AND LCCA

The previous section provided guidance on characterizing and analyzing uncertainty that is applicable to all LCAs and LCCAs, but it is worth discussing issues that are relevant specifically to pavement analyses. Following the structure of the previous section, there are three activities required to conduct a comparative analysis: define quantities, characterize uncertainty, and analyze comparative uncertainty.

3.1 *Define uncertainty quantities*

Figure 2 shows the scope of analysis that may be considered in pavement LCAs and LCCAs, although some elements may be excluded in analyses. Inventory data is required for each process in the life cycle and consists of background and foreground data. Although there is no clear differentiation between the two types, foreground data is generally relevant to a specific analysis (e.g., quantities of materials or fuel consumption), whereas background data represents upstream impacts or costs that may be relevant to many analyses (e.g., unit material costs or environmental impacts). Uncertainty quantities and characterization of background data need only be done once, but the process needs to be conducted for foreground data in each analysis.

Empirical, model domain, and value parameters must be defined for inventory parameters that represent collected data (e.g., emission quantities) and model parameters that are used to calculate inventory parameters (e.g., pavement thickness). A complete list of the 82 input parameters for a comparative pavement LCA is listed in Section S6 of the supporting information in (Gregory et al. 2016), but examples of parameters for each uncertainty type are listed in Table 3. Note that the first four empirical parameters are model parameters. That is,

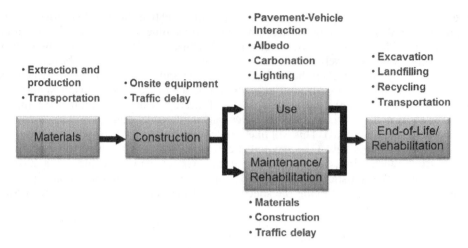

Figure 2. System boundary for pavement LCAs and LCCAs. Analyses may not include all elements.

Table 3. Examples of empirical, model domain, and value parameters from a pavement LCA (drawn from (Gregory et al. 2016)).

Empirical parameters	Model domain parameters	Value parameters
Fuel efficiency—cars	Design life	Salvage life allocation
Traffic growth factor	Analysis period	Maintenance strategy
PCC thickness	Scope*: albedo	
AC thickness, layer 1	Scope: PVI-deflection	
AC milling energy	Scope: PVI-roughness	

*Scope refers to whether or not the phenomenon (albedo or Pavement-Vehicle Interaction, PVI) is included in the analysis.

they will be used to calculate quantities, such as mass of aggregate or fuel, that will be used in an inventory. This is because it is typically more feasible to characterize uncertainty for model parameters because the sources of uncertainty can be decomposed.

Model domain parameters define the scope of the analysis and are chosen by the analyst—there are no true values, only appropriate values. It is important to note that it is possible to conduct analyses with and without elements of the life cycle if there are questions about data or models (these are the parameters listed with the "scope" prefix). This enables the analyst to determine the influence of the element on the comparative results.

Finally, value parameters represent preferences made by the analyst in conducting the LCA or LCCA where, once again, there are appropriate values, not true values. Allocation factors in LCA are a common example, as are discount rates in LCCA. The use of maintenance strategy as a value parameter is due to the fact that the analyst cannot know what the actual maintenance strategy will be. Rather, there may be several different potential maintenance strategies. This parameter enables the analyst to explore the impact of different strategies on comparative results.

3.2 *Characterize uncertainty*

Uncertainty for background data is already included in some databases, such as the ecoinvent and USLCI databases. Characterizing uncertainty in empirical parameters for foreground data is decomposed into the first three elements listed in Table 2: measurement uncertainty, Inventory Quantities Application Uncertainty (IQAU), and Intermediate Flows Application Uncertainty (IFAU). Obtaining primary data on the variation in application uncertainty is preferable, but estimates of measurement uncertainty for different types of quantities are provided by the ecoinvent center (Weidema et al. 2013). They also provide factors that can be used to quantify uncertainty associated with IQAU. We have demonstrated how those same factors can be used to quantify uncertainty associated with IFAU and then combined with a quantitative uncertainty for measurement uncertainty and IQAU (see section S2 of the supporting information of (Gregory et al. 2016)). This results in a weighted probability distribution (e.g., log-normal) for an empirical parameter.

Separating assessments of data quality for IQAU and IFAU is important. For IQAU, the pedigree of the data is related to the appropriateness of the input and output amounts in an inventory from a particular data source. For IFAU, the pedigree refers to the appropriateness of other life cycle inventory data sets to represent intermediate flows (i.e., cumulative LCI data for upstream processes) in the inventory under consideration.

As noted in Section 2.2, uncertainty in model domain and value parameters should be defined as a range of continuous or discrete values with equal likelihood (i.e., an unweighted or uniform distribution). The model domain parameters about scope will be binary (i.e., whether to include the activity or not). Model domain parameters about analysis period or design life will likely be uniform discrete because the choice of a different design life results in a different pavement design, and the choice of analysis period results in a different maintenance schedule. However, it is also possible to use uniform continuous distributions for model domain or value parameters, such as discount rate.

There is an opportunity to couple pavement LCA and LCCA with outputs from mechanistic-empirical pavement design software, such as Pavement-ME, as a means of quantifying inputs and characterizing uncertainty. In particular, the software can provide details on pavement mechanical response and deterioration over time, which are important for calculations of excess fuel consumption due to pavement-vehicle interaction. In addition, information is provided on the statistical confidence of the deterioration curves, thereby providing a quantitative characterization of the uncertainty in the curves. Noshadravan et al. (2013) provide details on how to translate the deterioration curves at several levels of reliability into uncertainty characterizations and quantitative assessments of roughness-induced excess fuel consumption emissions over time. Swei et al. (2013) provide details on how to translate the uncertainty in the deterioration curves into uncertainty in the timing of future maintenance activities on the basis of when predicted pavement distress exceeds a performance threshold.

Initial and future material and construction costs used in pavement LCCAs are highly influential and thus, uncertainty characterization for these parameters is particularly important. Swei et al. (2013) describe how to conduct a univariate regression analysis using historical bid data in order to test whether a statistically significant relationship between average unit cost and bid volume exists. Defining this relationship enables more accurate estimates of unit material and construction costs, and it also enables quantification of the uncertainty in the estimate. In addition, forecasting techniques have been used to probabilistically project prices and their uncertainty (Swei et al. 2013; Swei et al. 2016).

3.3 *Analyze comparative uncertainty*

The process for conducting comparative uncertainty analyses for pavement LCAs and LCCAs is outlined in Section 2.3, but it is worth highlighting specific results of interest. We give examples of the format of the results from studies published in the literature. The intent is to demonstrate the type of results that can be drawn from probabilistic studies and the benefits of the approach.

The first result of interest is a probabilistic distribution of results for the complete life cycle along with each life cycle phase. Figure 3 shows an example of results from a comparative pavement LCA. From this result one can gain insight into the relative magnitude and variation in impacts among the alternatives, and the life cycle phases that most influence the results.

The second set of results of interest are intended to assess the statistical significance of the difference among alternatives (determine whether they are resolvable) and quantify differences among alternatives, statistically significant or otherwise. Table 4 shows four metrics that can be used to assess and quantify differences between alternatives. They are depicted graphically in Figure 4. We calculate $\Delta\mu$ because it can be considered as the conventional metric for comparing costs or environmental impacts in deterministic LCAs and LCCAs. We calculate ΔZ_{90} because it can be viewed as the difference between the alternatives from a risk-averse perspective. The Cumulative Distribution Function (CDF) in Figure 4a is useful for viewing the impacts of taking different risk levels in quantifying the differences among alternatives. $\Delta\mu$, α_{90}, and β values for several dozen pavement LCCA cases can be found in (Swei et al. 2015).

While it is conventional to use $\Delta\mu$ and α_{90} to calculate differences between alternatives, they do not comment on the statistical significance among alternatives. For this reason we use β to determine whether there is a statistically significant difference among alternatives, and we use CI_* to quantify the maximum statistically significant difference between the two alternatives. The example in Figure 4b shows a case where $\beta = 0.95$, which indicates that the result is statistically significant (based on a threshold value of 0.9), and the statistically significant difference, CI_*, is 6%. $\Delta\mu$, β, and CI_* values for several dozen pavement LCA cases can be found in (Xu et al. 2014).

There are two additional results that are necessary to support the comparative uncertainty analysis approach described in Section 3.3: a ranking of influential parameters, and a categorization of resolvable scenarios. Table 5 shows the influential parameters in a pavement LCCA case.

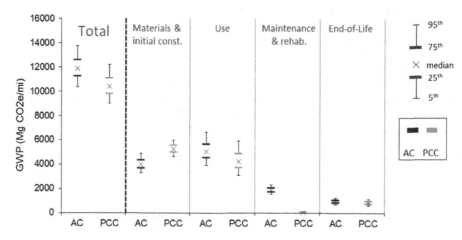

Figure 3. Probabilistic Global Warming Potential (GWP) results of a comparative pavement LCA of an urban interstate pavement in Arizona for the complete life cycle and each life cycle phase. Results from a case presented in (Xu et al. 2014).

Table 4. Metrics for comparative LCAs and LCCAs. Values may be costs or environmental impacts.

Metric	Meaning
$\Delta\mu = \dfrac{\mu_B - \mu_A}{\mu_A}$	The difference between the mean value of alternative A and the mean value of alternative B.
$\Delta Z_{90} = \dfrac{Z_{90,B} - Z_{90,A}}{Z_{90,A}}$	The difference between the 90th percentile value of alternative A and the 90th percentile value of alternative B.
$\beta = P(CI_L < 1)$	The frequency that alternative B has a lower cost or impact than alternative A. Values greater than 0.9 (or less than 0.1) indicate alternative A (or alternative B) have a statistically significant lower cost or impact. CI is defined in $CIL = \dfrac{Z_{L,B}}{Z_{L,A}}$ Equation 1.
$CI_* = \begin{cases} CI_{0.1}, & \beta < 0.1 \\ CI_{0.9}, & \beta > 0.9 \end{cases}$	The value of CI when β is 0.9 (or 0.1). This represents the maximum statistically significant difference between the two alternatives. it is only meaningful when $\beta > 0.9$ (or $\beta < 0.1$).

Influential parameters are listed for each of the two design alternatives and the difference between the two alternatives. These are calculated using sensitivity analysis techniques, such as regression or variance-based methods. These results are important because they inform the selection of parameters for data refinement. Influential parameters for LCCA scenarios have been calculated in (Swei et al. 2015) and for LCA scenarios in (Xu et al. 2014; Gregory et al. 2016).

Results from categorization analyses using CART algorithms show which scenarios have statistically resolvable results and the values of key parameters in those groups of scenarios. Figure 5 shows the results of categorization analyses in a comparative LCA. Meaningful scenario groups are isolated and designated by the parameter listed in the box, which indicates that all scenarios in that group have a fixed value of that parameter. Green boxes indicate scenarios that are statistically resolved. Figure 5a shows results after the initial probabilistic

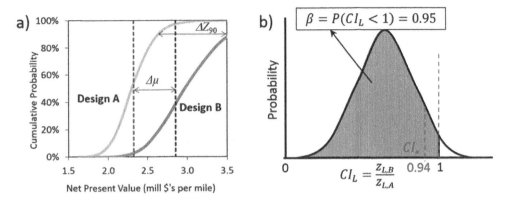

Figure 4. a) Cumulative probability density functions of life cycle cost for two design alternatives showing the $\Delta\mu$ and ΔZ_{90} metrics. b) Probability density function of the comparison indicator for global warming potential impacts of two alternatives. In this case, design B has a statistically significant lower impact than design A, i.e., $\beta = 0.95$. $CI_* = 0.94$, which means the maximum statistically significant difference is 6%.

Table 5. Influential parameters in a pavement LCCA using urban interstate cases in four climates. Cases and data from (Swei et al. 2015). Influential parameters are listed for the Hot Mix Asphalt (HMA) alternative, the Jointed Plain Concrete Pavement (JPCP) alternative, and the difference between the two alternatives.

Parameter	LTPP Climate Zone			
	Wet Freeze (Missouri)	Dry No Freeze (Arizona)	Dry Freeze (Colorado)	Wet No Freeze (Florida)
HMA	HMA	HMA	HMA	HMA
Pavement-ME Reliabiltity		0.08	0.06	0.10
Aggregate Price	0.16	0.07	0.01	0.00
AC Surface Price	0.25	0.13	0.06	0.31
AC Binder Price	0.38	0.10	0.16	0.11
AC Base Price	0.01	0.43	0.64	0.28
JPCP	JPCP	JPCP	JPCP	JPCP
JPCP Layer Price	0.82	0.96	0.95	0.99
Difference	Difference	Difference	Difference	Difference
JPCP Layer Price	0.19	0.56	0.32	0.85
Aggregate Price	0.13	0.01	0.01	
AC Surface Price	0.20	0.04	0.03	0.04
AC Binder Price	0.31	0.04	0.11	0.02
AC Base Price	0.02	0.23	0.55	0.04

scenario-aware analysis in which empirical quantities are at their full range of values. Figure 5b shows results after a second iteration that used more refined data for two parameters (rate of roughness evolution and the impact factor for bitumen) that were identified as influential in the sensitivity analysis.

There are 128 total scenarios analyzed (each one includes thousands of simulations) and in the first step 41 scenarios are resolvable, whereas the second iteration after data refinement leads to 83 scenarios that are resolvable. This highlights the effectiveness of the data refinement, but it is also interesting to observe the scenario subpopulations that lead to resolvable results. For example, 32 of the 41 resolvable scenarios in Figure 5a include albedo and exclude the impact of pavement roughness in the analysis scope. The other 9 resolved scenarios are distributed among the states examined, but all share the common feature of including the impact of pavement deflection. The subpopulation of scenarios which exclude albedo effects (the left half of tree), serves as a lesson on the importance of considering and isolating individual scenarios and scenario populations. This subpopulation seems irresolvable because it has a β of 0.6. Within this group, however, one can isolate six specific scenarios (labelled groups b and c in Figure 5a) that are, in fact, resolvable.

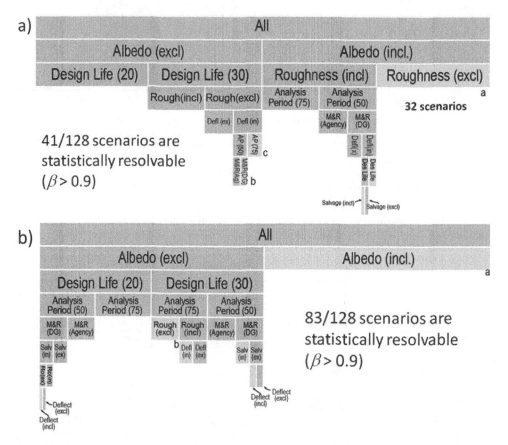

Figure 5. Categorization analysis results of resolvable scenarios from a comparative pavement LCA case in (Gregory et al. 2016). a) Results shown for all empirical quantities are at full range of values (iteration 1). b) Results shown for refined data analysis (iteration 2 – refined rate of roughness evolution and the impact factor for bitumen). Boxes indicate a collection of scenarios that share a fixed value for the parameter listed in the box. Green bars differentiate scenarios that are statistically resolved ($\beta > 0.9$). For binary model domain parameters the scope is either included (incl) or excluded (excl). Parenthetical numbers indicate the value of the parameter. Rough or Ro = roughness; AP = Analysis period; Defl = deflection; M&R = maintenance and rehabilitation; Des Life = design life; Salv = salvage.

The data refinement step successfully produced significant resolution in about 65% (83 of 128) of scenarios, but much of the scenario space remains unresolved. If those scenarios are of particular interest, then the analyst should iterate through the process again, identifying influential parameters and exploring whether resources are available to improve the fidelity of parameter estimates. Although an initial sensitivity analysis provides useful guidance for those iterations, that analysis should be repeated each time more refined information is introduced.

4 OPPORTUNITIES FOR FUTURE RESEARCH

There are several opportunities to improve uncertainty analyses in pavement LCA and LCCA, particularly uncertainty characterization. This includes the data used to quantify measurement uncertainty, which can be accomplished through data collection of the variation in inventory and model parameters used in LCIs. In addition, further research on the factors used to quantify uncertainty due to IQAU and IFAU would be beneficial. Another topic worthy of research is the development of models for how inventory data and environmental impacts will evolve over time and the associated methods to characterize uncertainty. This is particularly important given the long analysis periods used in pavement LCAs and LCCAs. Finally, additional research on uncertainty in impact assessment methods and their evolution over time would be an important contribution to the field of study.

If the pavement LCA and LCCA communities adopt a probabilistic approach in analyses, a body of work will be developed that can be used to establish priorities for data refinement and model improvement. It will also enable decision-makers to have a better understanding of the conditions that are most likely to lead to statistically significant results.

REFERENCES

FHWA Pavement Division, 1998. *Life-Cycle Cost Analysis in Pavement Design.*

Gregory, J.R. et al., 2016. A Methodology for Robust Comparative Life Cycle Assessments Incorporating Uncertainty. *Environmental Science and Technology*, 50(12).

Harvey, J.T. et al., 2016. *Pavement Life Cycle Assessment Framework*, Available at: https://www.fhwa.dot.gov/pavement/pub_details.cfm?id = 998 [Accessed October 21, 2016].

Ilg, P. et al., 2016. Uncertainty in life cycle costing for long-range infrastructure. Part I: leveling the playing field to address uncertainties. *International Journal of Life Cycle Assessment*, (1977), pp.1–16. Available at: http://dx.doi.org/10.1007/s11367-016-1154-1.

Lloyd, S.M. & Ries, R., 2007. Analyzing Uncertainty in Life-Cycle Assessment A Survey of Quantitative Approaches. *Journal of Industrial Ecology*, 11(1), pp.161–181.

Morgan, M.G. & Henrion, M., 1990. *Uncertainty: a guide to dealing with uncertainty in quantitative risk and policy analysis*, New York, NY, USA: Cambridge University Press.

Noshadravan, A. et al., 2013. Comparative pavement life cycle assessment with parameter uncertainty. *Transportation Research Part D: Transport and Environment*, 25.

Swei, O., Gregory, J. & Kirchain, R., 2016. Probabilistic Approach for Long-Run Price Projections : Case Study of Concrete and Asphalt. *Journal of Construction Engineering and Management*, In Press.

Swei, O., Gregory, J. & Kirchain, R., 2013. Probabilistic characterization of uncertain inputs in the life-cycle cost analysis of pavements. *Transportation Research Record: Journal of the Transportation Research Board*, 2366, pp. 71–77.

Swei, O., Gregory, J. & Kirchain, R., 2015. Probabilistic life-cycle cost analysis of pavements: Drivers of variation and implications of context. *Transportation Research Record: Journal of the Transportation Research Board*, 2523.

Weidema, B.P. et al., 2013. *Data quality guideline for the ecoinvent database version 3*, Available at: http://www.ecoinvent.org/database/methodology-of-ecoinvent-3/methodology-of-ecoinvent-3.html.

Williams, E.D., Weber, C.L. & Hawkins, T.R., 2009. Hybrid framework for managing uncertainty in life cycle inventories. *Journal of Industrial Ecology*, 13(6), pp. 928–944.

Xu, X. et al., 2014. Scenario Analysis of Comparative Pavement Life Cycle Assessment. In *International Symposium on Pavement LCA 2014*. pp. 13–26. Available at: http://www.ucprc.ucdavis.edu/p-lca2014/media/pdf/Papers/LCA14_Probabilistic Approach.pdf.

… Pavement Life-Cycle Assessment – Al-Qadi, Ozer & Harvey (Eds)

Functional unit choice for comparative pavement LCA involving use-stage with pavement roughness Uncertainty Quantification (UQ)

Mojtaba Ziyadi, Hasan Ozer & Imad L. Al-Qadi
Department of Civil and Environmental Engineering, University of Illinois at Urbana-Champaign, Urbana, IL, USA

ABSTRACT: An analysis of the use-stage for pavement Life-Cycle Assessment (LCA) is presented within a framework of an LCA tool developed for Illinois Tollway. Methodological choices that can significantly affect LCA results were evaluated in this study. The share of the use-stage in a comprehensive pavement LCA framework was evaluated with all life-cycle stages including materials, construction, use, maintenance, and end of life. The scope of the use-stage includes albedo, carbonation, and rolling resistance (including pavement roughness and texture components). Uncertainty of the pavement roughness and its effect on the results are investigated using Monte Carlo sampling technique. A discussion on the choice of functional unit and its effect in comparative LCA is presented and an appropriate functional unit is suggested. Four projects were selected as case studies to demonstrate the capabilities of the tool and proposed framework. A multi-point environmental performance evaluation was performed using four environmental indicators for comprehensive interpretation of results. The effect of each stage of LCA was evaluated with emphasis on the results of the use-stage. Additional fuel consumption and emissions, resulting from roughness and texture, constituted the largest share of use-stage impacts while the effect of carbonation was limited. As traffic reduced and the share of the materials and construction stage increased, the share of the use-stage could be decreased to 50% levels.

1 INTRODUCTION

Life-Cycle Assessment (LCA) is one of the sustainability measurement tools used to quantify life-cycle environmental impacts of a product. Recently, pavement LCA applications have been increased with an emphasis on sustainability for the construction industry aiming at reducing carbon footprint of pavements. In order to accurately characterize the environmental impact of pavements, Federal Highway Administration (FHWA) recently published pavement LCA framework following the International Standards Organization (ISO) 14040 and 14044 standards (Van Dam et al; 2015; Harvey et al., 2016; ISO, 2006).

There is certainly a need for an easy-to-use, and yet reliable, model capturing the nature of pavement–environment interaction that can be incorporated in pavement LCAs for the use of LCA experts as well as LCA users in agencies and industry. This can facilitate implementation of LCA with its most needed components related to the use-stage and improve sustainability of pavement systems. In general, some studies ignore the use-stage component completely (i.e., texture, albedo, and carbonation) or simplify it to a degree that renders the need to use other software platforms unnecessary (Wang et al., 2012). Finally, most previous studies focused on energy consumption and/or GHG only as part of the impact characterization and the sole outcome of the LCA (Wang et al., 2014, Shakiba et al., 2016). However, a comprehensive pavement LCA, similar to other products and services, should ultimately include environmental impact categories other than energy and GHG as outlined by characterization methods such as TRACI (Bare, 2011), CML (Guinée et al., 2002). This is provided by a multi-point environmental assessment opportunity, rather than relying on only energy and GHG.

Common challenges in performing pavement LCA were reviewed and discussed in details by Santero et al. (2011a, b) through a comprehensive review of existing pavement LCA literature and modeling tools as of 2010. Data and modeling gaps were identified in pavement LCAs and in particular for the use-stage, feedstock energy of bitumen, impact of traffic delay, maintenance phase, end-of-life stage, and inconsistencies for various methodological choices were summarized. Inyim et al. (2016) and AzariJafari et al. (2016) reviewed the current literature and pointed out similar gaps in more recent studies.

One of the challenges in performing comparative LCA is the choice of Functional Unit (FU). The functional unit is a unit of measurement of system components to which inputs and outputs of LCA are normalized. According to ISO 14044:2006, functional unit is "quantified performance of a product system for use a reference unit". ISO standards recommends to choose functional consistent with the goal and scope of the study where primary function and performance characteristics of the product are specified. The purpose of the functional unit is to quantify the services delivered by the product system. Therefore, the functional unit shall declare the relevant functions of the product with some performance characteristics. The flaws in some of the commonly used FUs are discussed in the following sections and a proper FU is proposed.

While comprehensive methodological choices are needed to implement the pavement LCA, there are many uncertainty sources associated with pavement LCA. Some of these uncertainties include, but are not limited to: input variability, human errors, source credibility, and parameter and model uncertainty. Studies have focused on the uncertainty analysis of pavement LCA (Noshadravan et al., 2013; Gregory et al., 2016). Uncertainty Quantification (UQ) methods are mathematical methods that help identify, propagate, quantify, and interpret the uncertainties in the system. Each stage of the pavement LCA may need different handling of uncertainty depending on the type of the source. Quantifying uncertainties would ultimately help understand the variations in the pavement LCA analysis and to what extent these variations affect the reliability of the outcome. Moreover, this would let decision-makers consider ranges of outcomes rather than deterministic values, henceforth, helping in informative decision-making with better understood consequences.

The main goal of this paper is to present the essential components needed to perform a complete LCA (cradle-to-grave analysis) to compare pavement systems with varying designs, and traffic characteristics. The main components targeted in this study include selection of an appropriate functional unit and uncertainty quantification. Uncertainty quantification of the use stage is performed by studying the variability in the pavement roughness. Both of these components are presented with illustrative examples through case studies.

2 METHODOLOGICAL CHOICES AND LIFE-CYCLE INVENTORY ANALYSIS

2.1 *Goal of the study*

The goal of the study is to develop an LCA tool and framework for the Illinois Tollway in order to perform a comparative LCA between projects constructed in the past and planned for future. The intended application is to evaluate the progress toward the agency's sustainability goal from environmental performance perspective by reporting and comparing environmental performance of past and present projects, and pavement type and design selection using environmental performance in addition to cost and performance.

2.2 *System boundary*

A system boundary in an LCA defines the processes and life-cycle stages included. The system boundary for the current study includes all stages of the pavement LCA. Material acquisition and production, construction, maintenance, use-stage, and end-of-life stages are included. Rolling resistance, carbonation, and albedo components are included in the use-stage.

Upstream (data from supply chain such as electricity production, extraction of crude oil and transportation, etc.) and downstream processes (typically processes starts with material stage) are included.

2.3 *Choice of functional unit for LCA involving the use-stage*

In comparative LCA, a consistent Functional Unit (FU) must be chosen in order to compare two or more pavement systems. The functional unit for pavements should represent physical dimensions and pavement performance. Performance requirements can include design life, traffic level, subgrade type, and pavement condition and should be reported or explicitly included in the functional unit. Typically, a functional unit for pavements is defined as *"entire project dimensions or lane-mile through a specified lifetime fulfilling specifications and performance requirements"*. This type of functions unit only reporting the physical dimensions of the project may not be sufficient for comparative LCAs where pavement performances are different.

Therefore, the functional unit should effectively reflect the characteristics of the pavement section under LCA study. Different functional units have been used in literature for pavement LCA studies. Three categories can be identified:

- Physical: Physical or geometrical functional units account for dimensions of the project such as, project-length (mile or kilometer), lane-length (lane-mile or kilometer), volume (material), etc. This is the most commonly used functional unit in the literature. Depending on the goal of the study, this functional unit can be appropriate to use. The following scenarios are among the typical examples: reporting total GHG emissions or energy consumption attributed to a pavement system or network of pavements, interpretation of life-cycle stage contribution, etc.
- Structural or performance-based: Structural or performance-related functional units account for the condition of pavement or factors that affect performance such as, traffic (ADT/AADT), load (Equivalent Single Axle Load (ESAL)), and performance-lane-length. For example, per pavement serviceability rating (PCR) per lane per mile (Santero & Horvath, 2011a). This type of functional units can be used for comparative evaluation of two or more pavement systems.
- Annualized: The results obtained from geometry-related functional units are normalized by the analysis period. This type of functional unit is commonly used for comparative LCAs to evaluate sustainability improvement between two or more types of pavements. For example, functional unit becomes lane-mile-year.

For a fair comparison of projects, FU should not penalize the projects by their total length, number of lanes, analysis period and traffic level. Instead, it should account for the performance. This means one would normalize LCA results by dimensions of the project, analysis period and traffic and directly account for the performance. In this regard, the categories in the above-mentioned list should not be considered as proper choices.

It should be noted that structural FUs are usually necessary when the goal is to conduct comparative LCAs for projects with different design characteristics e.g. different traffic or analysis period. These type of comparisons are necessary for benchmarking studies where current projects are meant to be compared to some baseline projects from the past while the design parameters can be different. While a true FU can never be sought for these types of projects, due to inherent differences in all aspects of the comparison, we argue that if the pavement performance is considered in the use-stage calculations, it would be unnecessary to explicitly include performance in the FU. Instead, FU should account for the missing component i.e. traffic or ESAL, along with physical dimensions of the project to avoid penalizing projects serving higher traffic levels.

It is clear that regardless of the chosen methodology, use-stage components somehow account for the performance e.g. through roughness, texture or deflection indicators, albedo change over time, or carbonation parameters for concrete sections. Meaning, poor performing section will result in more impacts. Therefore, this study proposes a new functional unit that accounts for traffic as well as physical characteristics of the projects being compared. Vehicle-Length Traveled (VLT) in terms of Vehicle-Miles-Traveled (VMT) or Vehicle-Kilometer-Traveled (VKT) is proposed as an alternative choice of functional unit choice when the goal of the study is to compare two pavements with design features common for benchmarking applications.

Although seem trivial, we expand the discussion here to explore flaws in the physical FUs as well as show sufficiency of the proposed FU when dealing with comparative LCAs involving use-stage.

Table 1 presents a comparison of some of the commonly used functional units with their potential consequences on the comparison of two hypothetical pavement systems. Varying scenarios are presented with the expected use-stage impact when three types of functional units are used. Performance of pavement is characterized by IRI and traffic level also affecting directly the use-stage impacts through excess vehicle fuel consumption. The higher the area under the IRI or traffic growth curves, the higher the use-stage impact is. When only physical units (i.e. lane-mile) or annualized functional unit (lane-mile-year) are used, in general longer living pavements and pavements serving for higher traffic volumes can be penalized.

As shown with the hypothetical examples, VLT choice results in more consistent comparison of two pavements favoring good performing sections serving for higher traffic volumes. Since VLT accounts for the "usage" of the system by users, it is applicable to all LCA stages when the system boundary includes use stage. Essentially, VLT is a proper choice if any aspect of the LCA include time variable in it. This is because a pavement section is designed to serve users over time usually represented by ADT/AADT. Thus, a proper FU should account for traffic in a similar way that should account for time and dimensions of the project.

Analysis period is another critical choice for a fair comparison of pavement systems. Analysis period should be chosen to capture initial pavement and through the life of at least the

Table 1. Functional unit scenarios and consequences for comparative LCAs.

Scenario Description for Comparison	Performance (in terms of IRI)	Traffic	Functional Unit Choice and Consequences on Use-stage Impact
Alternative pavement trials for the same traffic with same design lives. Different expected performance (poor performance for A) with same analysis period.	IRI curves A above B, $AP_A = AP_B$	Traffic A=B	LM: A > B LMY: A > B VMT: A > B *Consistent results when analysis period is the same.*
Different designs with different design lives accounting for traffic volume differences. Similar expected performance within the analysis period (shorter life for pavement A).	IRI curves B and A, AP_A, AP_B	Traffic B above A	LM: B > A LMY: most likely B > A VMT: most likely A = B *LM and LMY will penalize longer living pavement. Results depend on change in AP and differences in performance and traffic.*
Arbitrary selection from network with different performance and traffic. Same or different design lives. Assume A is the poor performing with higher traffic and same analysis period.	IRI curves A above B, $AP_A = AP_B$	Traffic A above B	LM: A > B LMY: A > B VMT: most likely A > B *- VMT will favor better performing pavement B if traffic volume is not too low. - If traffic volume is too low, indication of overdesign.*
Arbitrary selection from network with different performance and traffic. Same or different design lives. Assume A is the poor performing with different traffic.	IRI curves A and B, AP_A, AP_B	Traffic B above A	Total: can vary. LM: can vary. LMY: can vary. VMT: most likely A > B *- VMT will favor better performing pavement B.* *- Under higher traffic conditions LM and LMY may penalize pavement B.*

AP = analysis period
LM: lane-mile

next subsequent major rehabilitation treatment or next full reconstruction (Van Dam et al., 2015). The comparisons presented in Table 1 applies to the cases where analysis period is extended through the entire life-cycle of pavements till the end-of-life or subsequent major treatments (IRI performance curves are shown with linear curves for simplicity). In this study, analysis period is chosen as the time period from initial construction to the reconstruction.

2.4 *Material, construction, maintenance, and end-of-life stages*

Inventory data regarding the operation of equipment, production of materials, transportation of materials, and plant operations for pavement mixtures were considered using various primary and secondary data sources (e.g., literature, government and industry reports, commercial software and simulators, surveys). A life-cycle inventory database was developed combining operational or process activity data collected with processes available in commercial software and databases such as SimaPro and US-Ecoinvent. The inventory database was compiled in terms of pay items compatible with the agency's procurement procedures. A cost-based cut-off criteria was used including pavement pay items that contribute to the top 95% or 0.5% of the total pavement cost. In addition, a cut-off allocation was applied for recycled materials, by-products, and waste products where only processes directly related to the post-processing or preparation of secondary materials were considered. A similar cut-off allocation rule was used for the End-of-Life (EOL) stage, where the processing of any material destined to be recycled or reused after the current pavement system ends is cut-off and attributed to the future pavement system. Feedstock energy of bituminous material was also considered. Detailed information on material and construction, maintenance, and end-of-life stages can be found elsewhere (Yang et al, 2016, Kang et al., 2014).

2.5 *Use-stage*

Use-stage components include, rolling resistance, albedo, carbonation, lighting and leachate. Depending on the availability of data/models, and the importance of the components on overall impacts, the system boundary may include one or more of the use-stage components. The system boundary in this study includes all modules of the use-stage, except structural rolling resistance (deflection-related), leachate, and lighting. The modules were selected based on the availability of the methods and their significance on overall LCA results. The following sub-sections provide methodological choices for current study.

2.5.1 *Albedo*

Albedo is the main controlling property defining the measure of the ability of a surface to reflect solar radiation. It ranges from 0 (perfectly non-reflective) to 1 (perfectly reflective material). Albedo directly affects the radiative forcing, a measure of the balance of incoming and outgoing energy in the earth-atmosphere system. The effect of albedo can be used to quantify the heat island effect and radiative forcing using the following equation from the FHWA LCA framework (Harvey, et al., 2016).

$$m_{CO_2} = \sum_{n=1}^{N} 100 * \left(\alpha_{new}^n - \alpha_{ref}\right) * \left(f_{RF}\right) * A \qquad (1)$$

where,

m_{CO_2} = total CO_2 offset or gain in kg over the analysis period

α_{new}^n = the mean albedo value of original pavement construction and subsequent treatments as an average value of initial albedo right after construction and final albedo before the next treatment

α_{ref} = reference albedo value that can be taken as the albedo of old pavement surface replaced by new construction or network average albedo

f_{RF} = CO_2 offset per an increase of 0.01 rise in albedo in kg CO_2 m^{-2} (between 2.55 and 4.90 kg CO_2 m^{-2})

A = total surface area of new pavement construction in m^2

N = total number of treatment activities replacing surface layers during the analysis period

According to literature, the following albedo (α) ranges for new and weathered surfaces were used (Kaloush et al., 2008): 0.05 (new) and 0.15 (weathered) with an average of 0.1 for asphalt and 0.4 (new) and 0.2 (weathered) (average of 0.3) for Portland cement concrete surfaces, respectively. The first value is for the new surface, and the second value is for the weathered surface.

2.5.2 Carbonation

Carbonation is the process whereby carbon dioxide is reabsorbed and stored in concrete to form a bond with calcium oxide and calcium hydroxide, resulting from the calcination process during cement manufacturing. This process takes place only in concrete pavements. The majority of pavement LCAs that consider carbonation follow a model included in the works by Pommer and Pade (2006) and Lagerblad (2005). The model includes several factors that control absorption over time and uses a simplification of Fick's second law of diffusion as presented by Santero & Horvath (2009).

Following the work by Lagerblad (2005), the following equation was used to calculate carbon dioxide uptake during the life cycle of concrete:

$$CO_2(kg) = k \times t^{0.5} \times c \times CaO \times r \times A \times M \qquad (2)$$

where,

$k = 1.25 \times$ CF_lime $\times$ CF_FlyAsh (base k-value based on CEM type I half burried-half exposed). If CEM type II is used multiply the base of the equation (1.25) by 1.10 or (1.25×1.10)

CF_lime = 1.05 for 10% limestone, and 1.10 for 20% limestone
CF_FlyAsh = 1.05 for 10% fly ash, 1.10 for 20% fly ash, and 1.20 for 40% fly ash
c = Quantity of cement in mix (kg/m³)
CaO = 0.65, the amount of CaO content in Portland cement
r = 0.75, the proportion of calcium oxide that can be carbonated
A = the exposed surface area of concrete (m³)
M = 0.79, the chemical molar fraction (CO_2/CaO)
t = exposed carbonation time (years)

2.5.3 Rolling Resistance (RR)

It has been estimated that approximately 20% of transportation-related consumption is due to rolling resistance (IEA, 2005). Rolling resistance is defined in ISO 28580:2009 as the loss of energy or the energy dissipated per unit of distance traveled (ISO, 2009). Pavement-related factors that affect RR include unevenness (also called roughness), pavement structure (deflection based), and texture.

A recent study developed a model called the RSI (Roughness-Speed Impact) model to relate energy and emissions to pavement roughness and vehicle operating conditions using vehicle specific power model (Ziyadi et al., 2016):

RSI model for vehicle energy consumption:

$$\Delta \hat{E}(v, \Delta IRI) = (k_a + k_c \cdot v^2) \times \Delta IRI \qquad (3)$$

where,

$\Delta \hat{E}$ = Estimated additional energy consumption per vehicle distance (kJ/mile)
v = Average speed (mph)
k_a, k_c = model coefficients defined for each type of vehicle
ΔIRI = incremental changes in IRI between two consecutive analysis steps

This model uses a series of incremental tables (Ziyadi et al., 2016) to estimate the TRACI impact.

The main inputs of the RSI models are IRI and traffic composition and speed. Since LCA covers a long span of time, it is important to know how these inputs change over time.

IRI increases over time as pavement deteriorates and decreases (*improves*) upon application of any Maintenance and Rehabilitation (M&R) project. IRI progression curves and drop models were developed for the pavements in the Tollway network using historic data.

A calibrated HDM4 model (Chatti & Zaabar, 2012) is used to calculate the additional fuel consumption due to pavement texture for trucks only. Global Warming Potential (GWP) is calculated based on energy conversion. The following linear relationship is used to interpolate the percent change in fuel consumption due to 1 mm increase in Mean Profile Depth (MPD) and speed:

$$\delta E_{texture}(\%) = 0.02 - 2.5 \times 10^{-4} \times (v - 35) \quad (4)$$

where,
$\delta E_{texture}$ = percent change in fuel due to 1 mm increase in MPD
v = speed (mph)

Texture progression models were developed using literature model (Chatti & Zaabar, 2012) and calibrated with historic data from the Tollway network.

3 UNCERTAINTY QUANTIFICATION (UQ)

Pavement roughness measured and reported in terms of IRI is one of the main inputs to the pavement LCA that has significant effect on the results, yet it is prone to many uncertainties. Human and measurement errors are among those. Assuming there is a true value of IRI every year for each pavement section, variations in data can be modeled as normal probability distribution. Using this concept, uncertainties in the IRI were propagated throughout the LCA calculation. Monte Carlo sampling technique is used for this purpose. Statistical parameters of the distributions including mean and standard deviations for this study were collected from historical data and calculated using similar pavement types.

4 IMPACT ASSESSMENT

Four environmental indicators were selected to assess the impacts including Single Score (SS), Total Primary Energy (TPE), Global Warming Potential (GWP) and Primary Energy as Fuel (PEF). Results from different TRACI impact categories were normalized (Lautier et al, 2010) and weighted (Bare et al., 2006) based on National Institute of Standards and Technology (NIST) to a single score to simplify for external use. Care should be taken when interpreting SS because it is a highly simplified indicator of multiple complex impact categories. TPE includes all energy embodied as fuel (e.g., diesel, natural gas) and material (e.g., petroleum products such as plastics, asphalt binder), whereas PEF only includes the energy embodied as fuel.

5 CASE STUDIES

Four pavement-widening projects were selected for analysis using the developed LCA tool and methodologies presented in this paper. Projects were selected from the Illinois Toll highway (Tollway) network located in the Chicago metropolitan area. Two past (2008 and before) and two current (2012 and after) widening projects were selected from a total of 14 historical and current projects representing a wide range of cases. Table 2 summarizes these projects and presents relevant information. All projects are Jointed Plain Concrete Sections (JPCP). Traffic ranges from low AADT of 19,240 for to a high AADT of 148,200. The functional unit proposed in this study (VMT) was used to interpret the results from these projects.

Table 2. Case study project information.

Toll road	Year	Project code	Milepost	Length (mi)	Analysis period (yr)	AADT; % Truck	Description
Jane Addams Memorial I-90/I-39/ US 51	2012–2013	4077	49.7 to 53.6	3.9	62 yrs; 3 overlays	28,460 EB; 13.3%	Roadway widening (3 lanes 12-inch JPCP) and reconstruction
	2014	4133	24.9 to 33.5	8.6	62 yrs; 3 overlays	19,240 WB; 20.3%	Roadway widening (3 lanes 11.25-inch JPCP) and reconstruction
Tri-State I-94/I-294/ I-80	2007–2008	5228	15.84 to 13.24	2.6	62 yrs; 3 overlays	148,200 SB 14.6%	Roadway widening and reconstruction (with 12-inch JPCP) from 3 to 4 lanes
Ronald Reagan Memorial I-88	1999	723	133.7 to 138.8	5.1	44 yrs; 2 overlays	76,680 EB; 80,670 WB; 9.8%	Roadway widening and reconstruction to 3 (12-inch JPCP) lanes

6 RESULTS AND ANALYSIS

Four LCA projects were analyzed using the Illinois Tollway Roadway/Roadside tool and environmental impacts were calculated according to the system boundary as described above. The purpose of the analysis is to examine the ranges and share of each influential factor and perform a project-to-project comparison to evaluate environmental sustainability improvements over the years. Figure 1-a shows the share of each stage for each project per functional unit. The use-stage share is almost 50 to 90% of environmental impacts in terms of Single Score (SS), Total Primary Energy (TPE), Global Warming Potential (GWP), and Primary Energy as Fuel (PEF) (Figure 1-b). The figure presents a significant range of values each stage contribute to the total environmental performance calculated from four projects. The error bars on Figure 1-b show the range of values based on the four projects.

It should be noted that new projects have the lowest contribution from use-stage among all projects. This is partly because the new projects carry lower traffic. In addition, investigating the processes involved in the LCA stages reveals that new projects involve more material and construction processes and, therefore, the material and construction and maintenance stages account for around 30% of total impacts alone.

The breakdown of processes contributing to the use-stage is shown in Table 3. The table shows a comparison of the four projects with their corresponding total impacts and impacts per functional unit. First, it is clearly shown that a significant portion of GWP impacts are associated with the additional fuel consumption of passenger vehicles due to roughness. The total absolute impacts for the new projects (4077 and 4133) with relatively lower traffic are smaller than the others. Total absolute impacts reflected into the impacts calculated per the functional unit chosen with physical characteristics (lane-mile) and annualized lane-mile. These two functional units resulted in higher impacts for the two sections with lower traffic volume (Projects 4077 and 4133). However, impacts per functional unit (in terms of user-miles traveled, VMT) are in the range of the other two projects. When VMT was used as functional unit, the sections with higher traffic volumes are given a credit due to serving for higher traffic while performing satisfactorily. In the case of project 4133, GWP impacts are highest (11.1 tonne-CO_2-eq/mi VMT). This suggest that the project 4133 may not have an optimum design (in terms of structure) for the traffic condition it was built for.

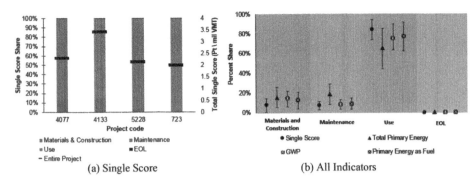

Figure 1. Environmental impact breakdown of LCA case studies for (a) single score and (b) all indicators.

Table 3. Use-stage breakdown for GWP.

| | | GWP (tonne-CO_2-eq) | | | |
| | | Project code | | | |
LCA Stage	Component	4077	4133	5228	723
Use-stage	Roughness-Related Passenger Vehicle	25,144	35,950	93,047	114,648
	Roughness-Related Small Truck	624	1,483	2,577	2,018
	Roughness-Related Medium Truck	824	1,957	3,400	2,662
	Roughness-Related Large Truck	2,988	7,095	12,325	9,651
	Texture-Related Medium Truck	1,413	3,263	5,604	5,457
	Texture-Related Large Truck	2,738	6,324	10,860	10,576
	Albedo Mainline	3,148	6,943	1,111	−8,321
	Albedo Shoulders	2,087	7,423	1,056	−3,645
	Carbonation Mainline	−55	−111	−46	−116
	Carbonation Shoulders	0	0	0	0
Total (tonne-CO_2-eq.)		*38,915*	*70,331*	*129,939*	*132,931*
Total VMT (millions)		*4,199*	*6,354*	*12,451*	*20,921*
Per Functional Unit of Lane-Mile (LM) *tonne-CO_2-eq./lane-mile*		3,326	2,726	12,494	8,688
Per Functional Unit of Annualized Lane-Mile (LMY) *tonne-CO_2-eq./lane-mile-year*		54	44	202	197
Per Functional Unit of Vehicles Mile Travelled (VMT) *tonne-CO_2-eq./million VMT*		9.3	11.1	10.4	6.4

According to the results albedo and carbonation only contribute to GWP and single score. In general, roughness effect governs use-stage impacts, followed by texture. The exception to this is GWP where the albedo effect is significant given the high surface exposure and change of albedo from new to weathered surface over time.

Different components of the use-stage contribute differently to the LCA indicators. This can result in the change of environmental rankings of comparable sections when different indicators are used. For example, since albedo and carbonation do not contribute directly to energy consumption, texture effect can be as high as 35% of energy indicators (TPE and PEF), however it is 20% (or less) of environmental indicators (SS and GWP) where albedo and carbonation contributions are involved.

Albedo can be a major contributor to GWP (~20%) when traffic is also relatively low. However, on overall environmental score (SS) of contributes less than 7%. Project 4133 in particular has the highest environmental (SS and GWP) impacts. This can be attributed to the relatively lower traffic, thus indicating a possible non-optimal design under current conditions.

Monte Carlo technique was used to run 1000 simulations of each project with probability distributions associated with IRI every year. The standard deviation of the measured IRI increases

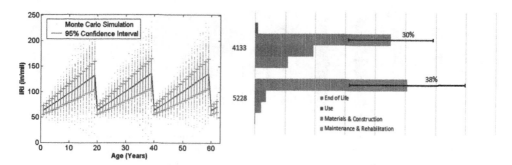

Figure 2. (a) IRI progression with uncertainty band and (b) resulting use stage output range.

linearly from the initial value of 4.2 to 15.0 before rehabilitation. Figure 2-a shows the IRI progression for projects with 95% confidence interval calculated using Monte Carlo simulation. These uncertainties were propagated throughout the LCA analysis. Figure 2-b shows the resulting use stage GWP range. It can be noted that such small variations in the measurements can lead up to 30% change in the impact of the use stage. This is in line with the fact that pavement roughness is the main contributor to the overall project environmental impacts. Hence, measurement of the IRI and using appropriate use stage models are crucial for implementing pavement LCA. Therefore, the case studies highlighted the significance of a complete LCA with multiple use-stage components included, multi-point environmental performance evaluation and selecting an appropriate functional unit during the interpretation phase of a pavement LCA.

7 CONCLUSIONS

The development of a complete pavement LCA tool with an emphasis on the use-stage life-cycle stage is presented. The introduced framework aimed to address some of the current challenges in the area of pavement LCA by performing a comparative cradle-to-grave type of LCA using multi-point environmental performance evaluation and a proposed functional unit allowing for fair project-to-project comparison.

A functional unit was proposed that can be used in comparative LCAs to compare pavement systems with different performance and traffic characteristics. The proposed functional unit is based on the total vehicles travelling the section during the analysis period and designated as Vehicle-Length Travelled (VLT), which can be implemented as Vehicle-Miles Travelled (VMT) or Vehicle-Kilometer Travelled (VKT). The unit provides a rational basis for comparative LCAs since it normalizes over years and traffic volume and does not necessarily penalize the longer living pavement systems carrying higher traffic volumes.

Four projects from the Illinois Tollway were selected to analyze the case studies. The projects were widening projects with jointed plain concrete pavement. In general, it was found that roughness and texture-related rolling resistance and resulting impacts constituted a significant portion of total use-stage impacts (between 50%-95% of total use-stage). Additional fuel consumption by passenger cars due to roughness was found to be responsible for the majority of the use-stage impacts. Albedo and carbonation only contribute to environmental indicators (GWP and SS) and not energy indicators (PEF and TPE). Albedo and the resulting radiative forcing can contribute significantly to GWP with relatively smaller effect on the overall environmental indicator (SS). GWP savings due to carbonation for concrete projects were less than 1%. Although the traffic level directly affected the total use-stage impacts, normalization of results by Vehicle-Miles Traveled (VMT) allowed for a more reasonable interpretation of the results, consistent with the agency's main goal to assess how much progress has been made over the years toward the sustainability goal by employing sustainable pavement strategies.

Also, uncertainty quantification of pavement roughness revealed that variations in IRI measurements can result in a significant change in the overall use stage results. Such significant impacts on the overall sustainability necessitate accurate measurements from the road network and the use of robust models in the LCA.

ACKNOWLEDGMENTS

Part of this work is funded by the Illinois State Toll Highway Authority through the Illinois Center for Transportation. The authors would like to acknowledge the input of Steve Gillen, Illinois Tollway, and other partners of the project: Applied Research Associates, Inc. and the Right Environment. The contents of this report reflect the views of the authors, who are responsible for the facts and the accuracy of the data presented herein. The contents do not necessarily reflect the official view or policies of the Illinois Tollway or ICT. This paper does not constitute a standard, specification, or regulation.

REFERENCES

Andersen, L.G., J.K., Larsen, E.S., Fraser, B., Schmidt & J.C., Dyre. 2014. Rolling resistance measurement and model development. *Journal of Transportation Engineering*, 141(2), 04014075.

AzariJafari, H., A., Yahia & M.B. Amor. 2016. Life cycle assessment of pavements: reviewing research challenges and opportunities. *Journal of Cleaner Production*, 112, 2187–2197.

Bare, J., T., Gloria & G. Norris. 2006. Development of the method and US normalization database for life cycle impact assessment and sustainability metrics. *Environmental science & technology*, 40(16), 5108–5115.

Bare, J. 2011. TRACI 2.0: the tool for the reduction and assessment of chemical and other environmental impacts 2.0. Clean *Technologies and Environmental Policy*, 13(5), 687–696.

Chatti, K. & I. Zaabar. 2012. *Estimating the effects of pavement condition on vehicle operating costs* (Vol. 720). Transportation Research Board.

Environmental Protection Agency (EPA). 2014. *MOVES (Motor Vehicle Emission Simulator)*. Environmental Protection Agency, Washington, DC.

Gregory, J.R., Noshadravan, A., Olivetti, E.A., Kirchain, R.E. 2016. A Methodology for Robust Comparative Life Cycle Assessment Incorporating Uncertainty. *Environmental Science and Technology*, 50, 6397–6405.

Guinée, J.B., M., Gorrée, R., et al. 2002. Handbook on life cycle assessment. Operational guide to the ISO standards. I: LCA in perspective. IIa: Guide. IIb: Operational annex. III: Scientific background. Kluwer Academic Publishers, ISBN 1-4020-0228-9, Dordrecht, 2002, 692 pp. cml.leiden.edu/research/industrialecology/researchprojects/finished/new-dutch-lca-

Hammarström, U., J., Eriksson, R., Karlsson & M.R. Yahya. 2012. *Rolling resistance model, fuel consumption model and the traffic energy saving potential from changed road surface conditions*. VTI rapport, ISSN 0347-6030; 748A

Harvey, J., Meijer, J., Ozer, H., Al-Qadi, I.L., Saboori, A., & Kendall, A. (2016). Pavement Life-Cycle Assessment Framework. Federal Highway Administration, Washington, DC.

IEA. 2005. *Energy efficient tyres: Improving the on-road performance of motor vehicles*. In IEA Workshop. International Energy Agency, November 2005.

Inyim, P., J., Pereyra, M., Bienvenu & A. Mostafavi. 2016. Environmental assessment of pavement infrastructure: A systematic review. *Journal of environmental management*, 176, 128–138.

ISO, I. 2006. 14040: Environmental management–life cycle assessment–principles and framework. *London: British Standards Institution*.

ISO. 28580. 2009. Passenger car, truck and bus tyres methods of measuring rolling resistance single point test and correlation of measurement results. The International Organization for Standardization.

Kaloush, K.E., J.D., Carlson, J.S., Golden & P.E. Phelan. 2008. *The thermal and radiative characteristics of concrete pavements in mitigating urban heat island effects* (No. PCA R&D SN2969).

Kang, S., R., Yang, H., Ozer & I.L. Al-Qadi. 2014. "Life-Cycle Greenhouse Gases and Energy Consumption for the Material and Construction Phases of Pavement with Traffic Delay", *Transportation Research Record: Journal of the Transportation Research Board*, 2428(1), pp. 27–34.

Karlsson, R., U., Hammarström, H., Sörensen & O. Eriksson. 2011. Road surface influence on rolling resistance—Coastdown measurements for a car and HGV—VTI notat 24A. *Swedish Road Administration (VTI)*.

Lagerblad, B. 2005. Carbon dioxide uptake during concrete life cycle–state of the art. *Swedish Cement and Concrete Research Institute CBI, Stockholm.*

Lautier, A., R.K., Rosenbaum, M., Margni, J., Bare, P.O., Roy & L. Deschênes. 2010. Development of normalization factors for Canada and the United States and comparison with European factors. *Science of the total environment*, 409(1), 33–42.

Noshadravan, A., Wildnauer, M., Gregory, J., Kircahin, R. 2013. Comparative pavement life cycle assessment with parameter uncertainty. *Transportation Research Part D.* 25, 131–138.

Pommer, K. & C. Pade. 2006. Guidelines: Update of Carbon dioxide in the Life Cycle Inventory of Concrete. Norden, Nordic Innovation Centre, Norway.

Santero, N.J. & A. Horvath. 2009. Global warming potential of pavements. *Environmental Research Letters*, 4(3), 034011.

Santero, N.J., E., Masanet & A. Horvath. 2011a. Life-cycle assessment of pavements. Part I: Critical review. Resources, *Conservation and Recycling*, 55(9), 801–809.

Santero, N.J.E., Masanet & A. Horvath. 2011b. Life-cycle assessment of pavements Part II: Filling the research gaps. *Resources, Conservation and Recycling,* 55(9), 810–818.

Shakiba, M., H., Ozer, M., Ziyadi & I.L. Al-Qadi. 2016. Mechanics based model for predicting structure-induced rolling resistance (SRR) of the tire-pavement system. *Mechanics of Time-Dependent Materials*, 1–22.

Van Dam, T.J., Harvey, J.T., Muench, S.T., Smith, K.D., Snyder, M.B., Al-Qadi, I.L., Ozer, H., Meijer, J., Ram, P.V., J.R. Roesler, J.R. & Kendall. A., 2015. *Towards sustainable pavement systems: a reference document.* Federal Highway Administration: Washington, DC.

Wang, T., Harvey J., & A. Kendall. 2014. Reducing greenhouse gas emissions through strategic management of highway pavement roughness. *Environmental Research Letters*, 9.

Wang, T., I.S., Lee, A., Kendall, J., Harvey, E.B., Lee & C. Kim. 2012. Life cycle energy consumption and GHG emission from pavement rehabilitation with different rolling resistance. *Journal of Cleaner Production*, 33, 86–96.

Yang, R., Ozer, H., Al-Qadi., I.L., and W., Yoo. 2016. Development and Application of a Roadway/Roadside Life-Cycle Assessment Software for the Illinois Tollway. Submitted to 96th Annual Meeting, Transportation Research Board of National Academies, Washington D.C.

Ziyadi, M., H. Ozer, S. Kang & I.L. Al-Qadi. 2016. Pavement Roughness Related Energy Consumption and Environmental Impact Calculation Model for Transportation Sector. *Environmental Research Letters* (submitted).

Role of uncertainty assessment in LCA of pavements

S. Inti, M. Sharma & V. Tandon
Department of Civil Engineering, The University of Texas at El Paso, El Paso, TX, USA

ABSTRACT: Life Cycle Assessment (LCA) is a tool to appraise the environmental impact of a pavement during its service life. Typically, the LCA assessor has to make some assumptions and judgments, employ data from different sources, use analytical tools and models, etc., in order to provide information to the transportation officials for decision making. Since assumptions and data from different sources have inherent uncertainty, it is essential to explicitly report the underlying uncertainties and their consequences in LCA outputs. This study recommends an approach for performing uncertainty assessment in pavement LCA. The first step of the recommended approach is to define a clear objective for the need of the assessment. Then establish the scope of assessment such that it is technically feasible and satisfies the assessment objective. Prioritizing the phases in LCA that influence the output and concentrating only on significant phases makes the uncertainty assessment viable. The next step is classifying the uncertainties that prevailed in prioritized phases. Classification supports in selecting the proper methods and techniques that can apprehend the uncertainty. Finally, present the uncertainty results clearly and effectively to the transportation officials. The recommended approach is demonstrated through an example.

1 INTRODUCTION

Life Cycle Assessment (LCA) tool has gained popularity in the pavement infrastructure because it comprehensively quantifies the emissions and energy flows of pavement in its service life. The output from LCA helps decision makers in choosing a design and/or material and/or construction practice, etc. to lessen the burden on the environment. The practitioners of LCA employ the guiding principles provided by various standards such as International Organization of Standardization (ISO) 14040 (2006), ISO 14044(2006), Environmental Protection Agency (EPA) document (Scientific Applications International Corporation (SAIC) 2006), etc. However, there is no US government authorized guidelines for performing LCA of pavement (U.S. Department of Transportation, Federal Highway Administration (FHWA) 2014). Additionally, lack of a comprehensive and official database for the construction of pavements impacts the reliability of LCA.

Various researchers (Nisbet et al. (2000), Mroueh (2000), Huang et al. (2009), Weiland and Muench (2010), Santero et al. (2011), Tatari et al. (2012), Vidal et al. (2013), Barandica et al. (2013), Liu et al. (2014), Thiel and Len (2014), Anastasiou et al. (2015)), have performed LCA of pavements using available tools and guidelines. These researchers attempted to perform a comprehensive assessment by making some assumptions, using models that were not exclusively developed for pavement LCA, employ data from other geographical location or technology, etc. These factors lead to uncertainty in LCA output at different phases. The existence of uncertainty is also acknowledged in the ISO standards. It is stated in ISO 14040 (2006) that "*LCA does not predict absolute or precise environmental impacts due to the inherent uncertainty in modeling of environmental impacts.*" LCA assessors need to focus on either reducing the uncertainty or develop an effective way of communicating the consequences of uncertainty (e.g., assumptions) in the assessment to transportation officials.

2 LITERATURE REVIEW

It is important to understand the true sense of uncertainty. Heijungs and Huijbregts (2004) stated that the uncertainty arises by using data that is inappropriate, unreliable, or possess some degree of inherent variability. Weidema et al. (2013) expressed uncertainty as any distribution of data within a population either caused by random variation or bias. The chances are high that every LCA assessor encounters certain type of uncertainty. Knowing the fact that uncertainty subsists in LCA, the questions that arise are:

1) What is the credibility of the LCA results? (Allaire and Willcox 2014), 2) Which part of the LCA processes leads to primary uncertainty? (Hung and Ma 2009), 3) How to present the uncertainty in LCA that can be easily comprehended by the policy makers?

A systematic uncertainty assessment answers to the above questions to a good extent. According to Baker and Lepech (2009), uncertainty assessment supports:

a. better decision-making
b. increases transparency in the data
c. increases the quality of analysis, and
d. helps in planning information gathering exercise(s).

May and Brennan (2003), and Helton et al. (2006) described that the uncertainty analysis could provide a possible range of outcomes rather than single outcome based on the individual data uncertainties. It is important to understand various chief sources of uncertainty and the best techniques to handle them.

2.1 Sources of uncertainty in LCA

Baker and Lepech (2009) classified the uncertainties as below:

- Database uncertainty (missing or unrepresentative data)
- Model uncertainty
- Statistical measurement error (data based on a limited set of sample)
- Uncertainty in preferences (uncertainty due to analyst's choices)
- Uncertainty in a future physical system, about the currently designed system (the degree to which the future conditions change in which present design will be subjected)

One can further classify these uncertainty sources, for example, database uncertainty can be due to outdated data, spatial or technological variability of available data, inconsistency between various available data sources, etc.

2.2 Handling uncertainty

Typically, the objective of uncertainty assessment can be either reducing the uncertainty or communicating the possible impacts of uncertainties on LCA outputs. If the focus is reducing the uncertainty, Heijungs and Huijbregts (2004) clearly mentioned three approaches:

1. Scientific Approach: Performing thorough research while developing the characterization models, increasing the data size, etc.
2. Constructivist Approach: Involving stakeholders, discussing and finally deciding on or voting for a consensus characterization factors.
3. Legal Approach: Relying on authoritative bodies reports, like ISO or the US EPA.

Out of the three ways, the first two are not practical at the level of an analyst because it needs the involvement of state or governmental bodies. Therefore, the analysts often attempt in using the data from authoritative bodies like the US EPA, the FHWA, etc. Additionally, some researchers have suggested handling uncertainty by using local databases containing site specific data (Cellura et al. 2011).

However, most attention in the published literature of uncertainty assessment in LCA of pavements is given in communicating the possible outcomes due to the variability in inputs or data quality. There are various analytical tools such as a first order approximation of the Taylor expansion of the underlying model, fuzzy set methods, neural networks, etc., but the principal tool employed is Monte Carlo Simulation (MCS) (Hung and Ma 2009) (Noshadravan et al. 2013) (Allaire and Willcox 2014) (Yu et al. 2016) followed by sensitivity analysis.

In MCS, the input parameter is expressed in the form of a probability distribution. Multiple simulations can be performed by varying the input parameters (as per probability distributions), and outputs can be collected for each simulation. From the pool of outputs obtained, statistical properties can be evaluated, and the possible outcomes are presented. The principal challenge in using MCS is assigning a probability distribution and defining statistical input parameters. Most common type of probability distributions employed are the normal, lognormal, uniform, and triangular (May and Brennan 2003), (Heijungs and Huijbregts 2004).

In general, if the experimental data is not available, the LCA analyst assumes the probability distribution and statistical parameters (variance, standard deviation, etc.) for performing MCS. In some cases, if the input LCA data is available from numerous sources or from different time periods etc., then the analyst can characterize the quality of the available data and develop the statistical parameters for performing simulations. The Data Quality Indicators (DQI's) approach developed by Weidema and Wesnes (1996) is a well-known method for data characterization. The data is characterized based on five parameters that are reliability, completeness, temporal correlation, geographical correlation, and technological correlation. A numerical score is assigned to each parameter, and an aggregated DQI score is estimated at the end. Data quality score can be converted into a probability distribution function or as an additional variance in the data. For example, Yu et al. (2016) employed DQI score to develop a modified beta distribution based on the works of Canter et al. (2002). Noshadravan et al. (2013) used DQI's scores to calculate the additional variance to the data as per the procedure presented in Weidema et al. (2013). Similar to MCS, sensitivity analysis, scenario analysis are other tools that can be used to include uncertainty assessment in LCA.

Previous studies employed uncertainty propagation tools for a particular input parameter or a single phase of LCA. For example, Yu et al. (2016) worked on the albedo effect of pavements, another study by Yu et al. (2016) presented a procedure to evaluate the data quality of inputs for asphalt layers in pavement construction, Noshadravan et al. (2013) accounted for the uncertainty in the prediction of pavement roughness. There is a lack of guidance for performing the overall uncertainty assessment for LCA of pavements. The efforts from the previous researchers are commendable; the employed methodologies and findings from their studies can be implemented in performing a comprehensive and systematic uncertainty assessment of LCA of pavements.

3 OBJECTIVE

The objectives of this study are to recommend a suitable approach for performing uncertainty assessment in LCA of pavements, and to demonstrate the proposed approach through an example.

4 PROPOSED APPROACH

This study proposes an approach which comprises of four basic parts

1. Goal of uncertainty assessment
2. Prioritization of major influencing emission sources
3. Classification of uncertainty and selection of uncertainty propagating or mitigating tools
4. Presentation of results

4.1 *Goal of uncertainty assessment*

The necessity of conducting uncertainty assessment needs to be clearly defined. The goal of assessment influences the whole assessment and helps in outlining the scope.

4.2 Prioritization of major influencing emission sources

After defining the goal, the next step is to choose a well-outlined approach performing the assessment. Uncertainty may happen at every phase of LCA. Even though performing uncertainty assessment at every phase is desirable, it may be neither practical nor necessary. Heijungs and Huijbregts (2004) cautioned about the darker sides of uncertainty assessment. Incorporating uncertainty assessment makes LCA a more tedious practice because additional efforts are required for data collection. Employing various assessment techniques on all uncertainty sources generates a swarm of results which may overshadow the LCA itself. Often too much information leads to pessimism. Therefore, the uncertainty assessment needs to strike a balance between the practicality to analysts and meaningfulness to transportation officials.

Prioritization helps in maintaining the desired balance. A feasible scope for the uncertainty assessments can be attained by prioritizing assessment only on the phases which significantly impact the overall LCA results, and on the judgments or assumptions made by analysts, which affect the LCA outputs or decisions.

4.3 Classification of uncertainty sources

Once the key phases are prioritized the next step is to classify the uncertainty existed in these phases. Classifying uncertainty supports in identifying the tools and techniques required to propagate or lessen the uncertainty, as each source of uncertainty needs to deal with an appropriate approach. Figure 1 shows the primary sources of uncertainty in each phase of LCA and the possible actions that can be implemented to mitigate or propagate the uncertainty.

4.4 Presentation of results

Since uncertainty assessment involves complicated calculation techniques like simulations, statistics, etc., the output may overwhelm the policy or decision maker. Therefore, it is essential that the results are presented in a lucid fashion that strengthens the understanding of risks in LCA and helps in decision making.

5 EXAMPLE

Three equally performing pavement designs are analyzed using Mechanistic-Empirical Pavement Design for 30 years of El Paso Texas climate conditions are shown in Figure 2. Type of pavement layers, thicknesses, and year of maintenance are also mentioned in Figure 2. Each

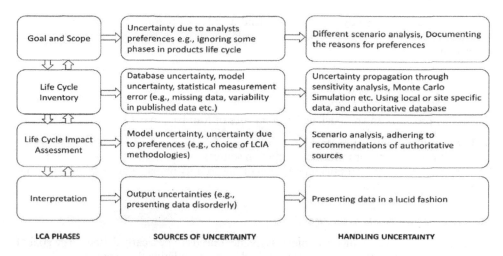

Figure 1. Sources of uncertainty and handling it at various phases of LCA.

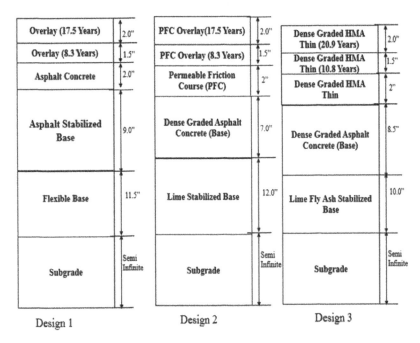

Figure 2. Pavement sections.

design is a six-lane highway (three lanes on each side) for an Annual Average Daily Traffic (AADT) of 60,000 at a growth rate of 0.75. Ten percent of the traffic comprises of trucks. The performance of each pavement for 30 years are assessed using the pavement design software.

Initially, LCA is performed to evaluate the environmental impacts due to each pavement design. Global Warming Potential (GWP) was estimated for the three designs for a one-mile length of the highway for 30 years. Five segments (material extraction, transportation, construction, use, and maintenance) of a pavement life cycle are considered in the GWP estimation. The elements considered in each segment, the models and data sources employed, and the assumptions made in the GWP estimation are shown in Table 1. For instance, during usage segment of pavement, emissions due to combustion of gasoline and diesel, and tire wear are considered. Other emission sources during highway usage like albedo, carbonation, and lighting are not considered because they are more beneficial when one compares an asphalt and concrete top surfaces (In this case study all designs have asphalt surface on top). The end of life of the pavements is not considered in LCA because as there is no well-defined end of life where the pavement would be demolished and thrown away (Weiland and Muench 2010).

The estimated GWP of three designs is summarized in Figure 3. The results suggest that the Design 3 has the lowest GWP compared to Designs 1 and 2. Design 3 bettered Design 1 and Design 2 by 6259, 8432 tons of GWP per mile of highway. However, by closely examining the models, data sources, and assumptions in the GWP estimate and associating them with types of uncertainty portrayed in Figure 1, it is evident that there is certain riskiness in the estimate. For example, the material extraction includes data from different geographical locations (Tire LCA from Japan, Asphalt production emissions from Europe, etc.,), construction phase has limited sample sizes (construction equipment data), transportation and maintenance phases have some critical assumptions (future maintenance timing 9AM-5 PM), and use phase has modeling errors (an average passenger car emissions used for estimating overall emissions).

5.1 *Uncertainty assessment*

Since uncertainty persists in the assessment, explicitly presenting the influence of uncertainty enhances the transparency of LCA and increases the confidence in decision making. In order

Table 1. Data sources for life cycle inventory.

Phase	Model used	Data sources	Assumptions
Material Production	Customized emissions model are developed using Greenhouse Gases, Regulated Emissions, and Energy Use in Transportation Model (GREET) (Argonne GREET Model).	Asphalt: Life Cycle Inventory Bitumen (Blomberg et al. 2011), Aggregate: Greenhouse Gas Emissions Inventory CEMEX Jesse Morrow Mountain Plant (Downey 2009), Asphalt Plant Operations: Hot Mix Asphalt Plants Emission Assessment Report (US EPA 2000) (Myers et al. 2000), Emulsion: Life Cycle Inventory Emulsion (Blomberg et. al. 2011), Fuel Combustion factors: Emissions Factors & AP42.	The mix designs for materials like asphalt concrete; lime stabilized base were considered as per Texas commonly used proportions.
Transportation	Customized emissions model are developed using Greenhouse Gases, Regulated Emissions, and Energy Use in Transportation Model (GREET).	A 20-ton capacity truck with full front haul and empty backhaul is assumed with fuel (diesel) consumption of 5.3 miles per gallon. Diesel is considered as the fuel used in trucks for transporting materials.	Transportation distance of raw materials like (asphalt, lime, etc.,) to plant or construction site 50 miles and 12 miles considered as the distance between asphalt plants to the construction site.
Construction	The emissions during construction from equipment and machinery were estimated using NONROAD 2008 database.	The type of construction equipment and working durations are estimated by using RSMEANS 2012. The equipment details were taken from common construction equipment manufacturers like Caterpillar, Dynapac, Bomag, Roadtech, Wirtgen, etc. The emissions data for equipment (machinery) available in NONROAD 2008 is matched with construction equipment by horsepower.	After estimating the emissions from each equipment per hour, the total impacts is calculated by multiplying machinery hours required for activity (i.e. asphalt concrete) and time efficiency.
Use	Models reported in NCHRP 720 for estimating vehicle operating costs were used to calculate the gasoline and diesel consumption, and tire wear.	Tire: Typical tire manufacturing LCA is developed by using Life Cycle Assessment of a Car Tire Continental (Kromer et al.), Emission Factor Documentation for AP-42 Manufacture of Rubber Products, Tyre LCCO$_2$ Calculation Guidelines. Emissions from Cars: Greenhouse Gas Emissions from a Typical Passenger Vehicle. Emissions from Trucks: Developed emissions from a truck with varying mileage using GREET.	The emissions for manufacturing a passenger car is estimated first, and later the emissions for trucks are estimated based on the proportions of materials used in manufacturing truck and a passenger car tire.
Maintenance	For materials production and transportation, the methods explained above are followed. For estimating the emissions due to traffic delays during maintenance is estimated using Motor Vehicle Emission Simulator.	For traffic delay emissions method proposed by Inti et al. 2016) is employed which used MOVES.	Traffic delay emissions depend on the timing of maintenance and in this study we considered maintenance from 9 AM–5 PM by closing one lane because this working time has a maximum impact on the environment.

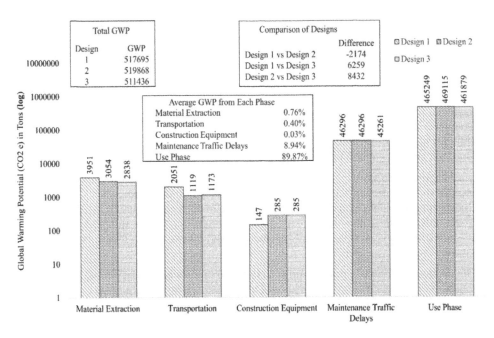

Figure 3. GWP of three pavement designs (deterministic approach).

to perform an uncertainty assessment on this example, the approach explained in the previous section is employed.

5.1.1 *Goal of uncertainty assessment*
The main goal of this assessment is to evaluate the influence of variation in input data quality and assumptions made in the LCA and finally present the range GWP for the three designs.

5.1.2 *Prioritization of major influencing emission sources*
The most affecting sections of pavement life cycle on the overall GWP need to be categorized. It is apparent from the LCA results in Figure 3 that the emissions from vehicles during highway usage, predominantly (90% of total GWP) impacts the overall LCA followed by the traffic delay emissions (9%) during maintenance of the highway. Close to 70% of the GWP difference between Design 1 and Design 3 can be attributed to use and maintenance phases, and in the case of Design 3 and Design 2, the percentage is around 98%. Hence, use and maintenance phases can be considered as a priority for uncertainty assessments as they are driving the overall LCA and the decision.

5.1.3 *Classification of uncertainty and selection of uncertainty propagating or mitigating tools*
Use Phase: Vehicle emissions, tire wear related to pavement-vehicle interaction are considered in the use phase of LCA in this study. Vehicle emissions and tire wear are a function of International Roughness Index (IRI) (a pavement distress). The performance of each pavement design is different and yields a distinct IRI. In the initial GWP estimate, the models suggested (for estimating user costs) in NCHRP Report 720 are used. These models were calibrated using vehicles driven on roads of the known condition in the US.

The IRI of the pavement, vehicle type, speed of the vehicles, etc., are the key input to these models. In the initial estimate, the speed of vehicles was considered as 55 Miles Per Hour (MPH) and calculated the emissions based on the IRI generated by pavement design software for three designs. Primarily, the mileage for different vehicles (cars, trucks, vans, etc.) were estimated at 55 MPH and later fuel consumed to travel one mile. Emissions were estimated based on the combustion of a gallon fuel (diesel and gasoline) for a typical passenger and trucks from the US EPA. The software estimated the IRI on an average monthly basis for 30 years. The emissions were estimated for a single day of the month and multiplied it by the total days in a month. The results in Figure 3 were the aggregated emissions for 30 years. Increasing IRI wears the tire quickly, resulting in a frequent change of tires than anticipated. Pavements with poor performance (higher IRI) causes more damage to tires and more emissions due to frequent tire replacement. For tire wear, the LCA of a typical passenger car tire is taken from the Continental Tire manufacturers of the Japan and integrated them with AP 42 emissions for rubber products of the US EPA.

There are various sources of uncertainty in the use phase estimate such as:

- Uncertainty in the model: NCHRP 720 models were calibrated at a geographical location which is different from the current location of study (climate variability)
- Uncertainty in assumptions: Speed of all vehicles for thirty years assumed as 55 MPH
- Uncertainty in the data sources: Tire emissions inventory from a different geographical location (the Japan), fuel consumption emission factors were taken from a typical car and truck in the US.

The impact of climate variability on NCHRP 720 models is not considered in this study. As calibrating the models through field tests in El Paso region (geographical location of the case study) is beyond the scope of this study. However, for another type of uncertainties (vehicle speed, tire emissions, car and truck emission factors), this study used MCS to present the possible range of emissions.

Due to the lack of guidance in choosing the probability distribution functions and parameters from an authoritative body for LCA of pavement, this study assumed probability distributions based on the works of May and Brennan (2003), Heijungs and Huijbregts (2004) and the author's experience. The following probability distributions and parameters are assumed: The speed of vehicles from 30–70 MPH (triangular probability distribution), ten percent variance in the fuel consumption of passenger car and trucks (normal probability distribution, and additional variance (geographical variation) was estimated in the tire emissions using the DQI's method as explained by Weidema et al. (2013) (used lognormal probability distribution). MCS with Latin Hypercube sampling method is used to run 20000 simulations for each design varying these input parameters (results discussed in a later section).

Maintenance Phase (Traffic Delay Emissions): The traffic delay emissions, during maintenance (additional fuel consumption), were estimated as per the procedure explained by Inti et al. (2016). This procedure uses the Motor Vehicle Emissions Simulator (MOVES 2014) and the FHWA's life cycle cost analysis method. Initially, a sensitivity analysis is performed on the various input parameters. It is observed that the major influencing parameter in this procedure is the maintenance timing and maintenance strategies (number of lanes open for traffic during maintenance). These two parameters greatly overwhelm the influence of other inputs. This study performed the scenario analysis as an uncertainty assessment by varying the maintenance timing.

5.2 *Presentation of results*

Use Phase: Figure 4 shows the results from MCS employed for the use phase uncertainty assessment. Figure 4a displays the cumulative probability distribution of three designs. Even though there is a difference between the three designs in Figure 4a, it is not easily noticeable, and it is hard to interpret such data. Since the designs are being compared, the results from MCS can be presented as shown in Figure 4b. The results indicated that Design 3 performs better than the other two designs and it generates less GWP during use phase at all input combinations. Design 3 excels Design 1 by a range of 2194–4907 (mean 3320) and Design 2 by 4737–10627 (mean 7181) tons of GWP at a probability of 0.9. This type of analysis gives confidence to evaluators as well as to policy makers that their decisions are based on sound assessments. The results also indicated that the speed of vehicles is the critical input factor, which influences the overall use phase emissions.

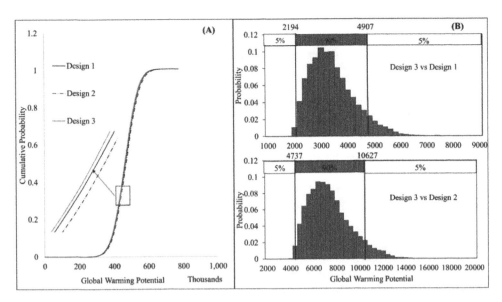

Figure 4. Monte carlo simulation results for use phase.

Table 2. Scenario analysis (maintenance timing) for traffic delay emissions.

Maintenance timing	Number of maintenance hours per day	GWP (1000 Tons)		
		Design 1	Design 2	Design 3
9 AM–5 PM	8	46.30	46.30	45.26
10 PM–6 AM	8	0.01	0.01	0.03
11 AM–3 PM & 8 PM–12 AM	8	12.23	12.23	12.44
8 AM–8 PM	12	86.11	86.65	71.43
24 Hours	24	61.57	61.96	51.08

Maintenance Phase: The results of scenario analysis are portrayed in Table 2. It is observed that maintenance timing influences the emissions predominantly. Maintenance during non-peak hours produced fewer emissions and the difference from the three pavements is negligible, and the situation is vice versa if maintenance happens during peak hours of traffic. Design 3 produced lowest GWP than the other designs when maintenance happens during the peak hours of traffic. This analysis shows that the LCA analyst's need a guidance of maintenance practices that the local and state bodies adapt on a regular basis. This type of scenario analysis is required for the phases which are very sensitive to the preferences of LCA analyst.

It is very clear from the two uncertainty analyses that the Design 3 is the one which causes lesser environmental impacts compared to the other designs. The uncertainty assessment bolsters the findings from an initial estimate that Design 3 generates lowest GWP than the other designs.

6 CONCLUSIONS AND LIMITATIONS

Uncertainty dwells at all phases of LCA, and there is a need to incorporate the uncertainty assessment to increase the transparency in LCA. Uncertainty assessment displays the possible range of outcomes and the risk involved in the LCA. This study proposed an approach for conducting uncertainty assessment by emphasizing on key steps like prioritization, classification, selection of uncertainty propagating or mitigating methods, and presentation of outputs.

The sources of the uncertainty are numerous and addressing every type of uncertainty is not practical. Prioritization directs the uncertainty assessment to focus on the critical phases of pavements life cycle and aids in keeping the scope of the assessment feasible. Classification of uncertainties helps in understanding the type of uncertainties prevailed in the LCA and guides in choosing an appropriate uncertainty appraising method. The presentation of uncertainty results is equally important, as they need to be presented in an understandable way.

A fictional example (three pavement designs) is presented to demonstrate the proposed approach. An initial estimate of GWP for three pavement designs is performed. The initial estimate shows that the Design 3 generated lower GWP than other designs. Use phase of pavement and maintenance traffic delay emissions contributed to the maximum GWP for three designs. So, uncertainty assessment is performed on these two phases by identifying the type of uncertainties. This study used different uncertainty propagation tools like Monte Carlo simulation, data quality index, sensitivity analysis, and scenario analysis. The MCS results for use phase indicates that Design 3 is better than the other two designs at all possible variation in inputs. The simulation results support the findings of the initial estimate. Sensitivity and scenario analysis were performed on the emissions due to traffic delay during maintenance. It is observed that the timing of maintenance and maintenance strategy greatly influences the emissions. If the maintenance happens during non-peak traffic hours, then the difference between the three designs is negligible, and if it happens during peak traffic hours, then Design 3 generated the lowest GWP.

The case study is used just like a numerical example but not to state the best design explicitly. In the case study, we presented only the uncertainty assessment related to input data for the models. A future study is required on handling the uncertainty in the models. Similarly, a future study is needed to include the uncertainty assessments for the other phases of LCA such as life cycle impact assessment.

REFERENCES

Allaire, D. & Willcox, K. 2014. Uncertainty Assessment of Complex Models with Application to Aviation Environmental Policy-Making. *Transport Policy* 34: 109–13.
Anastasiou, E.K., Liapis, A. & Papayianni, I. 2015. Comparative Life Cycle Assessment of Concrete Road Pavements Using Industrial by-Products as Alternative Materials. *Resources, Conservation and Recycling* 101: 1–8.
Argonne GREET Model. https://greet.es.anl.gov/. Accessed July 26, 2016.
Baker, J.W. & Lepech, M.D., 2009, September. Treatment of uncertainties in life cycle assessment. *In Proceedings of the ICOSSAR*.
Barandica, J.M., Fernández-Sánchez, G., Berzosa, A., Delgado, J.A., & Acosta, F.J. 2013. Applying life cycle thinking to reduce greenhouse gas emissions from road projects. *Journal of Cleaner Production* 57: 79–91.
Blomberg, T., Barnes, J., Bernard, F., Dewez, P., Le Clerc, S., Pfitzmann, M., Porot, L., Southern, M. & Taylor, R. 2011. Life Cycle Inventory: Bitumen. *European Bitumen Association*, Brussels.
Canter, K.G., Kennedy, D.J., Montgomery, D.C., Keats, J.B., & Carlyle, W.M. 2002. Screening stochastic life cycle assessment inventory models. *The International Journal of Life Cycle Assessment* 7(1): 18–26.
Cellura, M., Longo, S., & Mistretta, M. 2011. Sensitivity analysis to quantify uncertainty in life cycle assessment: the case study of an Italian tile. *Renewable and Sustainable Energy Reviews* 15(9): 4697–4705.
Curran M.A. 2006. Life Cycle Assessment: Principles and Practice. *Scientific Applications International Corporation (SAIC)*.
Downey B. 2009. Greenhouse Gas Emission Inventory CEMEX Jesse Morrow Mountain Plant. Project No. 03-23510A, Environ International Corporation.
Emission Factor Documentation for AP-42: Manufacture of Rubber Products. 2008. For U.S. Environmental Protection Agency, Office of Air Quality Planning and Standards, Measurement Policy Group.
Greenhouse Gas Emissions from a Typical Passenger Vehicle. http://www.epa.gov/otaq/climate/documents/420f14040a.pdf. Accessed July 27, 2015.
Harvey, J., Meijer, J., & Kendall, A. 2014. *Life Cycle Assessment of Pavements*. U.S. Department of Transportation, Federal Highway Administration No. FHWA-HIF-15-001.

Heijungs, R., & Huijbregts, M. A. 2004. A review of approaches to treat uncertainty in LCA. In *Pahl Wostl C, Schmidt S, Rizzoli AE, Jakeman AJ. Complexity and Integrated Resources Management. Transactions of the 2nd Biennial Meeting of the International Environmental Modelling and Software Society* 1: 332–339.

Helton, J.C., Johnson, J.D., Sallaberry, C.J., & Storlie, C. B. 2006. Survey of sampling-based methods for uncertainty and sensitivity analysis. *Reliability Engineering & System Safety* 91(10): 1175–1209.

Huang, Y., Bird, R., & Heidrich, O. 2009. Development of a life cycle assessment tool for construction and maintenance of asphalt pavements. *Journal of Cleaner Production* 17(2): 283–296.

Hung, M.L., & Ma, H.W. 2009. Quantifying *system uncertainty* of life cycle assessment based on Monte Carlo simulation. *The International Journal of Life Cycle Assessment* 14(1): 19–27.

Inti, S., Martin, S.A., & Tandon, V. 2016. Necessity of Including Maintenance Traffic Delay Emissions in Life Cycle Assessment of Pavements. In *Procedia Engineering 145: 972–979*.

Krömer, S., Kreipe, E., Reichenbach, D. & Stark, R. Life Cycle Assessment of a Car Tire. Continental AG, Hannover, Germany.

Liu, X., Cui, Q., & Schwartz, C. 2014. Greenhouse gas emissions of alternative pavement designs: Framework development and illustrative application. *Journal of environmental management* 132: 313–322.

Marceau, M., Nisbet, M.A., & Van Geem, M. G. 2007. *Life Cycle Inventory of Portland Cement Concrete.* Portland Cement Association.

May, J.R., & Brennan, D.J. 2003. Application of Data Quality Assessment Methods to an LCA of Electricity Generation. *The International Journal of Life Cycle Assessment* 8 (4): 215–25.

Mroueh, U. 2000. *Life cycle assessment of road construction.* Finland: Helsinki Finnish National Road Administration.

Myers, R., Shrager, B. & Brooks, G. 2000. Hot Mix Asphalt Plants Emission Assessment Report. No. EPA-454/R-00-019.

Noshadravan, A., Wildnauer, M., Gregory, J., & Kirchain, R. 2013. Comparative pavement life cycle assessment with parameter uncertainty. *Transportation Research Part D: Transport and Environment* 25: 131–138.

RSMeans Heavy Construction Cost Data. 2012. R.S. Means Company, Incorporated.

Santero, N. J., Masanet, E., & Horvath, A. 2011. Life-cycle assessment of pavements Part II: Filling the research gaps. *Resources, Conservation and Recycling* 55(9): 810–818.

Tatari, O., Nazzal, M., & Kucukvar, M. 2012. Comparative sustainability assessment of warm-mix asphalts: a thermodynamic based hybrid life cycle analysis. *Resources, Conservation and Recycling* 58: 18–24.

Thiel, C., Stengel, T., & Gehlen, C. 2014. Life Cycle Assessment (LCA) of road pavement materials. *Eco-efficient Construction and Building Materials: Life Cycle Assessment (LCA), Eco-Labelling and Case Studies 368*.

Tyre LCCO$_2$ Calculation Guidelines Ver. 2.0. 2012. The Japan Automobile Tyre Manufacturers Association, Inc.

U. E. Office of Air Quality Planning and Standards. *Emissions Factors & AP 42.* https://www3.epa.gov/ttnchie1/ap42/. Accessed July 26, 2016.

Vidal, R., Moliner, E., Martínez, G., & Rubio, M. C. 2013. Life cycle assessment of hot mix asphalt and zeolite-based warm mix asphalt with reclaimed asphalt pavement. *Resources, Conservation and Recycling* 74: 101–114.

Weidema, B. P., & Wesnæs, M. S. 1996. Data quality management for life cycle inventories—an example of using data quality indicators. *Journal of cleaner production* 4(3): 167–174.

Weidema, B. P., Bauer, C., Hischier, R., Mutel, C., Nemecek, T., Reinhard, J., Vadendo, C. O. & Wernet, G. 2013. Overview and methodology: Data quality guideline for the ecoinvent database version 3. *Swiss Centre for Life Cycle Inventories.*

Weiland, C. D., & Muench, S. T. 2010. Life Cycle Assessment of Portland Cement Concrete Interstate Highway Rehabilitation and Replacement No. WA-RD 744.4.

Yu, B., Liu, Q., & Gu, X. 2016. Data quality and uncertainty assessment methodology for pavement LCA. *International Journal of Pavement Engineering* 1–7.

Calculation method of stockpiling and use phase in road LCA: Case study of steel slag recycling

O. Yazoghli-Marzouk
CEREMA, Direction Territoriale Centre-Est, Département Laboratoire Autun, Autun, France

M. Dauvergne & W. Chebbi
LUNAM University, IFSTTAR, IM, EASE, Bouguenais, France

A. Jullien
LUNAM University, IFSTTAR, DAEI, Bouguenais, France

ABSTRACT: The environmental assessment of EAF-S recycling in road using LCA is investigated in this paper. Generally, in the LCA framework, the environmental assessment of recycling waste does not take into account the life cycle phases corresponding to stockpiling and use, as they are considered to have no impact, only waste processing is counted. We therefore propose to assess recycling of EAF-S in roads considering the phases of processing, stockpiling, transport and use in roads. The calculation of the use phase based on results from percolation test is compared to its calculation based on Ecoinvent data (total content and a transfer factor), in order to determine whether if the results are in the same range or not and propose recommendation for the use of Ecoinvent data in case of lack of data for waste materials. The impact indicators calculated are: energy, global warming potential, ecotoxic and toxic potentials using Ecorcem and OpenLCA softwares.

1 INTRODUCTION

The past two decades, show a great interest in life cycle assessment applied to pavement sector [Santero et al, 2011 (a) and (b); Yu and Lu, 2012; AzariJafari, 2016]. Recently, several symposiums and workshops dealing with this area have also been organized [Pavement LCA Workshop, 2010; 2012 RILEM Symposium on LCA for Construction Materials, Nantes, France; 2014 Pavement LCA Symposium, Davis, California and 2016 Development, standardization and implementation of LCA and integration with economics for transportation infrastructure and operations in SETAC Europe, Nantes]. These works showed that when LCA is applied to pavements, the use phase was rarely considered in comparative LCA, as the emissions and energy flows were considered to be the same as for the use phase. Moreover, many questions linked to the assessment of alternative materials actual engineering performances and their effects on the environment in the context of use for pavement construction, are left with no satisfactory answers for the potential user. In fact, material production processes (primary production) are usually assessed by LCA practitioners and producing them for recycling purposes (secondary production) is taken into account only through waste processing. The phases of the life cycle corresponding to stockpiling and use are in general considered to have no impact counted in LCAs. However, rainwater may leach chemicals elements from alternative (and natural) material, such as heavy metals, metalloids, polycyclic aromatic hydrocarbons and salts, either during handling and stockpiling [Yazghli-Marzouk et al., 2012 and 2015; Proust et al., 2014] before recycling or due to infiltration through the pavement surface containing recycled materials [Mroueh et al., 2001; Olsson, 2005; Birgisdòttir, 2005; Yazghli-Marzouk et al., 2012; Proust et al., 2014; Schwab et al., 2014] that contribute to different environmental impacts. Release to water from alternative materials is therefore the important flux to be examined in LCA of alternative materials for road construction.

In this paper, we propose to assess the recycling of electric arc furnace slag (EAF-S) in pavement as a case study of alternative material recycling, considering the phases of production of steel slag (for recycling purposes), stockpiling, transport and use in pavement. Calculation of the stockpiling and use phases based on experimental data (results from lixiviation and percolation tests) is compared to its calculation based on Ecoinvent data which take into account the total content (results from x-ray fluorescence) and a transfer factor (from the Ecoinvent data base), in order to determine whether the results are in the same range or not and propose recommendations for the use of Ecoinvent data in case of lack of data for waste materials.

2 METHODS AND MATERIAL PROPERTIES

2.1 *Methods*

Performing an LCA as initiated by SETAC [SETAC, 1993] involves two main types of underlying objectives, leading to:

- Comparing products (or processes). In this case, the chosen systems include only materials, processes and life cycle steps that may induce differences between compared products (or processes);
- Providing environmental information (for public and/or private organizations). Here the chosen systems may be much wider.

According to some authors, LCA is also a diagnosis tool that makes it possible to improve the global environmental profile of any system considered. It may be broken down into successive levels that depend upon the authors: 1/system description, 2/elementary process, 3/ flux calculations, 4/building the appropriate model, 5/analysing and interpreting the results and drawing up a report. The system usually includes all the elementary processes that are defined as the "smallest unit of the system" with inputs and outputs related to the industrial operation under consideration [AFNOR, 2006].

Figure 1 indicates the scenario of interest and the processes considered in this study for alternative materials life cycle impacts assessment. The lack of data on EAF-S primary production (EAF-S production in steel plant) conduces to not consider it. The phases investigated are waste processing for recycling purposes, stockpiling (temporary stocks), transport and use in roads.

The calculation of impact indicators, according to a model explained in a previous work by Sayagh *et al.* (2010), is described as:

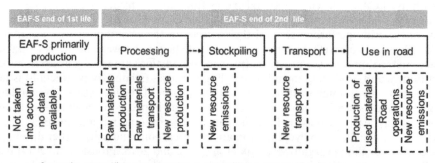

Figure 1. The scenario and processes investigated in this study.

$$Indj = \sum_i \alpha_{ij} \times C_{ij} \times m_i \qquad (1)$$

where Indj = indicator associated with impact category j; α_{ij} = classification coefficient (from Goedkoop, 2001); C_{ij} = contribution coefficient of inventory flow i to impact category j; m_i = mass of inventory flow i (kg).

Each indicator is expressed in specific units per kilogram or ton.

The contribution coefficients selected from the literature and used for the impact calculations, based on Equation (1), and the chosen impact categories (and indicators) derived from classical LCAs comprise all references given in [Sayagh et al., 2010]:

- Energy consumption: the specific energy consumption of each item of equipment (named CESP, the French acronym for Specific Production-related Energy Consumption); Global Warming Potential (GWP), from IPCC (2001),
- Toxic and Ecotoxic Potentials (TP and EP), from Huijbregts et al. (2000).

As a reminder, during stockpiling and use phases [Yazghli-Marzouk et al., 2012 and 2015; Proust et al., 2014] rainwater may leach chemicals elements from alternative materials. Therefore, in this LCA calculation, the output flux considered for EP and TP is release into water.

For stockpiling, release into water is simulated by leaching for six months. When a granular material is stockpiled, without compaction, its surface area in contact with water is substantial and close to the leaching test. A duration of six months is considered since alternative materials are generally stockpiled for a short period due to lack of space.

For the use phase, the structure simulated is a non-covered (by bituminous material) 10–15 cm thick road layer of EAF-S. The calculations proposed in this phase take into account construction operation, road-making equipment in operation and release of chemical substances from the alternative material into water for 100 years. Two calculation methods for EP and TP indicators in the use phase are compared: calculation based on experimental data and calculation based on Ecoinvent data. In the first case, the release of chemicals to water from the waste is based on results from percolation tests, as this test simulates the transfer of substances through a layer of compacted granular material due to water infiltration. In the second case, the release of chemical to water from the waste is based on calculation with a general Equation (2), where F_i (kg/kg of waste material) is the output flux of substance i, k_i is the transfer factor from Ecoinvent dataset and m_i is the total content of the substance i in the waste material determined by X-ray fluorescence.

$$F_i = k_i \times m_i \qquad (2)$$

Total content, leaching and percolation tests were therefore performed on alternative materials (EAF-S) to characterize water release, and toxic and eco toxic effects. The leaching test was performed on crushed aggregates according to NF EN 12457–2 (24 hours) [AFNOR, 2002]. The test consists of extractions of the material at liquid on solid ratio (L/S) equal to 10 by specific mixing. The percolation test was also performed on crushed aggregates according to NF CEN/TS 14405 [AFNOR, 2005] at different L/S (0,1; 0,2; 0,3; 0,5; 1; 3 and 5). The cumulative values correspond to L/S equal to 10. The leachant is demineralized water and the particle size is less than 4 mm. The solutions obtained are then filtered (filter pore size 0,45 µm) and analyzed to determine the concentrations in mg/kg of released substances. The total content of each element in the alternative material was determined on crushed aggregate (particle size < 80 µm) using NITON X-ray fluorescence spectrometer. This technique is used for elementary analyses. Under the effects of an X-ray beam, the sample resonates and emits its own X-rays. On the spectrum of fluorescent X-rays obtained, each peak (fluorescence emission) shows which element is present in the sample as the fluorescence emission is element-specific, and the peak height gives its quantity.

A distance of 30 km was considered for EAF-S transport as this is the mean distance in France for the aggregate market.

2.2 Material properties

The alternative material tested is steel slag produced by an Electric Arc Furnace (EAF-S). Its gap-grading analysis, according to EN 933-1 European standard (AFNOR, 2012), shows that it can be compared to a 0/1 mm sand with a large amount of fines (32% of fines passing through a 63 µm sieve). Its physical characteristics are presented in Table 1.

Chemical characterization of raw EAF-S shows (Table 2) that it is composed of 70% by weight of siliceous oxides of silicon, calcium, aluminum and iron. It also contains metals (copper (Cu), nickel (Ni), chromium (Cr), molybdenum (Mo), vanadium (V), zinc (Zn), etc.).

Table 3 shows some of the results of percolation tests on EAF-S after processing for recycling purposes. These data are used in EP and TP calculation.

Table 1. Physical analyses of raw EAF-S.

Elements	EAF-S
Real bulk density (t/m^3)	2.5
Natural water content (Wnat%)	18
Blaine specific surface area (cm^2/g)	5790

Table 2. Chemical composition of raw and processed EAF-S (X-ray fluorescence).

Elements	Units	Raw EAF-S	Processed EAF-S
Silicon dioxide (SiO$_2$)	% by weight	32.9	ND
Titanium dioxide (TiO$_2$)	% by weight	0.4	ND
Aluminium oxide (Al$_2$O$_3$)	% by weight	8.1	ND
Iron oxide (Fe$_2$O$_3$)	% by weight	14.1	ND
Manganese oxide (MnO)	% by weight	1.5	ND
Magnesium oxide (MgO)	% by weight	3.9	ND
Calcium oxide (CaO)	% by weight	15.8	ND
Sodium oxide (Na$_2$O)	% by weight	0.2	ND
Potassium oxide (K$_2$O)	% by weight	0.8	ND
P$_2$O$_5$	% by weight	0.2	ND
Arsenic (As)	mg/kg	<100	42
Cobalt (Co)	mg/kg	275	0
Chromium (Cr)	mg/kg	18000	10706
Copper (Cu)	mg/kg	1040	208
Molybdenum (Mo)	mg/kg	589	502
Nickel (Ni)	mg/kg	1510	849
Vanadium (V)	mg/kg	426	228
Zinc (Zn)	mg/kg	399	205
Zircon (Zr)	mg/kg	1350	ND

ND: not determined.

Table 3. Results of EAF-S percolation tests after processing (cumulative results; Liquid/Solid = 10)—ND means non determined.

Elements	EAF-S after processing (mg/kg)
Arsenic (As)	8,6 10^{-4}
Chromium (total)	4.29
Copper (Cu)	2.4 10^{-2}
Molybdenum (Mo)	9.94
Nickel (Ni)	ND
Zinc (Zn)	0.01

Table 4. Comparisons of impacts range for recycling of EAF-S (400,000 tons) in roads by calculation based on experimental data (1) and calculation based on Ecoinvent data (2).

Indicators	Energy MJ	GWP kg Eq CO_2	EP kg Eq 1,4 DCB	TP kg Eq 1,4 DCB
Transport (30 km)	54	3.12	2.43	0.47
Processing	34.4	0.4	2.68	0.79
Stockpiling	0	0	0.33	0.006
Use in roads (1)			49.9	1.08
Use in roads (2)			2640	60.3

GWP: global warming potential, EP: ecotoxic potential, TP: toxic potential.

3 RESULTS AND DISCUSSION

Table 4 presents the results of the LCA calculation for recycling EAF-S in roads, considering the processing of EAF-S for recycling purposes, stockpiling, transport and use in roads. Use in roads was calculated using experimental data (method 1), i.e. EAF-S percolation data. In method 2 Ecoinvent data was used, i.e. total content of EAF-S and transfer factor from the Ecoinvent database.

The contribution of the use phase to EP and TP is substantial, whereas the contribution of the stockpiling phase is negligible, when release to water is controlled.

Furthermore, the results (Table 4) show that EP and TP calculated with Ecoinvent data are fifty times higher than those calculated with experimental data.

Arsenic (As), cobalt (Co), chromium (Cr), copper (Cu), molybdenum (Mo), nickel (Ni) and zinc (Zn) contribute to EP and TP indicators. As and Mo are the main contributors, as their coefficients of contribution are close to 1 for Mo and 1 for As. In addition, their transfer coefficient at short term (100 years) is equal to 1. As the total content of Mo is higher (502 mg/kg of EAF-S) than that of As (42 mg/kg of EAF-S), EP and TP seem to be driven by the Mo content when they are calculated by method 2 (based on the Ecoinvent database). In this case, the calculation assumes that the entire Mo content is transferred to water.

However, when the calculation uses experimental data (percolation test) all the Mo is not transferred to water (9.94 mg/kg of EAF-S). The diffusion of Mo from its mineral bearing phases—iron silicates ($Fe_xSi_yO_4$) and melilite which is a solid solution of akermanite ($Ca_2MgSi_2O_7$) and gehlenite ($Ca_2Al[AlSiO_7]$)—identified in a previous study (Chebbi et al, 2016), is low in demineralised water.

Calculation using the Ecoinvent database appears to overestimate EP and TP. But studies of EAF-S percolation in other solutions, with pH values close to those encountered in the field, should be made to confirm this observation.

4 CONCLUSION

This paper presents the study of the environmental assessment of EAF-S (Electric Arc Furnace Slag) recycling using Life Cycle Assessment (LCA) by two methods. The first of these use experimental data and the second one uses the Ecoinvent database.

The life cycle phases corresponding to stockpiling and use are generally considered to have no impact counted in LCAs. In this study we therefore undertook the evaluation of stockpiling and use phase impacts on waste (EAF-S), considering that the release of chemicals from the waste occurs during these phases and could be evaluated by EP and TP indicators.

The results obtained show that the environmental assessment of EAF-S recycling should take into account the life cycle phase corresponding to use in roads as EP and TP exhibit important impacts, whereas the stockpiling phase need not be considered as its contribution to EP and TP is very low. In fact, even if recycling this alternative material is considered possible from the standpoint of local regulations, water may leach chemicals during the use phase by infiltration (percolation) through the road layer containing the alternative material.

Moreover, EP and TP seem to be driven by the Mo content. When these indicators are calculated using the Ecoinvent database, the entire Mo content is assumed to be transferred to water, whereas experimental data show that a very low fraction is leached by percolation in laboratory conditions. The results obtained using the Ecoinvent database therefore overestimate the environmental impacts. This result will be completed by other studies on different alternative materials, as part of the OFRIR database (http://ofrir.ifsttar.fr), where Life Cycle Assessment (LCA) is now its main objective.

REFERENCES

Afnor (2002). French standard NF EN 12457–2: Characterization of waste—Leaching—Compliance test for leaching of granular waste materials and sludges—Part 2: one stage batch test at a liquid to solid ratio of 10 l/kg for materials with particle size below 4 mm (without or with size reduction). AFNOR, 2002-12-01.
Afnor Cen TS 14405: Characterization of waste—Leaching behaviour test—Up-flow percolation test (under specified conditions), AFNOR, July 2005.
Afnor ISO 14040 (2006) Management environnemental—Analyse du cycle de vie—Principes et cadres, Environmental management -- Life cycle assessment—Requirements and guidelines, in: I.O.f. Standardization (Ed.), ISO 14040:33–46.
Afnor Nf EN 12457–2: Characterization of waste—Leaching—Compliance test for leaching of granular waste materials and sludges—Part 2: one stage batch test at a liquid to solid ratio of 10 l/kg for materials with particle size below 4 mm (without or with size reduction). AFNOR, 2002-12-01.
Afnor. (2012). Essais pour déterminer les caractéristiques géométriques des granulats—Partie 1: détermination de la granularité—Analyse granulométrique par tamisage.
AzariJafari H., A. Yahia, M. Ben Amor 'Life cycle assessment of pavements: reviewing research challenges and opportunities', Journal of Cleaner Production 112 (2016) 2187–2197.
Birgisdòttir Harpa "Life cycle assessment for road construction and use of residues from waste incinerator", PhD Thesis, Institute of environment and resources, Technical university of Denmark, July 2005, 60 pages.
Chebbi W., O. Yazoghli-Marzouk, A. Jullien, J. Moutte, "Valorization of steel slag in road construction—comprehension of molybdenum's repartition", International Journal of Advances in Mechanical and Civil Engineering, ISSN: 2394-2827, Volume-3, Issue-1, Feb.-2016, pp 61–64.
French standard CEN TS 14405. 2005. Characterization of waste—Leaching behaviour test—Up-flow percolation test (under specified conditions). AFNOR.
French standard NF EN 12457–2. 2002. Characterization of waste—Leaching—Compliance test for leaching of granular waste materials and sludges—Part 2: one stage batch test at a liquid to solid ratio of 10 l/kg for materials with particle size below 4 mm (without or with size reduction). AFNOR. France, Paris.
Frischknecht R., Jungbluth N., Althaus H.-J., Doka G., Dones R., Heck T., Hellweg S., Hischier R., Nemecek T., Rebitzer G. and Spielmann M. 2005. The ecoinvent database: Overview and methodological framework, International Journal of Life Cycle Assessment 10, 3–9.
Goedkoop M., Spriensma R. (2001) The Eco-indicator 99, a damage oriented method for Life Cycle Impact Assessment, methodology report. *Pré Consultants B.V.* 132.
Huijbregts M.A.J., Thissen U., Guinée J.B., Jager T., Kalf D., Van de Meent D., (2000) Priority assessment of toxic substances in life cycle assessment. Part I. Calculation of toxicity potentials for 181 substances with the nested multimedia fate exposure and effects model USES-LCA. Chemosphere, 41: 541–573.
Ifsttar. 2014. http://ecorcem.ifsttar.fr. Accessed 26 October 2015.
IPCC, 2007, Intergovernmental Panel on Climate Change. Fourth assessment report, climate change: the physical science basis. Chapter 2: changes in atmospheric constituents and radiative forcing. 106 pp.
Mroueh, U.M., Wahlström, M., 2001. By-products and recycled materials in earth construction in Finland—an assessment of applicability. Resour. Conserv. Recycl. 25 (1–2), 117–129.

Olsson Susanna "Environmental assessment of municipal solid waste incinerator bottom ash in road construction", September 2005, 34 pages. TRITA-LWR.LIC 2030; ISSN 1650-8629; ISRN KTH/LWR/LIC 2030-SE; ISBN 91-7178-151-X.

Pavement LCA Symposium, October 14–16, 2014, Davis, California, USA.

Pavement LCA Workshop, Davis, California; (Harvey et al. 2011). Pavement life cycle assessment workshop, may 5–7, 2010, Davis, California, USA. International journal of life cycle assessment, 16 (9) 944–946.

Proust C., Yazoghli-Marzouk O., Ropert C., Jullien A., "LCA of roards alternative materials in various reuse scenarios", International Symposium on Pavement Life Cycle Assessment, October 14–16, 2014, Davis, California, USA.

RILEM Symposium on LCA for Construction Materials, 2012, Nantes, France.

Santero NJ, E. Masanet, A. Horvath 'Life-cycle assessment of pavements. Part I: Critical review', Resources, Conservation and Recycling 55 (2011) 801–809.

Santero NJ, E. Masanet, A. Horvath "Life-cycle assessment of pavements Part II: Filling the research gaps", Resources, Conservation and Recycling 55 (2011) 810–818.

Sayagh S., Ventura A., Hoang T., François D., Jullien A., 2010. Sensitivity of the LCA allocation procedure for BFS recycled into pavement structures, Resources Conservation and Recycling.

SETAC (1993). Guidelines for Life-Cycle Assessment: a "code of practice". Ed. SETAC Foundation for Environmental Education, Florida 1993.

SETAC Europe, Development, standardization and implementation of LCA and integration with economics for transportation infrastructure and operations in SETAC Europe, 2016, Nantes, France.

Yazoghli-Marzouk O., Vulcano-greullet N., Cantegrit L., Friteyre L., Jullien A. (2014). Recycling foundry sand in road construction-field assessment. Construction and Building Materials, vol. 61, 69–78.

Yu B., Q. Lu, "Life cycle assessment of pavement: Methodology and case study", Transportation Research Part D 17 (2012) 380–388.

Concrete pavement life cycle environmental assessment & economic analysis: A manitoba case study

M. Alauddin Ahammed
Manitoba Infrastructure, Winnipeg, Manitoba, Canada

S. Sullivan
Cement Association of Canada, Toronto, Ontario, Canada

G. Finlayson, C. Goemans & J. Meil
Athena Sustainable Materials Institute, Ottawa, Ontario, Canada

Mehdi Akbarian
Department of Civil and Environmental Engineering, Massachusetts Institute of Technology, Cambridge, Massachusetts, USA

ABSTRACT: Life Cycle Assessment (LCA) is recognized as one of the most comprehensive ways to evaluate the environmental impacts of different strategies associated with a physical feature such as highway pavement structures, bridges and vertical structures. Alternatively, Life Cycle Cost Analysis (LCCA) is performed by different agencies including Manitoba Infrastructure (MI) to select the most cost-effective construction option or life cycle strategy. Highway agencies started to realize the need to combine the LCA and LCCA in the design, construction and management of roadway assets.

This paper presents comparisons of the environmental impacts and life cycle costs of various alternative strategies for a Portland Cement Concrete (PCC) pavement project in Manitoba to demonstrate the opportunity to optimize the cost, pavement performance and environmental impacts. A matrix of 10 different strategies that include alternative PCC mix, pavement design, and Maintenance and Rehabilitation (M&R) treatments have been used to contrast both the life cycle costs and environmental impacts with MI's past practice (base case). The presented analysis is expected to assist highway agencies to better understand and weigh not only the economics of alternative strategies, but also the environmental implications of alternative roadway materials, design, construction, and maintenance and rehabilitation practices.

1 INTRODUCTION

1.1 *Background*

Environmentally sustainable design, construction, preservation and maintenance practices can assist in preserving our natural environment. LCA is acknowledged as one of the most comprehensive ways to evaluate the environmental impact of activities associated with a physical feature or to compare the environmental impacts of different strategies for a given analysis period. The Athena Pavement LCA software allows users to analyze and compare the environmental implications of multiple roadway design scenarios taking into consideration materials production, construction, and M&R treatments. It is also the first commercially available software capable of modeling Pavement Vehicle Interactions (PVI), so designers have the option of considering roadway roughness and deflection on predicted roadway traffic fuel consumption and associated environmental impacts.

LCCA has long been performed by different agencies to select the most cost-effective construction option or life cycle strategy among the feasible alternatives. The life cycle approach to pavement management decisions has also continued to grow in asset financing, as agencies search for effective ways to allocate their budget to maintain an aging infrastructure network. Consequently, several new LCCA tools have been developed that can account for uncertainty in the decision-making process and the costs accrued to users of a roadway facility (Tighe 2001, Salem et al. 2003, Lee et al. 2011).

1.2 *Objectives and scope*

The objectives of this paper are to show, using the Athena Pavement LCA tool the environmental impacts of various PCC mixes and pavement design scenarios, and to explain the performance reasons for the current MI specification changes and future considerations. Furthermore, this study evaluates the life cycle costs associated with each material and design scenario. These include the agency cost of initial construction and future M&R activities as well as the user costs from excess fuel consumption due to PVI.

This case study uses a concrete pavement section constructed in 2015 on the northbound direction of Provincial Trunk Highway (PTH) 75 divided highway, in Manitoba. This project was the first project where the new MI specification was used (listed as Scenario 6 of this case study). The other analysis scenarios include a matrix of alternative PCC mix designs based on varying cement/aggregate, fly ash and slag compositions, varying reinforcing steel type, size and quantity, and rehabilitation practices based on the expected extension of pavement's initial service life from the usual initial service life due to the construction of more durable pavement. These variations, alone and in combination, facilitated the comparison of the life cycle environmental impacts and costs for 10 alternative scenarios relative to a base case. The environmental impacts and life cycle costs presented in this paper reflect the total for the entire project i.e., 11.02 km × 2 traffic lanes plus shoulders.

It should be noted that the empirical approach of linking PCC proportioning and slab design (10 alternative scenarios) to pavement performance, which in turns affect the M&R and PVI, is intended to demonstrate the potential impact of those concrete mixes and slab design. A more robust assessment will require actual long term field performance data for each of these scenarios which is out of the scope of this paper but could be investigated further in future studies.

2 PAVEMENT LCA SOFTWARE

2.1 *Overview*

In the Athena Pavement LCA tool, first released in 2011, users may enter their project specific roadway cross section designs (subbase, base and surface layers) and overall length or select from a library of sample regional PCC or Asphalt Concrete (AC) roadway designs and customize to their intended analysis scenario including the life cycle period. In May 2016, a web-enabled version of the tool was released (http://pavementlca.com). The Pavement LCA tool follows the guidance and frameworks as set out by the US Federal Highways Administration (FHWA) (see FHWA 2015 and FHWA 2016). The software has been made available free of charge to all with support from the Athena Sustainable Materials Institute's members, collaborators and supporters.

2.2 *Data sources*

Pavement materials, energy and transportation data were derived from proprietary Athena Institute databases as well as publicly and commercially available Life Cycle Inventory (LCI) databases and Environmental Product Declarations (EPDs). The materials data typically represent national or industry averages for the extraction, processing and manufacture of materials and products. Canadian regional energy grids and transportation distances are then applied to manufacturing process energies and material sourcing distances to arrive at regional data profiles. The current web-enabled version contains recent EPD results for Canadian portland cement, North American slag cement and updated Canadian regional energy and electricity grid profiles.

2.3 *Analysis outputs*

In the Athena Pavement LCA software, users can quickly describe roadway design, M&R and (optionally) PVI parameters through a few input screens and then can view a comprehensive set of life cycle impact assessment results comprised of: energy and raw material flows plus emissions to air, water and land. The software also enables side-by-side comparison of different options across life cycle stages. The life cycle impact assessment results are based on the US EPA's TRACI characterization model (https://www.epa.gov/chemical-research/tool-reduction-and-assessment-chemicals-and-other-environmental-impacts-traci) and cumulative resource use metrics.

The total operating energy may also be included in the LCA if the user inputs an estimate for annual operating energy consumption by fuel type. The software will calculate total energy, including pre-combustion energy (the energy used to extract, refine and deliver energy) and the related emissions to air, water and land over the life cycle of the roadway project, and can subsequently compare the life cycle operating and embodied energy and other environmental effects of various design options, allowing the user to better gauge construction, reconstruction and M&R trade-offs.

3 PAVEMENT LIFE CYCLE COST ANALYSIS (LCCA)

The LCCA model used in this study follows the guidelines of Federal Highway Administration (FHWA) for life cycle cost analysis of pavements (Walls & Smith 1998). The details of the model are presented in previous work by Swei et al. (2013) and Swei et al. (2015). The model was validated against Minnesota's LCCA tool by Akbarian et al. (2017). For the analysis presented in this paper, the deterministic LCCA model has been implemented without considering the sources of uncertainty. In addition, the road user costs are accounted for based on fuel consumption by vehicles. The initial and maintenance agency costs are calculated based on the quantity of each activity in the bill of materials and operations of each scenario, and according to the local (average) item costs for these activities. Moreover, the impacts of material and design decisions on the user cost due to PVI are evaluated for vehicles throughout the pavement life cycle. The deflection-induced excess fuel consumption is evaluated using previously developed model as a function of top layer thickness, stiffness, relation time, subgrade stiffness, pavement width, temperature, vehicle axle load, and speed (see Louhghalam et al. (2014) and Akbarian (2016) for details of model, experimental development, calibration, and validation). The impact of roughness-induced PVI on vehicle fuel consumption is evaluated through the calibrated Highway Development and Maintenance Management (HDM-4) model developed by the World Bank as a function of surface roughness in terms of the International Roughness Index (IRI) (Chatti & Zaabar 2002). The parameters for this analysis include annual IRI, vehicle type and traffic volume, and IRI after construction and maintenance. The cost of excess gasoline and diesel consumption are assumed to be equal, at a value of $0.95 CAD per liter.

4 CASE STUDY PROJECT DETAILS

4.1 *Project description*

As mentioned earlier, a PCC pavement constructed in 2015 on the northbound lanes of PTH 75 in Manitoba is used as a case study in this paper. The project is located approximately 30 km south of Winnipeg and has a total length of 11.02 km. For the 50-year analysis presented in this paper, distances from site to the stockpile, plant to the site and from equipment depot to the site were assumed to be 30 km. The current 1-way Annual Average Daily Traffic (AADT) is 3,900 inclusive of 650 heavy vehicles (trucks) and an annual growth rate of 2%. The existing roadway consisted of 100 mm AC (bituminous) surface layer over a 250 mm PCC and 125 mm granular base. The pavement reconstruction consists of reclaiming the existing bituminous layer, rubblizing the existing PCC, placing 100 mm drainable stable granular base, topped with 255 mm PCC and diamond grinding the new PCC surface (5 mm loss). Shoulder construction includes widening with granular base and bituminous (100 mm thick)/gravel surfaces). The cross section of the new roadway is shown in Figure 1.

4.2 Life cycle strategy

MI performs LCCA for a 50-year period to select a pavement strategy. MI's rigid (PCC) pavement life cycle strategy, which is based on local performance experience, is shown in Table 1. These periodic treatments and their lives have been used for the analysis presented in this paper. It should be noted that Pavement ME Design program may be used to predict the pavement performance and rehabilitation schedule. However, the globally calibrated performance and distress models in the Pavement ME Design program are found to be inadequate for Manitoba local conditions. MI is currently working on the local calibration of the Pavement ME Design performance and distress prediction models as well as the field verification of the LCCA strategy to determine if the life cycle rehabilitation activities are realistic.

4.3 Concrete pavement construction and concrete mix design

Recently, MI has revised the concrete mix design, dowel and tie bar configurations for the rigid pavement structure to make it more durable and economical. Changes include reduction

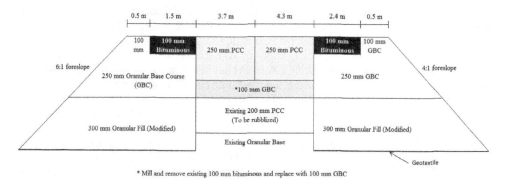

Figure 1. Roadway cross section (not to scale).

Table 1. Life cycle strategy for PCC pavements in Manitoba.

Item no.	Activities	Quantity	Activity year
1	New Construction or Reconstruction (Design pavement for 20 years accumulative traffic loading)	100%	0
2	Concrete Partial Depth Repairs	2% Surface Area	15
3	Concrete Partial Depth Repairs	5% Surface Area	25
4	Concrete Full Depth Repairs	10% Surface Area	25
5	Diamond Grinding	100% Surface Area	25
6	Concrete Partial Depth Repairs	5% Surface Area	40
7	Concrete Full Depth Repairs	15% Surface Area	40
8	Diamond Grinding	100% Surface Area	40
9	Salvage Value	5 Years of Service Life (1/3 of Items 7 plus 8)	50

of the cement content by implementation of optimized gradation, increase of fly ash content, use of Drainable Stable Base (DSB) under the rigid layer, use of non-corrosive (zinc clad or stainless) steel dowel and tie bars, use of smaller diameter dowels and tie bars, reduction of number of dowels per joint, and reduction of tie bars length and spacing. This paper presents the comparison of life cycle environmental impacts and costs of these changes, except the DSB whose specification is still under evaluation, with the base case (MI's past practice before adoption of these changes) scenario. Additional comparison includes environmental impacts of optimized concrete slab design using Thin Concrete Pavement (TCP) methodology, addition of slag in the concrete mix (to produce fly ash/slag cement ternary mix) and a hypothetical extension of initial pavement service life.

4.3.1 *Supplementary Cementing Materials (SCMs) and ternary mixes*

The aim of mixture proportioning is to find the combination of available and specified materials to ensure that a mixture is cost effective and meets all performance requirements. In the case of sustainable design, minimizing the pavement's environmental footprint over the life cycle must be one of the performance requirements. Cementitious content should be kept as low as possible without compromising mixture performance, both in the fresh and hardened states.

The two most commonly used SCMs in PCC paving are fly ash and slag cement. The amount of fly ash that can be used is often limited by concerns of delayed setting times and lower early strength gain. Its dosage varies from 15 to 40% by mass of cement. Slag cement may be used in pavements in dosages up to 50% but is limited by concerns of slow early strength gain, especially when placed during cooler ambient temperatures, and of scaling resistance.

Ternary concrete mixtures include three different cementitious materials. The optimum mixture proportions for ternary blends will depend on the final use of the concrete, construction requirements and seasonal restrictions. As with other concrete, cold weather will affect the early strength gain and mixture proportions may need to be adjusted to assure job-site performance. In low water-cementitious materials ratio (w/cm) applications such as paving, mixtures with 15% fly ash and 30% slag cement have been used successfully. For example, in 1998, airfield PCC pavements were constructed at the Minneapolis Airport using a ternary mix consisting of portland cement with 35% slag and 10% class C fly ash. The pavement has been performing very well.

MI previously allowed a maximum of 15% fly ash; this was recently increased to 20% (maximum). For ternary concrete mixes, agencies typically use a combination of 15% fly ash and 25% slag by mass of total cementitious (portland cement and SCMs) materials. This ternary mix scenario was added to the analysis to see if MI should consider ternary mixtures in the future, from environmental and performance standpoints, since it has already been shown that ternary mixtures improve concrete durability in the field (Taylor 2014).

4.3.2 *Optimization of aggregate gradation*

Another technique to reduce the total cementitious materials content in the PCC mixes is optimization of aggregate size and proportion. The tarantula curve, which provides an envelope for the desirable amount of material retained on each sieve, is the newest technique of mixture optimization (Ley et al. 2012). The curve has been independently validated by concrete pavement contractors and shows that over time, concrete mixtures have evolved to fit within the recommended limits of the Tarantula curve (Ley et al. 2014). This technique allows a reduction in paste content by use of an intermediate size aggregate, while still allowing sufficient paste volume to fill voids and provide workability. Benefits reported by contractors include: reduced shrinkage, lower cost, greater strengths and improved workability.

MI's PCC mix design for pavements had been based on the individual gradation of coarse and fine aggregates. In 2015, MI adopted the tarantula aggregate gradation band in the specification and increased the maximum size of the aggregate from 19 mm to 25 mm for gravel. The optimized gradation and reduced strength requirement allowed for a reduction of the total cementitious material content from 355 kg/m^3 to 307 kg/m^3 of concrete. Figure 2 shows how the pre-2015 and 2015 aggregate gradations fit into the tarantula aggregate gradation band (tarantula curve).

(a) Pre 2015   (b) 2015

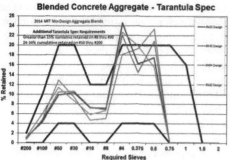
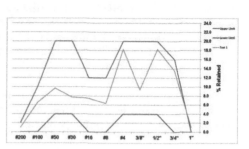

Figure 2. Aggregate gradation in tarantula curve.

4.3.3 Modifications to pavement structure (dowels and tie bars)

Dowel detail and joint design are new innovative concepts evolving in concrete pavement engineering today. Due to a decreasing transportation funding and a rising focus on sustainability, highway agencies are looking for ways to minimize initial construction costs and to use alternative construction materials.

Current trends in concrete pavement joint design include reducing the number of dowel bars and placing dowels directly under the wheel paths only as opposed to the conventional dowel placement at 300 mm interval throughout the transverse contraction joints. Placing dowels at 300 mm (or even 450 mm) off the outside edge of pavement and at 450 mm (or even 600 mm) off the longitudinal joint instead of typical 100–150 mm eases construction and can also help reduce interference from the adjacent tie bars.

MI had been constructing Jointed Plain Concrete Pavements (JPCP) using epoxy coated steel dowels and tie bars. Traditionally, 38 mm (diameter) × 450 mm (length) dowels were placed at 300 mm interval along the transverse contraction joints (26 dowels per 8 m wide transverse contraction joint). The size of tie bars was 19.5 mm (diameter) × 915 mm (length) and four tie bars were used along a 4.5 m long longitudinal joint. For the 11.02 km long case study road section, the total steel requirement was calculated to be 276 tonnes. MI has experienced corrosion of both dowels and tie bars resulting in deterioration of both transverse and longitudinal joints including the separation of longitudinal joints. Hence, MI has been exploring alternative reinforcing dowels and tie bar materials and configurations. MI recently revised the JPCP specifications for dowels and tie bars.

The new JPCP specifications require the use of stainless steel or zinc clad dowels and stainless steel tie bars. Based on economical and technical analysis of dowel size requirements, longevity and location of wheel paths (where about 90% of wheel load repetitions occurs), the size of dowels has been changed to 32 mm with a requirement to place only four dowels in each wheel path i.e., total 16 dowels per transverse contraction joint (two lanes). The new requirements for stainless steel tie bars are 16 mm (diameter) × 750 mm (length) and five dowels per 4.5 m long longitudinal joint. The estimated steel quantity is 126 tonnes for the 11.02 km long case study road section i.e., 54% reduction in steel quantity from that was used in the past. However, the LCA tool does not contain data for stainless steel. Therefore, all dowels and tie bars are modeled as galvanized steel (a limitation of the Pavement LCA tool) to estimate the effect of changes in steel quantity.

4.3.4 Optimized concrete slab design using Thin Concrete Pavement (TCP) methodology

This design procedure configures slab size so that each slab is loaded by only one wheel or a set of wheels (i.e., 50% of each axle load) at the same time. This significantly reduces the top tensile stresses of the slabs. Research shows that minimizing the critical top tensile stress will allow for thinner concrete slabs with smaller slab dimensions (length and width). Research also shows that fiber reinforced concrete slabs contribute to a longer service life (Covarrubias et al. publication date is unavailable). This concept was developed by a Chilean company,

TCPavements, in 2007. A mechanistic software, OptiPave, has been developed to design concrete pavement using this methodology. This design software predicts cracking, faulting and initial IRI, which is similar to the AASHTOWare Pavement ME Design program, and allows for the prediction of pavement performance over time. Typically, slabs are 70 mm thinner (on average) for higher trafficked roadways relative to traditionally designed pavements using the AASHTO 1993 method and no dowels or tie bars are used. Slab length and width are typically between 1.4 m to 2.5 m for the TCP (Covarrubias & Covarrubias 2008). In the analysis for this paper, a slab size of 2.0 m is used. The initial slab thickness is assumed to be 205 mm to obtain a net thickness of 200 mm after 5 mm loss due to diamond ground texturing of new concrete surface. The thickness designs for the JPCP and TCP using Pavement ME Design and TCPavements programs, respectively, are beyond the scope of this paper because of the inadequacy of Pavement ME Design performance/distress prediction models for the local condition and no local access to TCPavements software.

5 ANALYSIS AND RESULTS

5.1 *Alternative analysis scenarios*

Eleven cases (base case plus 10 alternative cases) are run in the Athena Pavement LCA software to estimate the environmental impacts of changes in one or more attributes. These alternative scenarios including their rationales are listed in Table 2. The base case represents a past standard mix design using 355 kg of cementitious material per m^3 of concrete, of which 15% was fly ash. The base quantity of steel is 276 tonnes. The proportion of coarse and fine aggregates is 61:39. MI's regular M&R strategy is used to estimate the post construction environmental impacts. Subsequent analysis cases include increase of fly ash content, addition of slag cement to produce a ternary mix, reduction of steel quantity, modification of M&R cycles and reduction of PCC slab size (TCP) from this base case.

Case 6 represents MI's current standard mix design, construction and M&R cycles. This mix design optimizes the aggregate blend to reduce cementitious material content with the

Table 2. Alternative analysis scenarios and their rationales.

Case #	Case description	Analysis rationale
Base	355 kg cementitious, 15% fly ash, 0% slag, 276 t steel, regular M&R	Impacts of past practice
1	355 kg cementitious, 20% fly ash, 0% slag, 276 t steel, regular M&R	Effect of additional fly ash
2	355 kg cementitious, 15% fly ash, 25% slag, 276 t steel, regular M&R	Effect of slag/ternary mix, if used
3	307 kg cementitious, 15% fly ash, 0% slag, 276 t steel, regular M&R	Effect of reduced cementitious material (tarantula optimization)
4	355 kg cementitious, 15% fly ash, 0% slag, 126 t steel, regular M&R	Effect of reduced steel
5	307 kg cementitious, 20% fly ash, 0% slag, 276 t steel, regular M&R	Combined effect of reduced cementitious and increased fly ash
6	307 kg cementitious, 20% fly ash, 0% slag, 126 t steel, regular M&R	Combined effect of reduced cementitious and steel, and increased fly ash (new spec.)
7	307 kg cementitious, 15% fly ash, 25% slag, 126 t steel, regular M&R	Combined effect of new spec. and slag/ternary mix, if used
8	307 kg cementitious, 15% fly ash, 25% slag, 126 t steel, extended M&R	Effect of extended M&R
9	355 kg cementitious, 15% fly ash, 0% slag, 0 steel, TCP, regular M&R	Effect of short concrete panel (TCP)
10	307 kg cementitious, 15% fly ash, 25% slag, 0 steel, TCP, extended M&R	Effect of reduced cementitious, TCP, ternary mix and extended M&R

use of larger maximum size and intermediate sized aggregates. However, there is a negligible change in the proportion of coarse and fine aggregates from the previous concrete mix design i.e., the base case. Cases 2 and 7 are intended to show the effects of ternary concrete mix as compared to the previous and new (current specification) mixes, respectively. Case 8 shows the effect of theoretical extension of initial pavement service life (extended M&R) due to the use of ternary concrete mix or other means (e.g., pavement designed for a longer service life). For this case, the timing of the M&R treatments are arbitrarily delayed by 5 years from the regular M&R treatments shown in Table 1. Case 9 shows the influence of short concrete panel i.e., TCP design as compared to the base case. Case 10 shows the influence of TCP, ternary cement and reduced cementitious with a theoretical extension of initial pavement service life (i.e., extended M&R).

To estimate the environmental impacts due to traffic use, the PVI analysis considered vehicle operating speed of 100 km/h, an initial IRI of 0.665 m/km (after diamond grinding of new pavement surface), a terminal IRI of 2.5 m/km that triggers a rehabilitation diamond grinding and a post-rehabilitation (diamond ground) IRI of 1.0 m/km. The subgrade resilient modulus is 30 MPa. The concrete density and elastic modulus are 2.320 t/m^3 and 28,600 MPa, respectively. The PVI analysis assumed 15 mm loss in concrete thickness per each periodic diamond ground treatment, except the diamond grinding of new concrete surface. The environmental impacts and life cycle costs presented in this paper reflect the total for the entire project (11.02 km).

5.2 Environmental impacts of base case

The LCA tool calculates the environmental impacts for several impact indicators and energy use metrics due to the manufacturing (material production and transportation), construction (equipment use and transportation), maintenance (M&R) and traffic use phase excess fuel consumption (due to roadway roughness and deflection). Table 3 summarizes the varying impact indicator and resource use metric results for the base case. A discussion of all these impact indicators is beyond the paper size (page) limitation for this symposium, and therefore the analysis focuses solely on the Global Warming Potential (GWP) or climate change effects (see EPA 2016).

Table 3. Base case environmental impacts.

Activities					Use phase excess fuel consumption due to PVI		Total life cycle
Impact category indicators and resource use metrics	Measurement units	Manufacturing	Construction	Maintenance & rehab	PVI effects (IRI)	PVI effects (deflection)	Total
Global Warming Potential	t CO_2 eq	8,086.9	6,358.5	9,500.6	1,632.7	3,701.6	29,280.3
Acidification Potential	t SO_2 eq	34.9	57.3	77.9	10.4	33.1	213.5
HH Particulate	t PM2.5 eq	13.9	3.4	5.6	0.5	1.9	25.3
Eutrophication Potential	t N eq	3.7	3.8	5.4	0.7	2.2	15.8
Ozone Depletion Potential	t CFC-11 eq	0.0	0.0	0.0	0.0	0.0	0.0
Smog Potential	t O_3 eq	653.5	1,922.9	2,515.9	357.8	1,111.9	6,561.9
Total Primary Energy	GJ	104,625.2	92,399.2	127,513.0	23,777.3	53,792.8	402,107.4
Non-Renewable Energy	GJ	95,722.9	92,360.2	126,125.1	23,772.2	53,770.4	391,750.9
Fossil Fuel Consumption	GJ	91,656.2	92,216.0	125,025.9	23,753.7	53,687.5	386,339.2

Figure 3 presents the base case percent contribution to GWP of the different activity stages. The percent contribution of material manufacturing, construction and maintenance (M&R) is a function of the roadway design, its construction and life cycle rehabilitation strategy, which together are categorized as embodied effects as opposed to the use phase effects of PVI. Figure 3 shows that the total embodied effect contributes 82% of the total life cycle environmental impact with the remaining 18% attributed to the effects of PVI. It should be noted that diamond grinding is usually considered a rehabilitation treatment and hence the effect of the initial diamond grind of the entire newly built roadway is reported in the maintenance effects instead of initial construction effects. Therefore, the M&R effects are somewhat overstated, which is another limitation of the software. Another reason that the M&R effects perhaps appear unusually significant for a PCC roadway is that MI's M&R schedule calls for a diamond grind of the entire roadway surface at year 25 and 40, which may not be a typical practice elsewhere.

5.3 *GWP impacts across scenarios*

Figure 4 contrasts the GWP impact across the various cases while Table 4 shows the variation (percentage change) from the base case within activity stages.

5.3.1 *Material manufacturing*

An increase in fly ash content of the base mix from 15% to 20% (Scn. 1) results in a 4.6% reduction in manufacturing GWP while 25% replacement of portland cement by slag (Scn. 2)

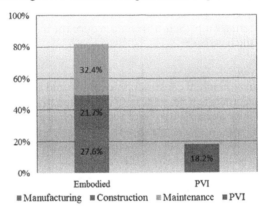

Figure 3. Percent contribution to GWP for the base case.

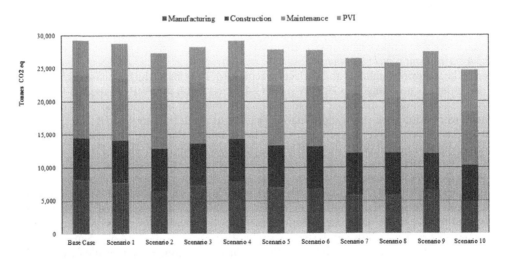

Figure 4. Comparison GWP of different scenarios.

Table 4. Variation of GWP of different scenarios from the base case.

Scenarios	Base	1	2	3	4	5	6	7	8	9	10
Manufacturing	N/A	−4.6	−19.6	−10.5	−1.3	−14.6	−15.9	−28.9	−28.9	−18.6	−40.7
Construction	N/A	0.0	0.0	0.0	0.0	0.0	0.0	0.0	0.0	−14.2	−14.2
Maintenance	N/A	−0.9	−3.9	−2.1	0.0	−2.9	−2.9	−5.4	−11.4	−4.7	−14.3
Embodied	N/A	−1.9	−8.1	−4.4	−0.4	−6.1	−6.5	−11.9	−14.3	−11.9	−23.2
PVI	N/A	0.0	0.0	0.0	0.0	0.0	0.0	0.0	−1.4	19.5	17.4
Life Cycle	N/A	−1.6	−6.7	−3.6	−0.4	−5.0	−5.3	−9.7	−12.0	−6.2	−15.8

in the base mix results in a 19.6% reduction in GWP from the base case. A reduction of cementitious material in the mix design from 355 kg/m³ of concrete to 307 kg/m³ (Scn. 3), due to the optimization using tarantula aggregate gradation, results in a 10.5% reduction in the GWP from the base case. The combined effect of concrete mix design change (Scn. 5) is 14.6% GWP reduction from the base case. The large reduction in steel quantity from 276 tonnes to 126 tonnes (Scn. 4) showed a net reduction of the GWP of only 1.3% as compared to the base case because of an equivalent and offsetting increase in concrete quantity use. The adoption of new pavement structure, concrete mix design and steel type and placement resulted in a total 15.9% reduction in GWP (Scn. 6). Use of ternary concrete mix could reduce the GWP by 28.9% (Scn. 7) while the use of the thin concrete panel could reduce the GWP by 18.6% (Scn. 9) as compared to the base case. Lastly, the use of thin concrete panel in combination with a ternary mix (Scn. 10) could reduce the manufacturing GWP by up to 40.7% as compared to the base case.

5.3.2 *Initial construction*

Since there was no change in initial construction quantity and equipment use for Scenarios 1 to 8 from the base case, there is no change in GWP related to construction for these cases. A 14.2% reduction in GWP as related to initial construction is possible with the use of TCP.

5.3.3 *M&R activities*

The new specification (Scn. 6) adopted by MI reduces the GWP during the M&R phase by 2.9% due to changes in portland cement content. Comparing the GWP results between Scenario 7 and 8 indicates that the five year extension (delay) in the M&R schedule (timing) could provide (11.4% − 5.4% =) 6.0% reduction in GWP as compared to the regular (standard) M&R schedule. When compared to the base case, there is a possible 11.4% GWP reduction by delaying the M&R treatments by five years. A five year delay in the M&R schedule in combination with use of TCP could provide a 14.3% reduction in GWP relative to the base case.

5.3.4 *Pavement vehicle interaction*

Since the roadway roughness and deflection remained the same as the base case for Scenarios 1 to 7, there is no change in GWP. A small (1.4%) reduction in GWP is expected for Scn. 8 due to fuel savings as the initial pavement remains smoother and experiences less deflection for longer period (five years additional service life to reach the terminal IRI) as compared to all other cases. However, employing the Thin Concrete Panel (TCP) resulted in a 19.5% increase in PVI related GWP due to higher expected deflection as compared to the base case.

5.3.5 *Overall GWP impacts*

As depicted in Table 4, MI can expect a 6.5% reduction of the embodied GWP; i.e., excluding PVI effects, with the adoption of the new concrete mix deign and steel configurations (Scn. 6) relative to the base case. Including the PVI, the overall life cycle GWP reduction is estimated to be 5.3% relative to the base case. Another (9.7% − 5.3% =) 4.4% (Scn. 7 − Scn. 6) reduction in GWP may be possible with the use of ternary concrete mix while further (12% − 9.7% =) 2.3% (Scn. 8 − Scn.7) reduction in GWP may be possible by extending the initial pavement service life by five years; i.e., by constructing more durable pavements. Overall, a 15.8% reduction in GWP is possible with the use of new concrete mix, ternary mix, thin concrete panel and extension of initial pavement service life by five years relative to the base case.

5.4 *Comparison of agency costs*

The initial construction costs, maintenance costs, and salvage values for the given cases are computed and the Net Present Values (NPV) are presented in Figure 5(a) together with the percentage reduction in total agency cost as compared to the base case scenario. Figure 5(b) provides a percent contribution of the agency costs for different scenarios. It is observed that initial construction drives the LCCA outcome and that each environmental impact reduction strategy has an associated cost benefit. A comparison of Scenarios 8 and 10 with other scenarios in Figures 5 (a) and (b) is interesting, where the five-year delayed maintenance strategy, in pair with the 3% discount rate, have decreased the relative contribution of the M&R costs as compared to other Scenarios, while resulting in considerable overall cost reduction levels of 14% and 23%, respectively from the Base Scenario.

5.5 *Comparison of user cost*

Figure 6(a) and Figure 6(b) present the user costs of roughness and deflection induced PVI impacts on passenger car and truck fuel consumptions for the old (regular) and new (extended) M&R strategies. The difference in the two cases is the five-year delay of the M&R treatments for the new M&R strategy, where they occur in years 30 and 45 instead of years 25 and 40, respectively. The upward trend of user costs are associated with increasing road IRI levels and the traffic growths of passenger car and trucks. The contribution of roughness induced PVI for passenger cars dominates the associated user costs, followed by the

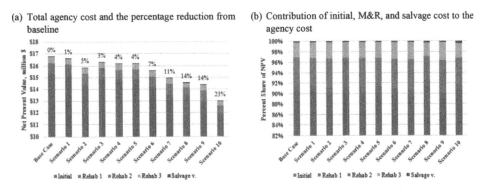

Figure 5. Agency LCCA results for the base case and the 10 scenarios.

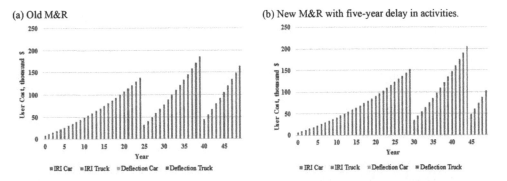

Figure 6. User cost associated with roughness and deflection induced PVI for scenarios.

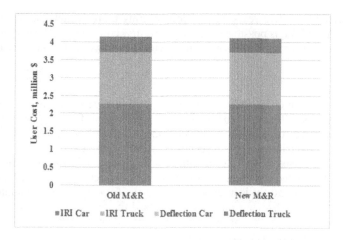

Figure 7. Total user cost associated with roughness and deflection induced PVI for passenger cars and trucks for the old and new M&R strategies.

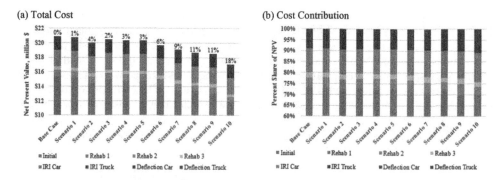

Figure 8. Total life cycle cost including agency and user associated costs.

same impact for trucks and deflection induced PVI for trucks. Figure 7 compares the overall old and new M&R user costs, which indicates that the five-year delay has almost no impact (<1%) in reducing the total user cost.

5.6 *Comparison of total cost*

The initial construction, M&R and PVI induced user costs are presented in Figure 8(a) together with the percentage reduction from the base case for each scenario. Figure 8(b) provides percentage contributions of total cost for the base case and the 10 alternative scenarios. Results indicate that the M&Rs contribute the least to the overall agency and user costs. Scenario 10, where several of the environmental impact reduction strategies are combined, shows an 18% reduction in the overall lifecycle cost. And, although the road section has a medium traffic volume, additional research can potentially justify more maintenance activities to mitigate user costs.

6 CONCLUDING REMARKS

This study demonstrates that the transportation sector can contribute significantly to the reduction of environmental impacts by adopting sustainable material, design, construction and maintenance practices and building durable infrastructures. Once a local or project specific scenario is populated, the Athena Pavement LCA software allows users to easily investigate additional alternative scenarios to understand environmental trade-offs and aid in decision making. When used in tandem with robust life cycle cost analysis software the user is additionally informed about the economic trade-offs associated with various roadway construction and life cycle management strategies.

The analysis of different scenarios showed that MI can expect about 5% reduction in overall life cycle GWP with the adoption of the new concrete mix design and dowel/tie bar configurations. Another 7% reduction in GWP may be possible by shifting to the use of ternary concrete mix and extending the pavement service life by five years. The adoption of the new concrete mix design and dowel/tie bar configurations showed a 6% reduction in total life cycle cost. With further enhancement to concrete mix and slab design, 18% reduction in the total life cycle cost is possible.

REFERENCES

Akbarian, M. 2015. Quantitative sustainability assessment of pavement-vehicle interaction: from bench-top experiments to integrated road network analysis. PhD Dissertation, Massachusetts Institute of Technology.

Akbarian, M., Swei, O., Kirchain, R. & Gregory, J. 2017. Probabilistic Characterization of Life-Cycle Agency and User Costs: Case Study in Minnesota. In Transportation Research Record: Journal of the Transportation Research Board, Transportation Research Board of the National Academics, Washington, D.C.

Chatti, K. & Zaabar, I. 2002. Estimating the effects of pavement condition on vehicle operating costs, Project 1–45. Report 720, National Cooperative Highway Research Program, Report 720.

Covarrubias Jr, J.P., Roesler, J. & Covarrubias, J.P. Publication date is unavailable. Design of Concrete Slabs with Optimized Geometry. Publication location is unavailable.

Covarrubias, T. & Covarrubias, V. 2008. TCP Design for Thin Concrete Pavements. In proceeding of the 9th International Conference on Concrete Pavements, San Francisco, CA, pp. 13.

Lee, E.-B., Kim, C. & Harvey J.T. 2011. Selection of Pavement for Highway Rehabilitation Based on Life-Cycle Cost Analysis. In Transportation Research Record: Journal of the Transportation Research Board, No. 2227, Transportation Research Board of the National Academics, Washington, D.C., pp. 23–32.

Ley, T., D. Cook & G. Fick. 2012. Concrete Pavement Mixture Design and Analysis (MDA): Effect of Aggregate Systems on Concrete Properties. National Concrete Pavement Technology Center, Ames, IA.

Ley, T., D. Cook, N. Seader, A. Ghaeezadeh, & B. Russell. 2014. Aggregate Proportioning and Gradation for Slip Formed Pavements. TTCC-National Concrete Consortium Fall 2014 Meeting presentation. September 9–11, 2014, Omaha, NE.

Louhghalam, A., M. Akbarian & F.J. Ulm. 2014. Scaling relations of dissipation-induced pavement-vehicle interactions. In Transportation Research Record: Journal of the Transportation Research Board, No. 2457, Transportation Research Board of the National Academics, Washington, D.C., pp. 95–104.

Salem, O., S. AbouRizk, & S. Ariaratnam. 2003. Risk-Based Life-Cycle Costing of Infrastructure Rehabilitation and Construction Alternatives. Journal of Infrastructure Systems, Vol. 9, No. 1, pp. 6–15.

Swei, O., J. Gregory, & R. Kirchain. 2013. Probabilistic Characterization of Uncertain Inputs in the Life-Cycle Cost Analysis of Pavements. In Transportation Research Record: Journal of the Transportation Research Board, No. 2366, Transportation Research Board of the National Academics, Washington, D.C., pp. 71–77.

Swei, O., J. Gregory, & R. Kirchain. 2015. Probabilistic Life-Cycle Cost Analysis of Pavements: Drivers of Variation and Implications of Context. In Transportation Research Record: Journal of the Transportation Research Board, No. 2523, Transportation Research Board of the National Academics, Washington, D.C., pp. 47–55.

Taylor, P.C. 2014. The Use of Ternary Mixes in Concrete, Ames, IA, National Concrete Pavement Technology Center.

Tighe, S. 2001. Guidelines for Probabilistic Pavement Life Cycle Cost Analysis. In Transportation Research Record: Journal of the Transportation Research Board, No. 1769, Transportation Research Board of the National Academics, Washington, D.C., pp. 28–38.

U.S. Environmental Protection Agency (EPA). 2016. Understanding Global Warming Potentials. [online]. (Web Link).
U.S. Federal Highways Administration (FHWA). 2015. Towards Sustainable Pavement Systems: A Reference Document. FHWA-HIF-15-002. Federal Highway Administration, Washington, DC. [online]. (Web Link).
U.S. Federal Highways Administration (FHWA). 2016. Pavement Life Cycle Assessment Framework. FHWA-HIF-16-014. Federal Highway Administration, Washington, DC. [online]. (Web Link).
Walls, J. III, & M.L. Smith. 1998. Life-Cycle Cost Analysis in Pavement Design: Interim Technical Bulletin. FHWA-SA-98-079. FHWA, U.S. Department of Transportation.

Life-cycle assessment tool development for flexible pavement in-place recycling techniques

Mouna Krami Senhaji, Hasan Ozer & Imad L. Al-Qadi
University of Illinois at Urbana-Champaign, IL, USA

ABSTRACT: The Federal Highway Administration (FHWA) has partnered with research teams at the Illinois Center for Transportation (ICT) at the University of Illinois at Urbana-Champaign (UIUC), the University of California Pavement Research Center (UCPRC), and Rutgers, the State University of New Jersey, in developing a tool to assess the environmental impacts and to predict the performance of preservation and rehabilitation treatments on flexible pavements. This tool can be used on a national scale. The Life Cycle Inventory (LCI) database has been developed to assess the environmental impacts of the materials and construction stages. The performance evaluation was conducted using both deterministic performance models and a qualitative decision matrix to determine the expected treatment life. The importance of this study lies in analyzing the capacity of the tool to help in the preservation and rehabilitation treatments selection, especially in-place recycling treatments, in terms of performance evaluation, and environmental assessment under a range of traffic, climate, structure capacity, and existing pavement conditions.

1 INTRODUCTION

In 2011, the National Cooperative Highway Research Program (NCHRP) Synthesis 421 compiled an extensive summary of the use of In-Place Recycling (IPR) techniques in the United States (Stroup-Gardiner 2011). An online survey about IPR techniques was conducted and included responses from 45 states to collect information about usage extensiveness, traffic, treatment service life, cost, and work zone conditions (opening time, lane closure time, reduction percent in lane closure). It was found that CIR (Cold In-place Recycling) is the most used IPR technique.

CIR involves milling and pulverizing the existing asphalt surface, mixing the recycled material with additives (as needed), and laying and compacting the mix into a new pavement layer to replace the existing one. A Hot-Mix Asphalt (HMA) overlay or surface treatment, such as a chip seal or cape seal, is then placed over the recycled pavement layer. The term *CIR* used in this study refers to a partial-depth recycling of only the existing HMA layers, while Full Depth Reclamation (FDR) penetrates to layers beyond the HMA. Typically, CIR is used with milling depths of 50–100 mm and is used to address distresses in pavements that are structurally sound (ARRA 2015). Stabilizing agents used in CIR can be mechanical (e.g. compaction), chemical (e.g. Portland cement, fly ash, calcium chloride, magnesium, lime), or bituminous (e.g. engineered emulsion or foamed asphalt) (ARRA 2015). The expected service life for CIR is estimated at 6–10 years with a surface treatment and 7–20 years with an HMA overlay (ARRA 2015).

Energy consumption and limited environmental impact characterization of in-place recycling technique were done to compare these techniques to conventional alternatives and validate environmental benefits as claimed. Schvallinger (2011) compared various types of CIR practices to conventional paving methods, finding savings up to 69% in energy when compared to HMA overlay and 76% in energy when compared to mill-and-fill methods (Schvallinger 2011). Finally, Santos et al. (2014) used LCA to compare IPR, traditional reconstruction, and corrective maintenance strategies by considering the entire life cycle, including material production, construction, work zone effects, and the pavement use (Santos et al. 2014). The IPR-based strategies resulted in a reduction of 1.5% in energy, as compared to traditional reconstruction and 30% in energy as compared to a corrective maintenance strategy. It should also be noted that studies concerning GHG and other emissions related to CIR have also been conducted (Liu et al. 2014; Alkins et al. 2008). None of these studies followed a life-cycle framework.

The focus of this paper is to present the methodology used in developing a Life-Cycle Assessment (LCA) tool to evaluate the energy consumption of in-place recycling techniques commonly used in the preservation program of flexible pavements and applying it in a case study. The tool analyzes techniques that local and state roadway agencies have been using as part of their preservation and rehabilitation programs, especially in-place recycling techniques, which include both Hot-In-place Recycling (HIR) and CIR. The research approach followed is based on the concepts of LCA and pavement life-cycle framework initiated by FHWA (FHWA 2016) that adheres to guidelines by the International Standards Organization (ISO) 14040:2006 standards for "Environmental management–Life-cycle assessment–Principles and framework" and the ISO14044:2006 standards for "Environmental Management–Life-Cycle Assessment–Requirements and Guidelines."

2 METHODOLOGY

The guidelines recommend four steps when conducting a LCA: (1) goal and scope definition, (2) inventory analysis, (3) impact assessment, and (4) interpretation. The first three steps are discussed below, while the fourth step is presented in the "Case Study" section.

2.1 *Goal and scope*

The goal of paper is to examine the upstream and downstream energy consumed by an IPR technique and the corresponding performance during the analysis period, using the LCA decision-making tool. This information is of interest to federal, state, and local transportation agencies that consider the environmental impacts of using IPR techniques in their pavement management strategies. The product system evaluated are IPR methods used in the United States. The functional unit is one lane-km of a flexible pavement upon application of a preservation treatment designed and expected to perform to satisfy average standards in the United States. A typical mainline lane of 3.65 m was assumed in the study, and only construction and materials related to the layers above the subgrade for the mainline pavement were included. Shoulders and supporting drainage, lighting, signage, and landscaping elements were not included and were assumed to be equivalent among alternatives.

Material production and hauling and construction stages of the pavement life cycle were considered, while the use phase, future maintenance or rehabilitation, and end-of-life stages were excluded. Thus, comparisons made between the alternatives assumed equal performance of the systems. Upstream materials and resources were considered for the major processes involved, including raw materials production, fuel production, and electricity generation.

As IPR techniques rely largely on recycled materials, a methodological choice for allocating recycled materials must be made. In this study, a cut-off allocation procedure was chosen for recycled asphalt pavement materials. Thus, only the processes related to the preparation of the previous pavement system were attributed to the recycled materials (i.e. scarifying, milling, crushing, and mixing of in-situ pavement); and processes related to the production of the original pavement were cut off.

2.2 *Life Cycle Inventory (LCI)*

In this LCA, a data-collection strategy has been developed using both primary and secondary data sources to build a comprehensive LCI database for the IPR preservation treatments, including material production, material hauling, and construction stages. The procedures described in the FHWA's pavement LCA framework were followed to develop the inventory database. The primary data collected included mainly process activity data associated with the construction of IPR treatments, equipment train, use of fuel, or other energy sources, and productivity that can be used to estimate total energy consumption and environmental impacts for specific on-site conditions. To complement the primary data, the secondary data were collected using various sources, including commercial LCI databases (e.g. Ecoinvent, US-Ecoinvent (Earthshift 2013)), software (e.g., EPA MOVES 2014 (US EPA 2015), eGRID (US EPA 2013)), governmental reports, material safety data sheets, and manufacturers' specifications.

2.2.1 *Primary data*

For data-collection purposes, customary U.S. units were used for all LCI data per unit area (i.e., in square yards) or unit lane-miles. Survey data collected from relevant local contractors in each region were compared with values reported in the literature for validation. As an example, process activity data from the CIR and FDR operations (fuel consumption, productivity, etc) were collected to estimate energy consumption. If collected data were outside the expected range, follow-up questions were asked to clarify the source of the discrepancy. After the survey data were benchmarked, the data were averaged where possible to achieve a representative value for each relevant region and to preserve the confidentiality of individual contractors.

2.2.2 *Secondary data*

Secondary data sources were used in this study. Upstream data for raw materials, resources, and transportation inputs were collected from various datasets (such as the commercial LCI database, the literature, and simulations) using publicly available software. Regional LCI data corresponding to the U.S. East and Midwest were obtained and modeled. Whenever regional data were not available, national LCI data representing the whole United States were used. Major unit processes that can be modeled to represent specific U.S. regions are listed below:

- Asphalt-binder production
- Diesel-fuel production
- Electricity generation
- Hauling trucks
- Construction equipment
- Vehicle operations (for use phase)

2.3 *Impact assessment*

The energy-use assessment in the case study was cumulative of all the energy consumed during material production, material hauling, and construction. Impact categorization was done using the EPA's TRACI method (Bare 2012).

2.4 *Interpretation*

The results of energy analysis and performance assessment are presented according to the life-cycle stages and major processes included in the scope. To allow for a comparative assessment, the results are presented for all alternatives. Interpretation of the final results are presented in the "Case Study" section.

2.5 *Performance estimation*

A two-pronged approach to evaluate the performance of preservation and rehabilitation treatments was developed and implemented in the comparative assessment tool. The two approaches were the deterministic performance models and the multi-criteria performance estimation.

The two approaches were used for calculating specific parameters to evaluate the pavement performance under certain project conditions. These parameters were the treatment life (i.e., the number of years until the next major treatment for the same pavement segment) and IRI (international roughness index) progression. The deterministic performance models approach predicts the wheelpath and fatigue cracking and time to estimate treatment lifetime under specific climate condition, traffic volume, and overlay thickness. The deterministic models were also developed to estimate IRI progression as a function of pavement variables.

As for the multi-criteria performance estimation, it calculates a performance score that varies from 1 to 5 and determines the performance of all rehabilitation and preservation treatments under various conditions (i.e., traffic, climate, existing pavement conditions, soil properties, and material characteristics). Furthermore, this approach estimates the treatment life and, with a specified IRI threshold, indirectly provides the IRI progression rate to describe how fast the pavement deteriorates.

2.5.1 *Deterministic performance models*

These models are obtained at a network or project level and may be used to predict the International Roughness Index (IRI) progression and cracking distresses performance over the entire life of the pavement. The output from these models is a function of continuous and/or categorical variables. This approach can be used to predict the performance of IPR pavements. Distress criteria considered to estimate treatment life include the load-associated wheelpath cracking, fatigue cracking, and IRI.

2.5.2 *Multi-criteria performance estimation*

The multi-criteria performance estimation process is used to estimate treatment life by collecting information related to various site-specific conditions and evaluating this information through a rating system to determine an expected treatment performance. This approach provides an estimate of treatment life based on reported lifespans in the literature or observed by the user in the region of interest and adjusted for site-specific conditions. The relationship of various site-specific factors to expected performance is compiled from multiple sources. These include existing literature for best practices and experimental data, agency and contractor surveys, decision trees adopted by local and state highway agencies, and expert opinions.

The performance estimation process is based on five categories: climate, traffic, existing pavement condition, soil properties, and pavement material properties. The process of integrating these into the performance estimation process is as follows:

1. Collect site-specific conditions under each criterion for a given treatment candidate from the user (e.g. traffic: Annual Average Daily Traffic (AADT) or/and Equivalent Single Axle Load (ESAL),% truck, road type; overall condition index representing existing pavement condition, distress severity and composition; soil properties; etc.)
2. Score the impact of each site-specific condition under each major category (or subcategory) by a rating score from 1 to 5. The interpretation of each rating score of a treatment under a certain condition is as follows:
1: High risk, 2: Medium-high risk, 3: Medium risk, 4: Medium-low risk, 5: Low or no risk.
For example, an HIR resurfacing treatment followed by a thin overlay of 50 mm or less is compared to a conventional treatment such as mill and overlay. If a high traffic level of >30,000 AADT is used, a score of "1" is assigned; the literature widely agrees that this treatment type is not suitable for high traffic levels. Assigning a rating of "1" in this case means there is a high risk for the selected design, and it expected to perform poorly.
3. Evaluate the score for each criterion and determine the final rating score for the application.
4. Based on the rating score, calculate the expected treatment life based on lifetime estimates compiled from the literature and surveys.

The treatment overall Performance Score (PS) is calculated based on the following formulation (1). It is assumed that all the factors are independent:

$$PS = \frac{\sum_{i=1}^{N_C} C_i + \sum_{i=1}^{N_T} T_i + \sum_{i=1}^{N_S} S_i + \sum_{i=1}^{N_E} E_i + \sum_{i=1}^{N_M} M_i}{N_C + N_T + N_S + N_B + N_M} \qquad (1)$$

where:
T_i: Rating for factor i related to each traffic
C_i: Rating for each climate condition i
S_i: Rating for each soil property i
E_i: Rating for factor i each existing pavement condition i
M_i: Rating for factor i related to material properties condition
N_C: Number of factors related to climate condition
N_T: Number of factors related to traffic
N_S: Number of factors related to soil properties
N_E: Number of factors related to existing conditions
N_M: Number of factors related to material properties

The interpretation of the PS and resulting impact on treatment life is explained as follows:

- PS is 4 or 5: Ideal on-site conditions for treatment; indicating very low risk for performance. Treatment life can be expected to be at the highest part of the range.
- PS is 2 or 3: Conditions are fair, carrying medium risk for the performance of treatment. Treatment life can be expected to be in the medium area of the range of expectations.
- PS is 1: On-site conditions are not appropriate for the treatment, with very high risk. Treatment life may be predicted at the lower end of the range of expected values.

3 FRAMEWORK OF THE TOOL

Both of the aforementioned approaches are implemented in the tool currently being developed. Once the project input parameters are entered, the user is provided with a list of treatment alternatives. The user has the option to select one or more in-place recycling treatments

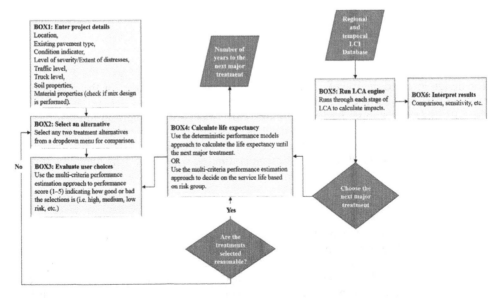

Figure 1. Flowchart of the comparative tool, illustrating the integration of the two performance-estimation approaches.

and compare them to one or more conventional methods. The selected treatments are initially screened for their applicability for the project conditions. If there are obvious and clear barriers against application of the selected treatment (e.g. geometric features impeding application of using a long, in-place recycling train), the user is warned. The selected treatments are evaluated by calculating treatment lifetime and developing IRI progression curves. Figure 1 presents a flowchart of the suggested tool, to illustrate the performance-estimation process integration to the overall flow of data and input flow.

4 CASE STUDY

The following case study was employed to make a comparative energy and performance assessment between the use of CIR and conventional mill-and-fill.

4.1 *Structural designs*

Two treatments were considered in the case study, namely, CIR with overlay (CIR/OL) and mill-and-fill. The existing structure consists of 100 mm of HMA, 200 mm of crushed aggregate base coarse layer, and 100 mm of crushed granular material on top of a subgrade soil having a California Bearing Ratio (CBR) value of 6%. The design was conducted based on chapter 46 of the Illinois Department of Transportation Bureau of Local Roads and Streets (IDOT BLRS) manual for pavement rehabilitation (IDOT 2012). The design procedure for both alternatives assumes typical values of structural coefficients for each material and calculates a Remaining Structural Number (SN_R) and a Final Structural Number (SN_F) for each layer, which are calculated based on inputs for traffic levels, subgrade strength, and existing layer thicknesses. The required thickness of the overlay is then calculated based on the difference between the two structural numbers.

In this study, the pavement design was carried out for a low-volume road with a traffic factor of 0.65, which is equivalent to an average daily traffic of 2,041 vehicles/day, with 7.5% single-unit trucks and 2.5% multiple-unit trucks. For the CIR with an Asphalt Concrete (AC) overlay design alternative, the existing 100 mm of HMA is recycled. The SN_R and SN_F were calculated to be 2.24 and 3.25, respectively, requiring an additional AC overlay thickness of 60 mm on top of the recycled asphalt. For the mill-and-fill design alternative, by contrast, the top 50 mm of HMA is milled, which resulted in calculated values of 1.84 and 3.25 for SN_R and SN_F, respectively, and an additional overlay thickness of 100 mm.

4.2 *Materials*

The design involved the use of 1% cement, 2.5% emulsion, and water for the cold in-place recycled-pavement surface course, and Hot-Mix Asphalt (HMA) for the overlay applied on the CIR and conventional paving methods. HMA was used for the conventional method overlay layer. The material inventory database described in the methodology section was used to calculate the energy use of the material production phase. Materials-hauling conditions were assumed to be 26°C for the air temperature, 60% for the relative humidity, and 0% grade (flat).

4.3 *Construction*

The CIR and conventional method treatments considered in the design used the same aforementioned single-pass equipment train and paving equipment set described in the construction-phase analysis. The number of teeth used in the CIR milling operation was estimated to be 15 per 100,000 kg. To evaluate the effect of pavement hardness on total energy savings, a harder pavement is considered by looking at the impact of using 30 teeth instead of 15 teeth per 100,000 kg. Pavement width was also varied from 3.65 to 4 m to assess its impact on energy savings using the aforementioned regression model. As for the chip seal, the aggregate spreader, binder distributor, roller and sweeper constitute the treatment construction checklist (FHWA 2013). The mill-and-fill operation consisted of milling part of or all of the existing HMA layer, replacing it with HMA laid down using a paver and compacted with a roller. The construction-equipment inventory was used to calculate the energy use during the construction practices of the two pavement designs.

5 RESULTS AND DISCUSSION

5.1 *Environmental impact assessment*

The environmental impact results for CIR and its equivalent conventional paving alternative with various hauling distances (i.e., 30 and 160 km) in this study are shown in Figure 2, separated by activity type, as well as material production (including mixing and raw material transportation), equipment operation, and materials hauling to site.

The results of the energy assessment of the case study (Figure 2) show that energy consumption associated with material production is 889 GJ for CIR/OL and 1186 GJ for mill-and-fill. CIR and mill-and-fill consume approximately the same amount of energy during the construction stage, with an average of 58 GJ, which makes the construction-stage contribution to the total energy savings very low, up to 0.95%. The construction stage for the two treatments are very similar because a similar set of equipment is used, with only a few exceptions. For the hauling stage, as the HMA hauling distance increases, the hauling energy increases from 56 GJ to 206 GJ for CIR/OL, and from 110 GJ to 312 GJ for mill-and-fill. Overall, the energy consumed during CIR/OL life stages showed a savings of 24% when the hauling distance is 30 km and a savings of 26% when the hauling distance is 160 km. The materials production stage contributes the most to energy consumption with 77%–87% to CIR/OL processes, while it contributes 76%–87% to mill-and-fill processes.

5.2 *Performance assessment and sustainability interpretation*

The decision-making process relies on the whole life-cycle environmental assessment, cost analysis, and future-performance prediction. As for the environmental assessment, while

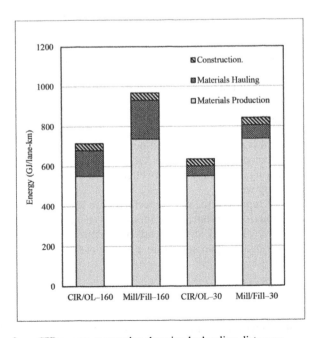

Figure 2. Energy from CIR versus conventional paving by hauling distances.

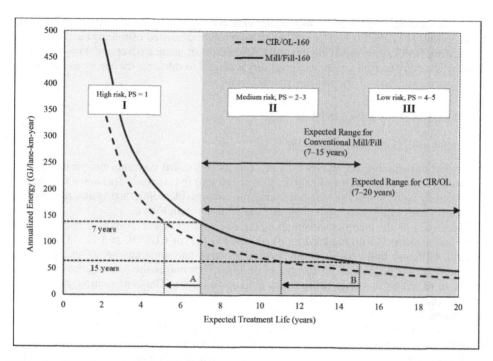

Figure 3. Scenario analysis for different expected treatment lives for each process using annualized energy.

the previous analysis may favor using CIR over conventional paving processes, a life-cycle approach requires that future performance, use, and end-of-life of the pavement be considered. As a full LCA is out of scope, a scenario analysis based on expected treatment life was used. The expected treatment life of a thin, dense-graded asphalt concrete overlay is typically 7–15 years, while that of CIR with asphalt concrete overlay is 7–20 years (ARRA 2015). Annualizing energy consumption over a range of expected treatment lives for both processes produces the relationships presented in Figure 3. It is important to note that the excess fuel consumption due to roughness and texture is not considered at this stage. However, the same concept of equivalency analysis will still be utilized with part of the use stage impact categories.

In Figure 3, the energy is annualized by expected treatment life for the 160 km hauling case. Arrow A shows that CIR/OL-160 can have a reduced treatment life as low as 5.8 years and still have the same annualized energy consumption as a poor-performing Mill/Fill-160 with a life of 7 years. By contrast, Arrow B shows that an CIR/OL-160 would need a minimum treatment life of 12.4 years to match a well-performing Mill/Fill-160 of 15 years for the same annualized energy consumption; the CIR/OL would need to perform beyond what is typically expected, indicating that it may be more effective to use Mill/Fill when the pavement is expected to perform very well.

The areas I, II, and III represent the risk in performance level zones, based on the multi-criteria performance estimation approach. The annualized energy decreases with the decrease of risk on performance level. The risk level depends on the project conditions where the treatment is applied. CIR is effective when applied in a local, low-traffic road having a good soil support (Caltrans 2008).

6 CONCLUSIONS

This study presents the functions of a new LCA tool being developed for FHWA. A case study was investigated using the tool, to perform a comparative study of two equivalent in-place recycling and conventional methods to demonstrate the capacity of the tool to perform LCA.

Deterministic performance models are viewed as the best approach to predict pavement performance. However, such models may not always exist, especially for the types of treatments this study focuses on. Therefore, an alternative multi-criteria performance estimation approach is proposed, to be implemented when such deterministic models are not available. The approach can be applied to a potential treatment by collecting on-site conditions consistent with the design variables.

This tool allows to systematically develop a broad baseline assessment of the preservation and rehabilitation treatments. Contractors and state and local agencies will be able to utilize this baseline to conduct their own environmental assessments.

REFERENCES

Alkins, A.E., B. Lane, and T. Kazmierowski. 2008. Sustainable pavements: environmental, economic, and social benefits of in situ pavement recycling. *Transportation Research Record: Journal of the Transportation Research Board No. 2084*. Washington, D.C.: Transportation Research Board of the National Academies, 100–103.

Asphalt Recycling and Reclaiming Association (ARRA). 2015. *Basic Asphalt Recycling Manual*. Annapolis, MD.

Bare, J. 2012. Tool for the Reduction and Assessment of Chemical and Other Environmental Impacts (TRACI, Version 2.1). [Software]. Cincinnati, Ohio: U.S. Environmental Protection Agency.

California Department of Transportation (Caltrans). 2008. Maintenance Technical Advisory Guide (MTAG), Volume I—Flexible Pavement Preservation. Second Edition, Chapter 13–In-Place Recycling. Sacramento, CA 95814: State of California Department of Transportation, Office of Pavement Preservation, Division of Maintenance.

Earthshift. 2013. US-Ecoinvent Database. Version 2.2. [Database]. St. Gallen, Switzerland: Swiss Center for Life-Cycle Inventories.

Federal Highway Administration (FHWA). 2013. Chip Seal Application Checklist. Pavement Preservation Checklist Series.

Federal Highway Administration (FHWA). 2016. *Pavement Life Cycle Assessment Framework*.

Illinois Department of Transportation (IDOT). 2012. Pavement Rehabilitation. In *Bureau of Local Roads and Streets Manual*. Springfield: IDOT.

International Organization for Standardization (ISO). 2006. Environmental Management—Life Cycle Assessment—Principles and Framework. ISO 14040:2006. Geneva, Switzerland: International Organization for Standardization.

Liu, X., Cui, Q., & Schwartz, C. 2014. Performance Benchmark of Greenhouse Gas Emissions from Asphalt Pavement in the United States. *ICSI 2014: Creating Infrastructure for a Sustainable World*: American Society of Civil Engineers.

Santos, J., J. Bryce, G. Flintsch, A. Ferreira, and B. Diefenderfer. 2014. A life cycle assessment of in-place recycling and conventional pavement construction and maintenance practices. *Structure and Infrastructure Engineering* (pp. 1–19).

Schvallinger, M. 2011. Analyzing trends of asphalt recycling in France. Master's Thesis, KTH Royal Institute of Technology, Stockholm, Sweden.

Stroup-Gardiner, M. 2011. *Recycling and Reclamation of Asphalt Pavements Using In-Place Methods—A Synthesis of Highway Practice* (NCHRP Synthesis 421). Washington, D.C: Transportation Research Board (TRB).

U.S. Environmental Protection Agency. 2013. Emissions and Generation Resources Integrated Database (eGRID).

U.S. Environmental Protection Agency. 2015. Motor Vehicle Emission Simulator (MOVES) 2014 [software]. Washington, D.C.

Life-cycle assessment of road pavements containing marginal materials: Comparative analysis based on a real case study

Marco Pasetto, Emiliano Pasquini, Giovanni Giacomello & Andrea Baliello
Department of Civil, Environmental and Architectural Engineering, University of Padua, Padua (PD), Italy

ABSTRACT: The Life-Cycle Assessment (LCA) is a standardized procedure generally used, in Italy, in industrial engineering to evaluate the economic-environmental efficiency of production processes. LCA is aimed at optimizing the design, with special emphasis on environmental sustainability. Also in the construction sector, LCA has recently gained a fundamental role as a quantitative measurement tool able to take into account correctly the environmental and economic benefits achievable adopting different alternatives (most of them uncommon) based on the entire service life, maintenance and end-of-life procedures included. As far as pavement engineering is concerned, the use of marginal materials (such as, for example, reclaimed asphalt pavement, crumb rubber, slags, etc.) is becoming of strategic importance due to the decreasing availability of virgin natural resources and the consequent increasing public consciousness addressed to environmental protection and preservation. In this regard, the LCA applied to road pavements constructed using marginal aggregates probably represents the only effective tool able to evidence the crucial aspects on which the design choices should be based, taking also into account long-term parameters. Given this background, the present research illustrates one real case study of LCA analysis applied to asphalt pavements of a motorway. The use of industrial by-products (i.e. steel slags) instead of natural mineral aggregates is considered. Comparative evaluation of different scenarios has been carried out using specifically developed spreadsheets. The research study demonstrates that LCA is able to highlight potentialities and issues related to the different analyzed scenarios, representing a valid tool for designers and decision-makers. Moreover, the obtained results contribute to enlarge the worldwide database about the implementation of LCA for pavements.

1 INTRODUCTION

1.1 *General overview of Life Cycle Assessment methodology*

The Life Cycle Assessment (LCA) is a methodology developed to support the decision-making process in environmental terms, implemented since the Seventies. LCA is usually utilized at industrial level since it allows the evaluation of the potential product impacts, considering all life cycle stages (from design to construction, from use to final disposal). This assessment includes all the interactions between product (or service) and surrounding environment, also considering the maintenance phases.

In the international field, Life Cycle Assessment is largely widespread to improve both industrial production and services. Thanks to the policies impulses, the use of LCA is strongly encourages also in European Union in order to achieve targets such as: i) reduction of energy and resource consumption; ii) health improvement; iii) environment saving. Moreover, since LCA allows to quantify the production processes and the environmental impacts indicating the possible strategies to reduce emissions, this methodology is becoming a necessary tool for the definition of public policies and industrial competitiveness.

1.2 The standards for LCA methodology

Life Cycle Assessment is defined and described in ISO 14040 and ISO 14044 (ISO 2006a, b) standards. LCA framework (Fig. 1) can be divided in the following steps: 1) goal and scope definitions; 2) inventory collection and analysis; 3) environmental impact assessment; 4) obtained results interpretation.

ISO standards describe in detail principles, framework, requirements and guidelines for the LCA, including: a) the goal and scope definition of the LCA; b) the Life Cycle Inventory analysis (LCI) phase; c) the Life Cycle Impact Assessment (LCIA) phase; d) the life cycle interpretation phase; e) reporting and critical review of the LCA; f) limitations of the LCA; g) relationship between the LCA phases; h) conditions for use of value choices and optional elements.

However, these standards do not state specific prescriptions to perform a LCA or defined methodologies for the specific LCA phase. Further, LCI phase can be performed separately from a specific LCA study, because LCI inputs/outputs introduction and quantification are not closely linked with the specific product or service evaluated with the LCA.

1.3 LCA in road engineering

The attention addressed to the minimization of impacts related to the construction inclusion in the environment defines a general trend concerning the study of ecological characteristics throughout the different infrastructure project hypothesis (from design to maintenance, from use to end-of-life). LCA can be applied to several civil engineering sectors, such as the transport infrastructure one. In this sense, many researchers already ventured in LCA implementation to evaluate the environmental impacts connected to transport infrastructures, facing the most important problems related to the material type and its transport, that strongly represent onerous items in road construction and maintenance (Jullien et al. 2009, Santero et al. 2011, Azarijafari et al. 2016).

The increasing expensiveness of road design (in terms of energy request and environmental impacts) involves the need to reduce work emissions and costs during its lifetime. In this sense, even more researchers, management and construction companies develop sustainable project and utilize LCA in decision procedures which affect several environmental aspects (such as impacts of different pavement types and materials). Also the decision-makers, with an accurate life cycle cost analysis, can use LCA to evaluate the project or policy impacts (Santero et al. 2011).

Many studies evaluate the environmental issues and the effects on road construction, management/maintenance and rehabilitation. Amini et al. (2012) compared conventional and perpetual pavement (i.e. not requesting structural maintenance) and their different effects on environment and costs. Others researchers suggested to take into account the effect on the environment and infrastructure caused by road traffic (Zaabar & Chatti 2010, Santos et al. 2015a). For example, Bryce et al. (2014) evaluated through LCA the possibility to reduce the road maintenance activities considering the pavement damage and the related tire vehicles consumption (the surface type affects the vehicles pollution). Yang et al. (2015) assessed the use of Reclaimed Asphalt Pavement—RAP and Recycled Asphalt Shingle—RAS in partial substitution of virgin aggregates to check the different environmental impacts, considering also the vehicles fuel consumes as a function of the IRI index.

Others authors studied LCA applied to road infrastructure materials. DeDene & Marasteanu (2012) examined the possibility to reduce production costs and harmful emissions of asphalt pavement with 15% of RAP. Butt et al. (2014) applied LCA to bituminous mixtures assessing the energy consumption and the environmental sustainability. In this case,

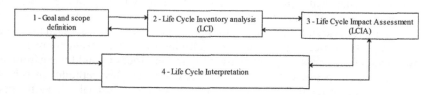

Figure 1. Life Cycle Assessment framework (from the ISO 14040 (ISO 2006b)).

the analysis of significant aspects such as the asphalt concrete production temperature, the vehicles type, the material transport and the location of the work site demonstrate the need to review the road construction method.

Software implementing the LCA to evaluate the environmental, social and economic impacts of a road pavement can be found. As examples, PaLATE developed by Californian researchers in the 2004 (Horvath et al. 2007), a model elaborated at Michigan University in the 2007 (Zhang et al. 2010), OPTIPAV proposed by Portuguese researchers in the 2011 (Santos & Ferreira 2013), DuBoCalc created by the Netherlands national public works agency in the 2013 (Harvey et al. 2014), SimaPro developed by consultants agency in the Netherlands (Anthonissen et al. 2015).

1.4 *The PaLATE tool*

The present work utilizes the Excel-based tool PaLATE (Pavement Life cycle Assessment Tool for Environmental and Economic Effects), a LCA applicator useful to assess the pavement and road life cycle (considering extraction, production, construction, maintenance and end-of-life phases). PaLATE provides many results in terms of environmental effects (e.g. energy consumption, CO_2, NO_x, PM_{10} emissions, etc.) and costs (e.g. during construction and maintenance phases).

The software is divided into different parts, each one able to collect input data and information (engineering, environmental and economic-based) in order to return a wide range of output results (air quality, amount of harmful emissions, etc.). Thanks to its analytical structure and flexibility, this tool is often used for research and commercial purposes. Utilizing PaLATE, Celauro et al. (2015) checked the influence of different virgin and recycled aggregates during construction and maintenance operation of a typical Italian road pavement. Similarly, Nathman et al. (2009) analyzed road pavement construction and maintenance activities quantifying the works feasibility and the economical, environmental and social sustainability with PaLATE.

2 CASE STUDY

2.1 *LCA phases*

Based on the scheme of Figure 1, the paper herein presents a Life Cycle Assessment performed through the overall road life cycle. In this sense, five phases (Fig. 2) can be cited (Celauro et al. 2015): 1) material production, 2) construction, 3) use (in this case it is not included the emission and the delay caused by traffic), 4) maintenance and rehabilitation (discard old material, production of new material, transport and laying processes), 5) end-of-life (pavement demolition and material disposal).

Data utilized in the LCI phase were obtained from different sources, depending on the information type (materials, transport, cost, etc.). The main sources are: Italian company of motorway (called "Autostrade per l'Italia"), asphalt concrete producer companies, transport companies, typical Italian price lists and specifications as well as Italian Minister of Public Works.

2.2 *Research objective*

The presented study is based on a real road infrastructure designed in Italy. Traditional construction/maintenance techniques were compared with recycling technologies since the reutilization of RAP and industrial by-products such as steel slags, fly ashes, municipal solid waste incinerator ashes or glass involves lower environmental impacts with respect to traditional techniques. Thus, the study is aimed at assessing the validity of LCA in order to promote the mitigation of environmental impacts, decrease costs and control energy (fuel, electrical energy, etc.) consumption when recycled materials are utilized in road construction. Moreover, in view of the importance of the maintenance policies and future investments,

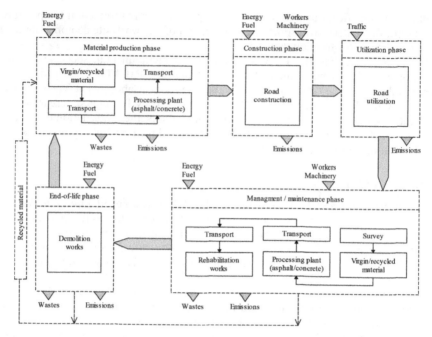

Figure 2. LCA phases and system boundary for the case study proposed.

LCA could be a promising instrument definitely assessing costs and environmental impacts. In this sense, since pavement construction and maintenance require significant amount of materials and energy levels for treatments, the paper focuses on the comparison between alternative pavement materials with respect to traditional (high-quality and expensive) ones.

2.3 Functional unit description

As mentioned above, the case study is based on a real Italian road construction project, the A3 motorway from Salerno to Reggio Calabria. LCA was applied to a motorway section of 1000 m having the geometry described in Fig. 3 (the left shoulder was considered not paved).

The semi-rigid pavement was designed using AASHTO method and it was composed by the materials and layers showed in Fig. 4 (the porous asphalt was prepared with polymer-modified bitumen, whereas the asphalt concretes for binder and base courses contained traditional bitumen).

2.4 Assumption and system boundary

The evaluation was performed considering a motorway section length of 1000 m and comparing different scenarios: A) transport distance equal to 100 km and only virgin aggregates for construction, B) transport distance equal to 100 km and by-products reuse (steel slags and RAP) in partial substitution of virgin aggregates, C) transport distance equal to 10 km (materials available in situ) and only virgin aggregates, D) transport distance equal to 10 km (materials available in situ) and by-products reuse (steel slags and RAP) in partial substitution of virgin aggregates.

Subgrades, drainages, tack coat, road markings and other particular street furniture were not included in the evaluation.

Fig. 2 illustrated the system boundary. This system took into account different processes (both from internal and external), the inbound flows (natural resources, materials and energy) and the outbound flows (products, emissions and wastes). As example, the different machinery type, i.e. on-site construction equipment (e.g., asphalt paver) and off-site processing

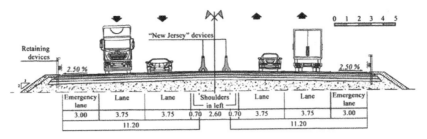

Figure 3. Scheduled maintenance activities during in-service life (20 years) for the Italian motorway A3 case study and pavement thickness description.

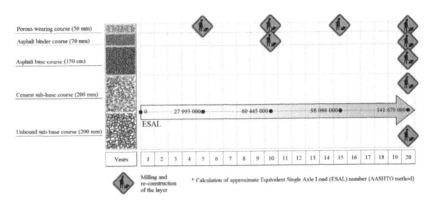

Figure 4. Scheduled maintenance activities during in-service life (20 years) for the Italian motorway A3 case study and pavement thickness description.

equipment (e.g., rock crusher), were considered from internal processes, whereas fuel and energy consumption or waste production were supposed from external processes. Differently, the feedstock energy was not examined.

The maintenance strategy is reported in Fig. 4 as found in the detailed project documents and assuming that maintenance activities will not be affected by the use of the recycled materials. The analysis period was set to 20 years after the initial construction. Every 5 years, the wearing course resurfacing (milling and laying new layer operation) was planned. Alternatively, the maintenance on the binder layer was designed after 10 years (milling and laying new asphalt operation), whereas after 20 years was scheduled a deep rehabilitation with total repaving operation (removing and reconstruction of the total pavement thickness).

3 LIFE CYCLE INVENTORY

3.1 *Materials*

The materials utilized in the road construction must demonstrate adequate physical and mechanical properties, complying with prescribed range specifications. Therefore, typical Italian road materials with known characteristics were considered for the construction. The information included in the Life Cycle Inventory concerned virgin aggregates, bitumen, recycled materials (RAP) and industrial by-products (steel slags). Materials data and costs were recovered from quarry in the proximity of the construction site, asphalt concrete producers and typical Italian price list. Wearing course was composed by polymer modified porous asphalt mix, whereas traditional bituminous mixtures were taken into account for binder and base layers. Two type of mixes were used in subbase layer: the first one on the top, composed by cement bound granular mix, and the second on the bottom, composed by granular unbound mix.

During binder layer maintenance phase (milling and reconstruction), 30% of RAP coming from the milling of the same layer was used. RAP re-use was carried out with a hot in-place recycling method: RAP was sent in plant and mixed with virgin aggregates and bitumen.

3.2 Consumption, transport and emissions

The data of energy consumption, transport and emission are inserted in the data sheet of PaLATE tool, which contains machinery performance, fuel and energy consumptions (from manuals of construction machinery and producers), water use, truck capacities, material densities, emissions to air (from plant and machinery for transport or work), etc.

It was assumed that materials (virgin or recycled) and asphalt concrete transports, from extraction sites (quarries or stockpiles) or asphalt plants towards the construction site, were developed by road only. In this case, considering the distance from the sites and the production plants equal to 100 or 10 km (depending on the scenario), the total length (in kilometers) that had to be covered by different truck type and numbers of trips was calculated.

Emissions from paver machine was not taken into account according to other studies indicating their negligibility if compared with other impacts (Hanson et al. 2012, Giani et al. 2015).

3.3 Cost analysis

The prices of raw materials, related to the material production, construction and maintenance phases, were obtained from a typical Italian price list: aggregates (virgin or recycled), binders (bitumen or cement), construction and maintenance techniques (pavement laying, full depth reclamation, asphalt plant, etc.).

The discount rate to calculate the total actual cost of road infrastructure was set in two different ways: a rate equal to 1% for the base scenario and one of 4% for the alternative scenario (in order to consider a less favorable context). Discount rates were both referred to construction and maintenance costs.

Then, the total construction, maintenance and disposal costs for pavement (indicated in scenarios, for the two discount rates) were calculated including 23% of overall costs and company profit.

4 LIFE CYCLE IMPACT ASSESSMENT

4.1 Environmental and costs evaluation

The environmental impact for scenarios (A, B, C and D) is reported in Table 1 and Table 2.

The assessment was based on energy, Water Consumption (WC), carbon dioxide (CO_2), nitric oxide (NO_x), Particulate Matter (PM_{10}), carbon mono-oxide (CO), sulfur dioxide (SO_2), heavy metals (Pb and Hg), Hazardous Waste Generated (HWG), Human Toxicity Potential cancer (HTPC) and Human Toxicity Potential Non-Cancer (HTPNC).

Fig. 5 and Fig. 6 show construction, maintenance and total costs for B, C and D scenarios, with discount rate of 1% or 4% comparing data with those of reference scenario A (Table 3).

4.2 Results interpretation and discussion

In general, it could be observed a typical decrease of different impact categories related to the substitution of virgin aggregates with recycled materials (see Table 1 and Table 2).

Otherwise, impact reductions found in this study were slightly lower with respect to other literature reporting because innovative production techniques of asphalt concrete, such as warm mix asphalt (Al-Qadi et al. 2015, Anthonissen et al. 2015, Pasetto et al. 2016), or in-place recycling practices, (Giani et al. 2015, Santos et al. 2015b) were not considered.

In this regard, in fact, many researcher authors demonstrated the strict correlation between parameter values and asphalt production technique (e.g. nitric oxide, according with Celauro et al. 2015).

Table 1. Environmental results from PaLATE tool for the four scenarios proposed.

Scenario	Energy [MJ]	WC [kg]	CO_2 [Mg]	NO_x [kg]	PM_{10} [kg]	SO_2 [kg]
A						
MP*	824,766,458	275,762	45,118	289,093	170,190	6,293,294
MT**	43,568,017	7,421	3,257	173,527	33,560	10,412
P***	3,904,838	367	29	6,750	710	446
Total	872,239,312	283,550	8,670	469,370	204,460	6,304,152
B						
MP*	784,998,830	261,100	42,465	277,573	155,479	6,283,626
MT**	46,761,742	7,965	3,496	186,248	35,857	11,175
P***	3,904,838	367	294	6,750	710	446
Total	835,665,409	269,432	46,255	470,570	192,046	6,295,247
C						
MP*	824,766,458	275,762	45,118	289,093	170,190	6,293,294
MT**	4,451,723	758	333	17,731	3,574	1,064
P***	3,904,838	367	294	6,750	710	446
Total	833,123,018	276,887	45,745	313,573	174,474	6,294,804
D						
MP*	781,227,517	261,100	42,564	280,836	155,510	6,544,939
MT**	4,006,379	734	322	17,152	3,400	1,029
P***	3,904,838	367	294	6,750	710	446
Total	789,138,734	262,200	43,180	304,737	159,620	6,546,414

*MP = materials production. **MT = materials transportation. ***P = processes (equipment).

Table 2. Environmental results from PaLATE tool for the four scenarios proposed.

Scenario	CO [kg]	Hg [g]	Pb [g]	HWG [kg]	HTPC	HTPNC
A						
MP*	159,794	1,065	53,168	10,596,575	174,183,222	150,792,239,042
MT**	14,461	32	1,478	314,074	1,569,208	1,954,641,048
P***	1,455	2	82	17,540	0	0
Total	175,710	1,099	54,728	10,928,190	175,752,430	152,746,880,089
B						
MP*	151,244	1,010	50,364	10,037,005	165,784,422	138,441,537,092
MT**	15,521	34	1,586	337,097	1,677,640	2,089,706,829
P***	1,455	2	82	17,540	0	0
Total	168,219	1,046	52,031	10,391,642	167,462,062	140,531,243,921
C						
MP*	159,794	1,065	53,168	10,596,575	174,183,222	150,792,239,042
MT**	1,478	3	151	32,092	241,144	300,374,143
P***	1,455	2	82	17,540	0	0
Total	162,727	1,070	53,401	10,646,207	174,424,366	152,092,613,184
D						
MP*	151,338	1,011	50,372	10,038,761	166,126,114	139,049,945,940
MT**	1,429	3	146	31,044	236,209	294,227,423
P***	1,484	2	84	18,109	0	0
Total	154,251	1,016	50,602	10,087,914	166,362,324	139,344,173,364

*MP = materials production. **MT = materials transportation. ***P = processes (equipment).

Construction and maintenance materials contribute to the greatest percentage of impacts (contribution by the materials phase varies with the type of pavement); however, the processes and equipment related to construction and maintenance phases constitute the smallest amounts (Al-Qadi et al. 2015). In general, materials production (bitumen, cement and mixtures) owns a marginal weight in life cycle of road pavement with respect to the transport processes (Anthonissen et al. 2015, Huang et al. 2009), which give a contribution about of the 20% of total impact. As far as the materials use concerns, the initial construction impact

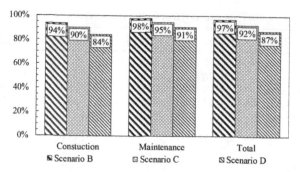

Figure 5. Costs of B, C and D scenarios compared to A scenario with discount rate of 1% (base scenario).

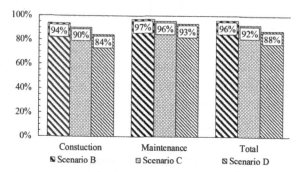

Figure 6. Costs of B, C and D scenarios compared to A scenario with discount rate of 4% (alternative scenario).

Table 3. Scenario A costs with discount rate of 1% and 4%.

Scenario	Costs [Euro]
Base (1%)	
Initial construction	21,511,453
Maintenance	19,599,351
Total	41,110,805
Alternative (4%)	
Initial construction	21,511,453
Maintenance	13,488,463
Total	34,999,917

ranges from 20% to 25%, the maintenance impact varies from 75% to 70% and the end-of-life impact is around 5%, according to literature (Anthonissen et al. 2015).

Especially during maintenance, RAP replacement (rejuvenated with the additive) reduces total energy consumption and environmental impact, requiring less virgin bitumen and allowing less landfilling material. Moreover, mechanical characteristics and durability of RAP mixtures are comparable with those of traditional ones, i.e. prepared with virgin aggregates.

Also by-products reuse (e.g. steel slags) prevents from conferring in dump, thus promoting environment protection (Mladenovic et al. 2015, Ferreira et al. 2016).

Otherwise, it has to be noticed that the higher specific gravity of steel slags burdens on the transportation costs; basing on this consideration, this project implied only the use 30% of slags by the total aggregate weight (to prevent the increasing of total impact). On the other hand, steel slags demonstrated excellent mechanical properties (which makes them suitable to be used as aggregate in asphalt pavement, for many traffic typologies), also improving skid resistance. Moreover, requested energy consumption for steel slags asphalt mixture production could be considered almost the same of that needed in the case of traditional mixes (with virgin aggregates only), even if slightly higher bitumen content, connected to steel slag surface porosity could be recommended (Pasetto & Baldo).

During construction phase, material costs are greater with respect to the equipment and worker ones. Vice versa, during maintenance phase, equipment and worker costs prevail.

Costs related to the marginal material rehabilitation operations were slightly lower with respect to those involving the virgin aggregate supply; thus B, C and D scenarios requested lightly reduced investments (in comparison with A scenario). Since A and C scenarios implicated the use of natural aggregates only, costs were higher because of the onerous activities connected to waste materials disposal towards landfills. Conversely, these items were not significant in B and D scenarios (where recycled material reutilization was planned). Figs. 5 and 6 do not clearly represent the differences between discount rates of 1% and 4%: in this sense, it is worth noting that a light decrease in maintenance costs (thus in total investments) with the highest discount rate can be detected.

5 CONCLUSIONS

The LCA seemed to be a useful tool to check the availability of projects, and, in general, a successful way to support decision-making processes considering environmental impact assessment and costs evaluation.

The specific case study proposed in the paper evaluated four scenarios (A, B, C and D) for a motorway section 1000 m long over a maintenance period of 20 years utilizing a specific software tool (PaLATE). Initial construction, maintenance, service life and end-of-life phases, as well as material transportation distances and modes were taken into account.

Different scenarios showed that the transport distance and the substitution of virgin aggregates with recycled materials (RAP or steel slags), played a key role in costs and environment impacts. Since the consumption of natural materials (bitumen and aggregates) represented another important factor for the impact assessment, recycling practices resulted in a suitable activity to promote resources preservation due to reduced needs of waste disposal.

In conclusion, an advanced LCA analysis, performed during the designing phases of road infrastructures, is definitely able to objectively and univocally state the best design and construction alternative, evaluating different costs and determining the most (social and environmental) sustainable constructive and maintenance technologies for the correct integration of the work.

However, performed LCA represents a first-step general analysis, thus further evaluations considering supplementary variables could be promoted.

REFERENCES

Al-Qadi, I.L., Yang, R., Kang, S., Ozer, H., Ferrebee, E., Roesler, J.R., Salinas, A., Meijer, J., Vavrik, W.R. & Gillen, S.L. 2015. Scenarios Developed for Improved Sustainability of Illinois Tollway. Life-Cycle Assessment Approach. *Transportation Research record: Journal of the Transportation Research Board*. 2523: 11–18.

Amini, A., Mashayekhi, M., Ziari, H. & Nobakht, S. 2012. Life cycle cost comparison of highways with perpetual and conventional pavements. *International Journal of Pavement Egineering*. 13(6): 553–568.

Anthonissen, J., Braet, J. & Van den Bergh, W. 2015. Life cycle assessment of bituminous pavements produced at various temperatures in the Belgium context. *Transportation Research Part D*. 41: 306–317.

AzariJafari, H., Yahia, A. & Amor, M B. 2016. Life cycle assessment of pavements: reviewing research challenges and opportunities. *Journal of Cleaner Production*. 112: 2187–2197.

Bryce, J., Katicha, S., Flintsch, G., Sivaneswaran, N. & Santos, J. 2014. Probabilistic Life-Cycle Assessment as Network-Level Evaluation Tool for Use and Maintenance Phases of Pavements. *Transportation Research record: Journal of the Transportation Research Board*. 2455: 44–53.

Butt, A.A., Mirzadeh, I., Toller, S. & Birgisson, B. 2014. Life cycle assessment for asphalt pavements: methods to calculate and allocate energy of binder and additives. *International Journal of Pavements Engineering*. 15(4): 290–302.

Celauro, C., Corriere, F., Guerrieri, M. & Lo Casto, B. 2015. Environmentally appraising different pavement and construction scenarios: a comparative analysis for a typical local road. *Transportation Research Part D*. 34: 41–51.

DeDene, C.D. & Marasteanu, M.O. 2012. Life Cycle Assessment of Reclaimed Asphalt Pavements to Improve Asphalt Pavement Sustainability. *Proceedings of 5th International SIIV Congress SIIV 2012—Sustainability of Road Infrastructures; Rome, 29–31 october 2012*.

Ferreira, V.J., Vilaplana, A.S.D.G., García-Armingol, T., Aranda-Usón, A., Lausín-González, C., López-Sabirón, A.M., & Ferreira, G. 2016. Evaluation of the steel slag incorporation as coarse aggregate for road construction: technical requirements and environmental impact assessment. *Journal of Cleaner Production*, 130: 175–186.

Giani, M.I., Dotelli, G., Brandini, N. & Zampori, L. 2015. Comparative life cycle assessment of asphalt pavements using reclaimed asphalt, warm mix technology and cold in-place recycling. *Resources, Conservation and Recycling*. 104: 224–238.

Hanson, C.S., Noland, R.B. & Cavale, K.R. 2012. Life-cycle greenhouse gas emissions of materials used in road construction. *Transportation Research Record: Journal of the Transportation Research Board*. 2287: 174–181.

Harvey, J., Kendall, A., Santero, N. & Wang, T. 2014. Use of Life Cycle Assessment for asphalt pavement at the network and project levels. *Proceedings of the 12th International Conference on Asphalt Pavements, ISAP 2014; Raleigh, 1–5 June 2014*: 1797–1806.

Horvath, A., Pacca, S., Masanet, E. & Canapa, R. 2004. PaLATE: Pavement Life-cycle Assessment Tool for Environmental and Economic Effects. Available from University of California, Berkley. Website: http://www.ce.berkeley.edu/,horvath/palate.html.

Huang, Y., Bird, R. & Heidrich, O. 2009. Development of a life cycle assessment tool for construction and maintenance of asphalt pavements. *Journal of Cleaner Production*. 17: 283–296.

International Organization for Standardization. 2006a. ISO 14040 Environmental management—Life cycle assessment—Principles and framework. ISO 14040:2006(E). International Organization for Standardization, Geneva. http://www.iso.org/iso/catalogue_detail?csnumber=37456.

International Organization for Standardization. 2006b. ISO 14044 Environmental management—Life cycle assessment—Requirements and guidelines. ISO 14044:2006(E). International Organization for Standardization, Geneva. http://www.iso.org/iso/catalogue_detail.htm?csnumber=38498.

Jullien, A., De Larrard, F., Bercovici, M., Lumière, L., Piantone, P., Domas, J., Dupont, P., Chateau, L. & Leray, F. 2009. The OFRIR project: a multi-actor approach to recycling in road infrastructure. *Bulletin des Laboratoires des Ponts et Chaussées*. 275: 65–84.

Mladenovic, A., Turk, J., Kovac, J., Mauko, A. & Cotic, Z. 2015. Environmental evaluation of two scenarios for the selection of materials for asphalt wearing courses. *Journal of Cleaner Production*. 87: 683–691.

Nathman, R., McNeil, S. & Van Dam, T.J. 2009. Integrating Environmental Perspectives into Pavement Management. Adding the Pavement Life-Cycle Assessment Tool for Environmental and Economic Effects to the Decision-Making Toolbox. *Transportation Research record: Journal of the Transportation Research Board*. 2093: 40–49.

Pasetto, M. & Baldo, N. 2008. Comparative performance analysis of bituminous mixtures with EAF steel slags: A laboratory evaluation. *Proceedings of the 2008 Global Symposium on Recycling, Waste Treatment and Clean Technology, REWAS 2008*: 565–570.

Pasetto, M., Baliello, A., Giacomello, G. & Pasquini, E. 2016. Rheological Characterization of Warm-Modified Asphalt Mastics Containing Electric Arc Furnace Steel Slags. *Advances in Materials Science and Engineering*. 2016: 1–11.

Santero, N.J., Masanet, E. & Horvath, A. 2011. Life-cycle assessment of pavements. Part I: Critical review. *Resources, Conservation and Recycling*. 55: 801–809.

Santos, J. & Ferreira, A. 2013. Life-cycle cost analysis system for pavement management at project level. *The International Journal of Pavement Engineering*. 14(1): 71–84.

Santos, J., Bryce. J., Flintsch, G., Ferreira, A. & Diefenderfer, B. 2015b. A life cycle assessment of in-place recycling and conventional pavement construction and maintenance practices. *Structure and Infrastructure Engineering*. 11(9): 1199–1217.

Santos, J., Ferreira, A. & Flintsch, G. 2015a. A life cycle assessment for pavement management: road pavement construction and management in Portugal. *International Journal of Pavement Engineering*. 16(4): 315–336.

Yang, R., Kang, S., Ozer, H. & Al-Qadi, I. 2015. Environmental and economic analyses of recycled asphalt concrete mixtures based on material production and potential performance. *Recources, Conservation and Recycling*. 104: 141–151.

Zaabar, I. & Chatti, K. 2010. Calibration of HDM-4 Models for Estimating the Effect of Pavement Roughness on Fuel Consumption for US Conditions. *Transportation Research Record*. 2155: 105–116.

Zhang, H., Keoleian, G.A., Lepech, M.D. & Kendall, A. 2010. Life-Cycle Optimization of Pavement Overlay System. *Journal of Infrastructure Systems*. 16(4): 310–322.

Integrated sustainability assessment of asphalt rubber pavement based on life cycle analysis

Ruijun Cao, Zhen Leng, Mark Shu-Chien Hsu, Huayang Yu & Yangyang Wang
Department of Civil and Environmental Engineering, The Hong Kong Polytechnic University, Hong Kong

ABSTRACT: This study aims to quantify and compare the life-cycle environmental, economic and social impacts of two pavement materials: namely 10 mm Asphalt Rubber Stone Matrix Asphalt (ARSMA10) and 10 mm Polymer Modified Stone Matrix Asphalt (PMSMA10). To achieve this objective, a comparative sustainability framework was developed by: 1) combining the Life Cycle Assessment (LCA) with pavement maintenance plan developed based on the pavement conditions predicted by the Mechanistic-Empirical Pavement Design Guide (MEPDG) software; 2) examining the additional fuel consumption and emissions based on predicted pavement condition; 3) and integrating the Cost-Benefit Analysis (CBA) to estimate the environmental damage cost and noise reduction benefit. The major findings of this study include: 1) in the 56-year analysis period, the dominant contributions to environmental impact are made by the extra vehicle Greenhouse Gas (GHG) emission and fuel consumption caused by pavement roughness change; 2) PMSMA10 has better environmental performance than ARSMA10 in terms of energy consumption and GHG emissions; 3) the long-term accumulated tire-road noise impact is considerable, which can almost offset its higher economic cost and environmental damage cost assuming ARSMA10 is able to decrease noise by 4%.

1 INTRODUCTION

Pavement, as one of the major components of the infrastructure system, plays an important role in the development of civilization and economic prosperity. Currently, the booming transportation and the increasing awareness of sustainable development have attracted significant research interest in reducing the negative environmental and social impacts of pavement during its construction and service. In 2005, the Greenhouse Gas (GHG) emissions from road transportation accounted for 12.5% to 13.0% of global total emissions (ASTAE, 2009), while pavement condition heavily affects vehicle emissions and fuel consumption. For instance, rough pavement surface has been reported to negatively affect vehicle fuel economy (Chatti and Zaabar 2012). Though the environmental impact variables in a pavement lifecycle are complicated, some sustained efforts have been made to minimize the environmental burdens from pavement projects, for example, many recycled materials, such as Reclaimed Asphalt Pavement (RAP) and waste tire rubber, are now being used in pavement construction to conserve raw material resources without compromising pavement performance.

Asphalt Rubber (AR), which is composed of raw asphalt and at least of 15% of waste tire rubber as modifier, is one good example of using waste materials in pavement. However, AR has received different popularities in different areas around the world, because on one hand it provides various benefits, such as recycling waste tires (Caltrans, 2013), enhancing pavement performance, and reducing tire-road noise (Lo Presti, 2013; SCDER & ABB. Inc., 1999; RPA, 1999), while on the other hand, it requires higher construction temperature and cost. To achieve a comprehensive sustainability assessment of AR, this study quantified the economic, environmental, and social impacts of 10 mm Asphalt Rubber Stone Matrix Asphalt (ARSMA10) through an extended Life Cycle Assessment (LCA) approach, using 10 mm Polymer Modified Stone Matrix Asphalt (PMSAM10) as a reference.

LCA provides a systematic procedure to quantify the environmental performance of products throughout their life cycles. The report of FHWA (2014) divided the life cycle of pavement into five phases: production, construction, use, maintenance, and disposal. The pavement LCA studies have been documented in many literatures since 1996 (Häkkinen & Mäkelä, 1996), but the research focuses and results differ in various studies. Recent studies have combined LCA with other analysis methods to achieve results that are more comprehensive. Yu et al. (2013) integrated LCA and Life Cycle Costing Analysis (LCCA) to estimate the evidential damage cost to optimize the pavement maintenance plan. Yang et al. (2015) developed an International Roughness Index (IRI) progression model in the LCA to compute the GHG emissions and energy consumption of vehicles. Chong and Wang (2016) developed an LCA framework based on the pavement design and management decisions.

However, noise, on which AR pavement can provide significant benefit, is often considered as one of the social-economic impacts, and has not been included and defined in the staple Life Cycle Inventories (LCI). European Commission (2014) has assessed the noise impact by computing the corresponding noise cost introduced by the Cost-Benefit Analysis (CBA) method. In this study, the comparative sustainability framework extended the use of LCA by: 1) combining the Mechanistic-Empirical Pavement Design Guide (MEPDG) software to predict pavement conditions and determine the maintenance plan; 2) integrating the pavement conditions to examine the additional fuel consumption and emissions; and 3) integrating the CBA to examine the environmental damage cost and noise reduction benefit.

2 METHODOLOGY

The methodology adopted in this study incorporated the MEPDG software, LCA, and the calculation of Environmental Damage Cost (EDC) and Noise Reduction Benefit (NRB). For the two compared pavement materials, i.e., ARSMA10 and PMSMA10, all calculations were conducted under the same traffic, project and climate condition. The calculation process can be further divided into five steps:

1. determine the maintenance plans for the two comparison materials according to the modeling results of the MEPDG;
2. perform LCA according to ISO 14044 (2006a) and ISO 14040 (2006b) to acquire the environmental impacts;
3. estimate the investment costs involved in the initial construction and life cycle maintenance;
4. convert the GHG emission and noise reduction to the Environmental Damage Cost (EDC) and Noise Reduction Benefit (NRB) following the method described in the Cost Benefit Guidelines (European Commission, 2014) and the Handbook on Estimation of External Costs in the transport sector (Maibach et al., 2008); and
5. calculate the overall cost based on the results of the former four steps.

2.1 Goal and scope definition

The goal of this study is to express the life-cycle environmental, economic and social impacts of two pavement materials (ARSMA10 and PMSMA10) into monetary values, and compare the overall sustainabilities of the two mixtures based on their estimated performance in the life cycle view. The performances of the two materials were predicted by the MEPDG software using the experimental data of the material properties as input. The only variable in this comparison study is the asphalt mixture and other parameters (i.e., traffic loading, climate, construction and maintenance methods) were fixed and assumed same.

2.2 Functional unit and system boundaries

The functional unit of this study is the square meter (m^2) of 40 mm wearing course throughout the 56-year analysis period.

The system boundary determination will have significant influence on the results when considering the environmental impacts. In this study, four life-cycle phases were considered, including: material production, construction, usage and End-of-Life (EOL). The examining processes of ARSMA10 in each phase are illustrated in Figure 1. The processes of the two asphalt mixtures are almost same except for the difference in the material production phase, where ARSMA10 includes an additional process of producing crumb rubber modifier from EOL tires.

2.3 *Maintenance strategies determination based on the MEPDG*

The MEPDG methodology was adopted to predict the pavement performance based on the traffic loading, material properties, and environmental data. The responses were used to predict incremental damage over designed life time (Baus & Stires, 2010). The MEPDG software applied in this study was AASHTOWare Pavement ME Design (Version 2.2, 2015), and its design methodology is documented in the Mechanistic-Empirical Pavement Design Guide, Manual of Practice, Interim Edition (AASHTO, 2008). There were two objectives of employing MEPDG: to make sure the pavement designs of the two materials can meet the performance criterion, and to predict the distress development of pavement material.

To predict the accumulated deterioration of the two materials, the laboratory measured material properties were used as the input variables. Specifically, the property inputs of ARSMA10 and PMSMA10 included unit weight, effective binder content, dynamic modulus of mixture, and Superpave performance grade of asphalt. Table 1 summarizes the general pavement design information, which was assumed according to the common practice in Hong Kong.

The IRI and rut depth were selected as the two measurement indexes in this study. The first index represents a standardized pavement unevenness (Sayers et al., 1986), while the second one refers to the accumulated pavement deformation (Simpson, 2003). Based on these two indexes, an effective pavement maintenance strategy was developed accordingly.

Previous studies have shown that preventive maintenance may prevent a pavement from requiring corrective maintenance, and can be six to ten times more cost-effective than a "do nothing" maintenance strategy (Johnson, 2000). Hence, preventive maintenance strategy was selected in this study. Furthermore, among various candidates for the preventive mainte-

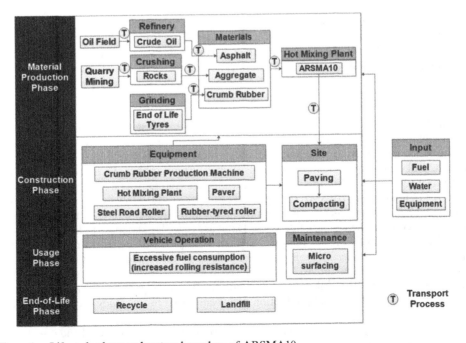

Figure 1. Life cycle phase and system boundary of ARSMA10.

Table 1. The summary of pavement design information.

General information		
Pavement type	Flexible Pavement	
Design life (years)	20	
Layer Thickness (mm)	40	
Length (m)	1000	
Lane width (m)	3.5	
Number of Lanes	4	
Discount Rate	4%	
Traffic information		
AADT	10000	
Growth Rate	3%	
Growth Function	Linear	
Vehicle Distribution	37.3%	Passenger car
	23.2%	Single-unit, short-haul truck
	37.2%	Single-unit, long-haul truck
	1.8%	Combination short-haul truck
	0.5%	Combination long-haul truck,
Operation Speed (km/h)	90	
Climate information*		
Climate Station	Hongkong, HK (99998)	
Mean annual Wind speed (kph)	9.31	
Mean annual Air temperature (deg C)	23.41	
Mean annual sun radiation	80.91%	
Mean annual precipitation (mm)	100.8	
Annual depth to water table (m)	6	

*The climate data is time-related and dynamic, the values of climate information in the table were calculated based on the hourly climate data from Jan/2000 - Jan/2010 to reflect the average level.

nance treatments, microsurfacing was selected in this study, considering that it is documented to be appropriate for most of the common distress types, such as roughness, rutting, cracking and raveling (Wilde et al., 2014). The maintenance intervals of the two asphalt mixtures were determined based on the accumulated deterioration predicted by the MEPDG software.

2.4 *Life Cycle Inventory Analysis*

LCI analysis is the second step of LCA method (ISO, 2006). This step is dedicated to present the major unit processes and relative calculation procedures within the considered life-cycle phase.

The material production phase included the extraction and initial processing of aggregates, asphalt, and other supplementary materials such as crumb rubber (Wang et al., 2012). The unit processes considered in this study included asphalt refinery, aggregate production, and mixture hot mixing. For ARSMA10, the rubber powder production was also considered. The data with respect to the energy consumption factor and GHG emission factor of aggregate were obtained from the Chinese Life Cycle Database (CLCD) Database (CLCD, 2010), and the relative factors of polymer modified asphalt were provided by the European Bitumen Association (Eurobitume, 2011). Besides, the process of processing asphalt rubber included crumb rubber production and asphalt rubber production. The data for energy consumption and GHG emissions were calculated according to the survey results of Zhu et al. (2014). The reference data of material production phase employed in this study are summarized in Table 2.

The construction phase consisted of two parts: transportation of material and on-site construction. The 30t diesel truck was selected in the transportations of raw materials and hot

Table 2. GHG emission and energy consumption of pavement life cycle inventory.

Life Cycle Inventory			ARSMA10	PMSMA10	Data reference
Material Production	Asphalt	Energy consumption (MJ/t)	189.33	311.20	CLCD (2010), Eurobitume (2011) and Zhu et al. (2014)
		Emissions (CO_2-e kg/t)	21.42	18.57	
	Aggregate	Energy consumption (MJ/t)	29.99	29.99	
		Emissions (CO_2-e kg/t)	2.29	2.99	
Transportation of Raw Materials (50 km)		Energy consumption (MJ/t)	40.20		
		Emissions (CO_2-e kg/t)	3.74		
Asphalt Mixture Hot Mixing		Energy consumption (MJ/t)	353.50	336.67	
		Emissions (CO_2-e kg/t)	29.67	28.26	
Transportation of Hot Mixture (20 km)		Energy consumption (MJ/t)	16.08		
		Emissions (CO_2-e kg/t)	1.50		
Pavement Construction	Paving	Energy consumption (MJ/t)	15.86		
		Emissions (CO_2-e kg/t)	1.18		
	Compaction	Energy consumption (MJ/t)	18.60		
		Emissions (CO_2-e kg/t)	1.38		
Pavement Usage	Preservation (Micro Surfacing)	Energy consumption (MJ/m²)	6.5		Chehovits & Galehouse et al. (2010)
		Emissions (CO_2-e kg/m²)	0.3		
	Additional energy consumption and GHG emissions caused by IRI changes	Energy consumption* (MJ/vehicle mile)	Passenger car 0.15, Single-unit, short-haul truck 0.12, Single-unit, long-haul truck, Combination short-haul truck 0.25, Combination long-haul truck		Yang et al. (2015) & Kang et al. (2014)
		GHG Emissions** (CO_2-e kg/vehicle hour)	0.004		Kalembo et al. (2011)

*Under the condition that per 63.4 in/mi (1 m/km) increase of IRI.
**Under the conditions that IRI increases from Good condition (95 in/mi) to poor condition (>150 in/mi).

mixtures, and the corresponding energy consumption and emissions were calculated according to CLCD (2010). The construction schedule and equipment activities of ARSMA10 were formulated following the Asphalt Rubber Design and Construction Guidelines (Hicks, 2002), by only considering the paving and compacting. Furthermore, the construction activities of PMSAM10 were assumed to be the same as ARSMA10. The emission factors of the paving and compacting processes of the two materials were calculated according to the power of machines and production efficiency (Zhu et al., 2014).

The usage phase primarily focused on the roughness effect on the additional fuel consumption and GHG emissions. The relationship between pavement smoothness and extra fuel consumption of vehicles has been studied in various literatures. It was reported that from 60 to 123.4 in/mi (0.95 to 1.95 m/km), a 63.4 in./mi (1 m/km) incremental change of IRI would increase the energy consumption by 3.7% for passenger cars, 1.2% for small trucks, 1.3% for medium trucks, and 0.9% for large trucks (Yang et al., 2015). According to Kalembo et al. (2011), the GHG emissions can increase 35,010 kg annually for the traffic volume of 1,000 vehicles per hour when the IRI changes from the good (<95 in./mi) to poor condition (>150 in./mi). The IRI changes of ARSMA10 and PMSAM10 were predicted by the MEPDG, so the calculation of the GHG emissions and energy consumption in the functional unit (Table 2) should be consistent with the AADT, growth rate and the vehicle distribution used in the MEPDG. Furthermore, the pavement maintenance work is necessary during the operation of the pavement system, and the corresponding emissions and energy required were also counted into the usage phase in this study.

When a road pavement reaches its service life, it can remain in place serving as support for a new pavement structure or be removed. By adopting a "cut-off" allocation method, no environmental impacts were assigned to the EOL phase in this study.

2.5 *Total cost and benefit accounting*

The final step was to convert all the impacts into monetary value for the purpose of more accessible comparison of the two pavement materials. The report by the World Conservation Union (2006) suggests that the three dimensions of sustainability: environmental, social and economic, are the mainstream sustainability thinking, which needs to be balanced and better integrated. In this study, all three dimensions were considered and expressed as the cost and benefit. The economic impacts were represented by the agency investment cost; the environmental impacts computed from LCA were converted to the environmental damage cost; and the social impacts of pavement materials mainly focused on the noise impact to people, which were presented as the noise reduction benefits.

The agency investment costs consisted of the material cost, construction cost and the maintenance cost, which were investigated according to the Asphalt Rubber Usage Guide (State of California Department of Transportation, 2003), Rubber Asphalt Industrialization Feasibility Report in Guangdong Province (Guangzhou Municipal Industries Ltd., 2011), and Handbook on Asphalt Pavement Maintenance (Johnson, 2000).

The Environmental Damage Costs (EDC) are the costs for unit of air pollutants that people need to pay to offset the effects on environment (Yu et al., 2013). A statistical analysis was conducted by Yu et al. (2013) to find the mean values (50$/t in 2010) of the EDC for CO_2 among the wide range from 5$/t to 1667$/t.

The value of noise was calculated as the unit marginal cost per person exposed to a certain noise level (European Commission, 2014), which are provided in the Handbook on Estimation of External Costs in the Transport Sector (Maibach et al., 2008). The noise

Table 3. Summary of life cycle cost.

Life Cycle Cost*			ARSMA10	PMSMA10	Data reference
Construction Cost**	Material ($/t)		87	68	State of California Department of Transportation (2003)
	Equipment ($/m²)			0.17	Guangzhou Municipal Industries Ltd. (2011)
	Fuel ($/m²)			1.37	
	Labor ($/m²)			0.07	
	Management ($/m²)			0.04	
Maintenance Cost*** ($/m²)				3.92	Johnson (2000)
Environmental Damage Costs**** ($/t)				63.27	Yu et al. (2013)
Noise Costs ($/person year)			510.86 (>73dB)	593.97 (>76dB)	Maibach et al. (2008), RPA (2003) and Sacramento County Department of Environmental Review and Assessment (1999)

*The costs have been converted to the Present Value (PV) according to the cost in the reference, and 4% was selected as the discount rate.
**The thickness of pavement is 40 mm.
***The maintenance refers to the pavement preservation treatment, microsurfacing.
****The air pollutant item considered is the mass of CO_2 in the whole life cycle.

reduction benefit of asphalt mixtures is the difference between the noise costs of ARSMA10 and PMSAM10. According to various global rubberized asohalt studies, the avergar noise reduction of asphalt rubber would be 2–3 dB and the noise of asphalt rubber overlay is measured and documented as 73.7dB. Consequently, 3dB was selected as the noise reduction when employing ARSMA10 in this evaluation, and 200 persons were assumed to be directly exposed to the highway every year.

The details of the life cycle costs are listed in Table 3.

3 RESULTS AND DISCUSSION

3.1 *Maintenance plan determination from MEPDG*

The results of the accumulated change of IRI and rut depth in 20-year design life predicted by the MEPDG software are illustrated in Figure 2.

It is evident that the IRI values for the two materials stay within the permissible range of MEPDG during the design life.

For ARSMA10, when it reaches approximate seven years, the predicted rut depth exceeds the threshold value, while for PMSAM10, it takes eight years. This indicates that the rut-resistance of the selected polymer modified asphalt mixture is better than that of the asphalt rubber mixture in this study. The maintenance strategies were then determined according to the estimated development of rut depth. The analysis period (56 years) is determined as the Least Common Multiple (LCM) of maintenance intervals (7 & 8 years) for the two materials based on the rut depth prediction. As illustrated in Figure 3, in the 56-year analysis period, the preservative maintenance (microsurfacing) would be conducted every eight years for PMSMA10, and every seven years for ARSMA10.

3.2 *Environmental performance*

The breakdowns of GHG emissions and energy consumption for the four major phases of the life cycle of the two mixes are shown in Figure 4 and Figure 5, respectively. In the 56-year analysis period, the dominant contributions were made by the extra impacts caused by the

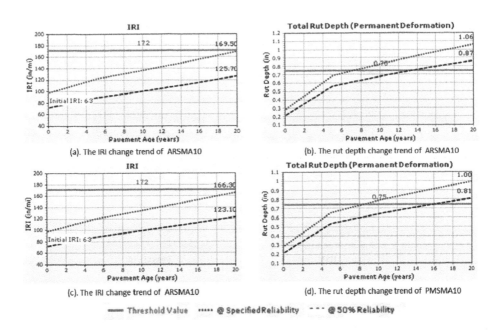

Figure 2. Performances of the two materials predicted by the MEPDG software.

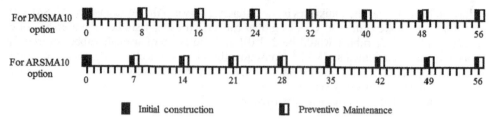

Figure 3. Maintenance plans for the two pavement material designs.

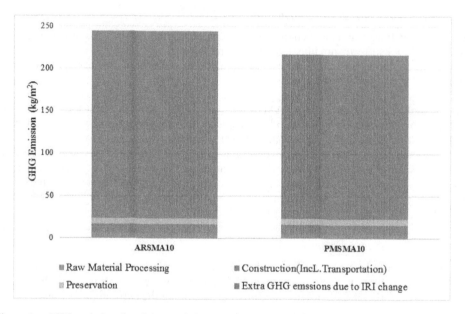

Figure 4. GHG emissions breakdown of the two pavement materials.

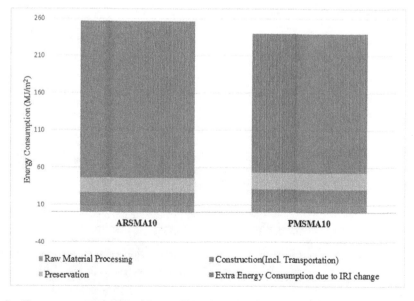

Figure 5. Energy consumption breakdown of the two pavement materials.

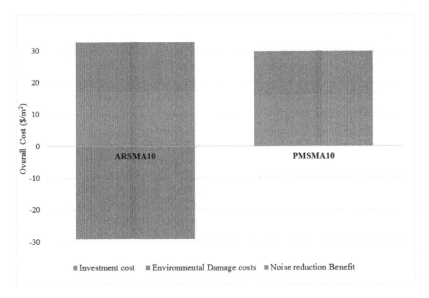

Figure 6. Overall cost breakdown of the two pavement materials.

pavement IRI change. In general, PMSMA10 has better environmental performance than ARSMA10 with regarding to both emissions and energy consumption.

As in Figure 4 shows, the percent distribution of the extra GHG emissions due to roughness change in the usage phase is especially overwhelming, approximately 242 and 214 kg/m^2 for the two mixes, respectively, since the sum of emissions in the other life-cycle phases of ARSMA10 and PMSAM10 (3.37 and 3.35 kg/m^2) are negligible in comparison.

As Figure 5 illustrates, when the impact of IRI in the usage phase is ignored, the energy consumption by the function unit of ARSMA10 (46.5 MJ/m^2) is less than that of PMSMA10 (53.5 MJ/m^2). Because of the better serviceability performance of PMSAM10 predicted by MEPDG, the extra energy consumed by vehicles on ARSMA10 pavement (209.8 MJ/m^2) is estimated to be more than that by vehicles on PMSMA10 (185.7 MJ/m^2), which leads to more overall life-cycle energy consumption of ARSMA10.

3.3 Overall sustainability performance

When the environmental impact is considered as the only performance measurement, PMSAM10 is estimated to perform better because of its lower EDC (13.5 \$/m^2). In addition, the investment cost of ARSMA10 (17.2 \$/m^2) is slightly higher than that of PMSAM10 (16.2 \$/m^2), as the asphalt rubber material had higher price than the polymer modified asphalt.

Nevertheless, when assuming that the ARSMA10 can contribute 3dB noise reduction from 76dB noise of PMSMA10, the noise reduction benefit (29.1 \$/m^2) of ARSAM10 in the 56-year analysis period is estimated to be able to offset the other cost expenditure (32.7 \$/m^2) in economic and environmental dimensions.

4 CONCLUSIONS

In this study, a comparative sustainability assessment was conducted on two asphalt mixtures: ARSMA10 and PMSAM10, by converting their corresponding economic, environmental and social impacts to the monetary values. The following points summarize the main findings of this study:

- In the 56-year analysis period, the dominating contributing factor for environmental impact is the extra GHG emissions and energy consumption of vehicles due to the pave-

ment roughness change. Overall, PMSMA10 has better environmental performance than ARSMA10, in terms of both emissions and energy.
- When the noise impact is ignored, the overall long-term performance of PMSMA10 is better with lower agency investment cost and environmental damage cost.
- The long-term accumulated noise impact is considerable assuming that ARSMA10 is able to decrease the noise by 4%, which can almost offset its higher agency investment cost and environmental damage cost.

Based on the findings of this study, further research is recommended on the cost and benefit analysis of installing noise barrier to PMSMA10 pavement to achieve the same amount of noise reduction as the ARSMA10 pavement. It is also worth to mention the land use saving due to the recycling of EOL by ARSMA10 was not considered in this study, which is a factor that will further improve the environmental performance of ARMSA10.

Furthermore, this research followed many standards: the cost benefit calculation technique was adopted from European standards; Pavement material design information and performance prediction based on Hong Kong traffic and climate data; and the data references were accessed from California and China. For simplicity, the potential effects from this multi-standard employment would be stated through compare Hong Kong with other regions (e.g. China, U.S., and Europe). For the life cycle assessment part, assuming the same decisions in the selection of construction equipment and techniques, the emission factors would largely rely on the energy source and distribution. According to Hong Kong Energy Statistics (2015), Hong Kong derives its energy supplies almost entirely from external sources. Energy is either imported directly, or produced through some intermediate transformation processes using imported fuel inputs. In the contrary, both China and U.S. can self-supply to satisfy the domestic energy demands (China Energy Group, 2014; Ratner & Glover, 2014). In this sense, when the energy demands are same, the energy costs and environmental impacts in Hong Kong would evidently larger than the other two regions. In terms of the monetary transformation of noise impacts, the cost factors are the investigated Willing-to-Pay (WTP) for reducing annoyance based on stated preference studies and quantifiable costs of health effects in Europe. It is hard to say that the WTP in Hong Kong would be higher than Europe. Nonetheless, the uniqueness of Hong Kong (e.g. hot climate, topography, dense population, high-rise buildings and intensive bus traffic) is most likely to create more serious noise impacts in Hong Kong. In general, the usage variety of standard would indeed cause some effects on the overall impacts, however, from the perspective of comparison, simultaneous increase or decrease would not greatly affect the relative results from the two comparative objects.

Finally, this study only investigated these two materials as a case study, sensitivity analysis is recommended in the future study to take into consideration of the effects of uncertainties in various variables, such as the material composition and performance, time period, system boundaries, transportation distance, and treatment of refinery allocation.

REFERENCES

American Association of State Highway and Transportation Officials (AASHTO). 2008. Mechanistic-Empirical Pavement Design Guide: A Manual of Practice. AASHTO Designation: MEPDG-1. Washington, DC.
Asia Sustainable and Alternative Energy Program (ASTAE) 2009 Greenhouse gas emission mitigation in road construction and rehabilitation.
Baus, R.L. & Stires, N.R. 2010. MEPDG Implementation University of South Carolina.
Census and Statistics Department of Hong Kong Special Administrative Region. 2015. Hong Kong Energy Statistics: 2015 Annual Report.
Chatti, K., Zaabar, I., 2012. Estimating the Effects of Pavement Condition on Vehicle Operating Costs. NCHRP Report 720. Project 01–45. Transportation Research Board, Washington, DC.
China Energy Group. 2014. Key China Energy Statistics 2014.
Chinese Life Cycle Database (CLCD), 2010. Sichuan University & IKE.
Chong, D & Wang, Y.H. 2016. Impacts of Flexible Pavement design and management decisions on life cycle energy consumption and carbon footprint. *International Journal of Life Cycle Assessment*.

Eurobitume. 2011. Life Cycle Inventory: Bitumen. Brussels, Belgium: European Bitumen Association.
European Commission, 2014. Guide to Cost-benefit Analysis of Investment Projects: Economic appraisal tool for Cohesion Policy 2014–2020. Brussels, Belgium: European Commission.
Federal Highway Administration, 2014, Tech Brief: Life Cycle Assessment of Pavements, FWHA-HIF-15-001, February.
Guangzhou Municipal Industries Ltd. 2011. Rubber asphalt industrialization feasibility report in Guangdong Province.
Hicks, R.G. 2002. Asphalt Rubber Design and Construction Guidelines [online]. Available from: https://rma.org/sites/default/files/MOD-043-Asphalt_Rubber_Design_and_Construction_Guidelines.pdf [Accessed Access Date February 28, 2017].
Häkkinen T, Mäkelä K. 1996. Environmental adaption of concrete: environmental impact of concrete and asphalt pavements. VTT, Tech. Res. Center of Finland.
ISO, 2006a. ISO 14040: Environmental Management—Life Cycle Assessment, Principles and Guidelines. International Organization for Standardization, Geneva.
ISO, 2006b. ISO 14044: Environmental Management—Life Cycle Assessment, Life Cycle Impact Assessment. International Organization for Standardization, Geneva.
Johnson, A. 2000. Best Practice Handbook on Asphalt Pavement Maintenance. University of Minnesota.
Kalembo, C., Jeihani, M., & Saka, A. 2011. Evaluation of the Impact of Pavement Roughness on Vwhicle Gas Emissions in Baltimore County. Paper presented at the TRB 2012 Annual Meeting.
Lo Presti, D. 2013. Recycled Tyre Rubber Modified Bitumens for road asphalt mixtures: A literature review. *Construction and Building Materials* 49: 863–881.
Maibach, M., Schreyer, C., Sutter, D., Essen, H.P.V., Boon, B.H., Smokers, R., Schroten, A., Doll, C., Pawlowska, B. & Bak, M. 2008. Handbook on estimation of external costs in the transport sector-Internalisation Measures and Policies for All External Cost of Transport (IMPACT). CE Delft.
Ratner, M., & Glover, C. 2014. US Energy: Overview and Key Statistics. Washington: Congressional Research Service.
Rubber Asphalt Association (RPA). 1999. Noise Reduction with Asphalt-Rubber.
Sacramento County Department of Environmental Review, & Assessment, B.B., Inc. (Scder & Abb. Inc.) 1999. Report on Status of Rubberized Asphalt Traffic Noise Reduction [online]. Available from: http://cupertino.org/inc/pdf/SR85/Exhibit%20D%20-%20Report%20on%20Status%20of%20Rubberized%20Asphalt%20Traffic%20Noise%20Reduction.pdf [Accessed Access Date February 28, 2017].
Sandberg, U., Haider, M., Conter, M., Goubert, L., Bergiers, A., Glaeser, K., Schwalbe, G., Zoller, M., Boujard, O., Hammarstrom, U., Karlsson, R., Ejsmont, J.A., Wang, T., Harvey, J.T., 2011. Rolling Resistance—Basic Information and State-of-the-Art on Measurement Methods. Deliverable #1 in MIRIAM SP 1.
Sayers, M.W., Gillespie, T.D., Paterson, W.D.O., 1986. Guidelines for Conducting and Calibrating Road Roughness Measurements. World Bank Technical Paper #46. The World Bank, Washington, DC.
Simpson, A.L. 2003. Measurement of a Rut. TRB 2003 Annual Meeting.
State of California Department of Transportation (Caltrans). 2003. Asphalt Rubber Usage Guide [online]. Available from: http://www.dot.ca.gov/hq/esc/Translab/ormt/pdf/Caltrans_Asphalt_Rubber_Usage_Guide.pdf [Accessed Access Date February 28, 2017].
The World Conservation Union 2006. The future of sustainability: Re-thinking environment and development in the twenty-first century.
Van Dam, T., Harvey, J., Muench, S., Smith, K., Snyder, M., Al-Qadi, I., Ozer, H., Meijer, J., Ram, R.V., Roesler, J.R., Kendall, A., 2015. Towards Sustainable Pavement Systems: A Reference Document. FHWA Report FHWA-HIF-15-002. Federal Highway Administration, Washington, DC.
Wang, T., Lee, I.-S., Kendall, A., Harvey, J., Lee, E.-B. & Kim, C. 2012. Life cycle energy consumption and GHG emission from pavement rehabilitation with different rolling resistance. *Journal of Cleaner Production* 33: 86–96.
Wilde, W.J., Thompson, L. & Wood, T.J. 2014. Cost-Effective Pavement Preservation Solutions for the Real World. Mankato: Center for Transportation Research and Implementation, Minnesota State University.
Yang, R., Kang, S., Ozer, H., & Al-Qadi, I.L. 2015. Environmental and economic analyses of recycled asphalt concrete mixtures based on material production and potential performance. *Resources, Conservation and Recycling*. 104: 141–151.
Yu, B., Lu, Q., & Xu, J. 2013. An improved pavement maintenance optimization methodology: Integrating LCA and LCCA. *Transportation Research Part A: Policy and Practice*. 55: 1–11.
Zhu, H., Cai, H., Yan, J. & Lu, Y. 2014. Life cycle assessment on different types of Asphalt Rubber in China. International Symposium on Pavement LCA 2014.

Environmental assessment and economic analysis of porous pavement at sidewalk

X. Chen, H. Wang & H. Najm
Rutgers University, New Jersey, USA

ABSTRACT: This study aims to evaluate economic and environmental benefits of porous pavement surface using Life-Cycle Cost Analysis (LCCA) and Life-Cycle Assessment (LCA). The research will focus on application of porous surface for sidewalk. The life-cycle inventory data were collected from literature search, online database, project records, and contractor survey. The comparison study was conducted between pervious paving systems with storm water treatment function and the conventional impermeable pavement with Best Management Practices (BMPs) in storm water management. Three pavement alternatives are considered in this study, including conventional concrete, porous asphalt, and porous concrete. The study results can help quantify the costs and benefits of using porous pavement surface at light traffic condition and select the sustainable pavement option at the same performance level.

1 INTRODUCTION

Porous pavement is one of the green infrastructures that reduce the negative environmental impact from storm water runoff in the urban area. In natural process, the rainfall will soak into the ground and gets filtered by vegetation and soil. However, in urban area with sufficient impermeable surface such as roofs and pavement, the storm water runoff releases contaminants into the nearby water bodies such as bacteria, trash, heavy metals and other pollutants making them one of the most important sources of water pollution (EPA, 2008). The major benefit of porous pavement is the storm water quality control and the potential to lower the requirement of Best Management Practices (BMPs), curbs, underdrainage system such as pipes, manhole, inlet and outlet, and gutters (FHWA, 2010). Recently, studies indicate that porous pavement can reduce the urban heat island effect under certain whether conditions (EPA, 2008). The cooling effect is mainly caused by absorbing heat for moisture evaporation. Studies also show that porous pavement can reduce tire noise by 2 to 8 decibels and increase surface friction by reducing hydroplaning risk at pavement surface (Lebens, 2012; EPA, 2008).

Porous pavement surfaces could be asphalt, concrete or interlock pavers. These pavements have been mainly used in light traffic areas such as parking lot, highway shoulder, sidewalk, and residential street and driveways (FHWA, 2010). Application of porous pavement on heavy traffic highway is also possible but less preferable due to the insufficient successful practice in the U.S. (Ting Wang, 2010). In the cold region, winter maintenance of porous pavement may be problematic. Porous pavement experiences decreased durability due to the damage caused by freeze-thaw cycle (Lebens, 2012). At the same time, it is commonly acknowledged that less dense, open-graded porous pavement has inadequate structural support compared to traditional asphalt and concrete pavement (Dave Rogge, Elizabeth A. Hunt, 1999). Other disadvantages of porous pavement include the potential clogging and raveling issues.

Although porous pavement as one of the storm water management practice has been widely accepted as an environmentally preferable, less expensive alternative to conventional pavement in light traffic areas, there is limited studies focus on the life cycle environmental impacts and the life cycle cost of porous pavement.

The objective of this paper is to conduct environmental assessment and economic analysis of porous pavements at sidewalk. The life-cycle assessment includes the definition of goal and scope, life-cycle inventory analysis, and impact assessment and interpretation. Due to the time and resource limitation, this study only focused on the environmental impact related to material acquisition and production. In addition, life-cycle cost analysis was conducted to compare different alternative pavement designs at sidewalk.

2 REVIEW OF PERTINENT STUDIES

Many states' and local Departments of Transportation (DOTs) have pavement design and construction requirements of sidewalk, trail, and bike lanes in their specifications and guidelines. Table 1 summarized examples of sidewalk design requirements and costs from case studies in Florida, Pennsylvania, and Washington. Most of previous researches about sidewalk focus on the safety, completeness, and accessibility of sidewalk, and the evaluation of satisfaction of pedestrian with different environmental elements of sidewalk. However, there are few studies on engineering design and construction of pavement at sidewalk, resulting in the lack of data of mix design and properties of sidewalk pavement material for both porous concrete and porous asphalt.

Wang, Harvey and Jones (2010) performed a preliminary life-cycle cost analysis and life-cycle assessment framework of different storm water management options based on the fully permeable pavement designs. Their study was to evaluate the present economic life-cycle cost of fully permeable pavement as compared to the conventional pavement with treatment BMP for two scenarios: 1) shoulder retrofit for high speed highway and 2) low-speed highway or parking lot. The analysis period used in their study was 40 years in which two constructions of BMPs and two to four replacements of fully permeable pavements were included.

Table 1. Summary of porous pavement design requirements and costs.

	Structure design	Drainage system	Construction cost	Source & Location
Pervious Concrete Shoulder (Highway Traffic)	Thickness: 10" Reservoir layer:12"	Two collection slotted pipes were placed to document runoff volume and water quality analyses	1.5 times conventional paving method due to skilled labor to install the concrete layer	Wanielista & Chopra, 2007 Florida
Permeable Asphalt (Light Traffic)	Thickness: 3"	N/A	$110 / sq. yard; full 8" including choker stone, Storage stone not included	Basch et.al, 2012 Philadelphia, PA
Pervious Asphalt (Minimal Traffic)	N/A	Combination of traditional curbs channel to receiving storm sewer and pervious asphalt	Cost of designing and installing slightly higher than traditional pavement due to innovative engineering	Basch et.al, 2012 Salem, OR
Permeable Concrete for Sidewalk	Length: 1500' Width: 5 1/2'	N/A	$20/ sq. yard (bid price lower than expected), $10000 additional engineering cost	Olympia, WA, 1999 Olympia, WA
Permeable Concrete for Bicycle Lanes Sidewalk	Bike lane 5' wide; Sidewalk 6' wide	Gutter slope and overflow basin to remove standing water into storm water facility	Pervious concrete lane: $140 / sq. yard, Pervious Concrete Sidewalk: $92.25 /sy	Tosomeen, 2008 Olympia, WA

The typical discount rate used is 4% and the salvage value at the end of the analysis period was assumed to be zero. The result indicates that the fully permeable pavement is more cost-effective than the scenarios of shoulder retrofit and parking lot with BMPs.

Life-Cycle Assessment (LCA) is a method to quantify the life-cycle environmental impact of a product from cradle to grave with flexibility and comprehensiveness (ISO 14044, 2006). The phases generally start from the raw material production to the end of life of a product. Vares and Pulakka (2015) preformed LCA and LCCA for comparison of the conventional and permeable pavement walkways in Finland. The cost analysis used the present value including maintenance and repair costs over a period of 30 year. The conventional pavements include asphalt and concrete surfaces with built-in sewage system with pipes and manholes for storm water treatment. The pervious surface includes partly permeable pavement structure of concrete and fully permeable pavement structure of permeable concrete and porous asphalt. The report also provides a scenario for operation and maintenance periods of surfaces. Sensitivity analysis was conducted with 25% higher and lower cost in case of permeable structures. The result shows the life cycle costs of permeable structure pavement are more cost-effective, which is also supported by the sensitivity analysis. The life cycle assessment of pavement structures shows that pavement surface has the highest contribution of the GHG emission compared to other sub-bases. All permeable pavements have less carbon footprint than conventional pavement. The reason is that the two conventional pavement use drainage manholes with concrete tube and reinforcement have the biggest contribute to the difference of carbon footprint between permeable pavement and conventional pavement.

3 POROUS PAVEMENT DESIGNS

Three pavement alternatives are considered in this study, including conventional concrete, porous asphalt, and porous concrete design 1 and 2. Table 2 shows the structure design of conventional concrete sidewalk. For conventional sidewalk, underdrain system with pipes is needed. According to the New Jersey Standard Roadway Construction 2007, the conventional concrete sidewalk required 4-inch thick cement concrete surface with a slope 1/4" per foot. Detail information about high-density polyethylene drainage pipe including sizes (like inside and outside diameters), thickness, and weight were found from a manufacture's product note.

Porous paving sidewalk designs used in the study include porous concrete and asphalt surface course. The reservoir layer of uniformly graded coarse aggregate is placed on soil subgrade with certain drainage requirement. One of the important benefits of porous pavement is the potential to reduce or eliminate the use of underdrain pipes. With the reservoir layer, more rainfall amounts can be temporally stored under the pavement until fully infiltrated into the ground. The thickness of storage layer depends upon required runoff storage volume and the typical rainfall amount of the area. The design of porous pavement follows New Jersey Storm Water Best Management Practices Manual—Standard for pervious paving system. Table 3 shows the structure design of porous paving sidewalk.

The porous asphalt mixture designs were obtained from the technical report of Porous Asphalt Pavements (Wisconsin asphalt pavement association, 2015). The porous concrete mixture design 1 and design 2 were obtained from lab experiments currently conducted at Rutgers University. The conventional concrete mixture design is from a sidewalk project of New Jersey Department of Transportation. The mix designs of different materials are shown in Table 4.

Table 2. Conventional concrete sidewalk structure design.

Conventional concrete sidewalk structure		Layer thickness (in)	Quantity (ton/mile)
Surface Layer	Cement concrete	4	41.85
Sub grade	Compacted earth	N/A	N/A
Drainage Pipes	4 (in.) in diameter	N/A	1.484

Table 3. Porous pavement sidewalk structure design.

Porous pavement sidewalk structure		Thickness (in)	Area (sq. in)	Quantity (ton/mile)
Surface	Porous Concrete Surface	4	–	38.00
	Porous Asphalt Surface	4	–	35.2
Bedding	Choker Course—AASHTO No. 57	1	–	7.10
Reservoir Layer	Coarse Aggregate—AASHTO No. 2	12	–	88.00
Filter Layer	Non-Woven Geotextile	–	253440	–
Subgrade	Un-compacted Subgrade	–	–	–

Table 4. Mixture design of conventional concrete, porous concrete and porous asphalt.

Material (lbs/cu.yd)	Conventional concrete	Porous concrete design 1	Porous concrete design 2	Porous asphalt
Cement	405	635	465	N/A
Slag	175	N/A	155	N/A
Fine Agg./sand	1314	224	N/A	N/A
Coarse Agg.	1850	2430	2500	N/A
Water	283	209	165	N/A
Asphalt	N/A	N/A	N/A	227
Fine Agg.	N/A	N/A	N/A	635
Coarse Agg.	N/A	N/A	N/A	1904

4 LCA CASE STUDY

4.1 LCA goal and scope

The LCA attempts to quantify the environmental impacts of three sidewalk pavements using a cradle-to-grave approach. The three pavements that are being studied include porous asphalt, porous concrete, and traditional concrete pavement. The functional unit is defined as one-mile sidewalk with four feet width. The system boundary covers material, initial construction, and maintenance stages. Due to the time and resource limitation, the construction and end-of-life stage is not considered in this study. Future study will include inventory of construction stage with detail information about construction equipment and working productivity.

The analysis period used to inventory the inputs and outputs is set to be 40 years in order to create a fair comparison between alternatives with different design life. Since there is limited data in the literature of the life span of sidewalk pavements, an effort was made in this study to conduct sensitivity analysis of pavement life with an assumption that reconstruction will take place at the end-of-life of pavement. The minor pavement repairs are neglected in this study.

The impact assessments include energy consumption and Greenhouse Gas (GHG) emission from the upstream and direct combustion processes of the activities. The greenhouse gases considered in this study include Carbon Dioxide (CO_2), Methane (CH_4) and Nitrous Oxide (N_2O), which are converted to CO_2 equivalent using the Global Warming Potential (GWP) in 100-year time horizon (Levasseur et al., 2009).

4.2 Life-cycle inventory data

The inventory data of energy consumption and GHG emission of material and manufacturing process were collected from published article and government reports. Upstream or indirect energy consumption and green house gas emission associated with the processes of different stages are considered. The combustion (direct) energy and GHG Emissions are due to the straight use of fuels and electricity; while the upstream (indirect) components are generated from processing fuel that is consumed during various processes from material extraction to production.

To consider the upstream values, GREET 2013 model was used to extract the unit upstream energy consumption and GHG emission. GREET 2013 model is a life-cycle modeling tool to evaluate the impact of fuel use including all fuel production processes from oil exploration to fuel use (from well to wheels) (Wang, 1999). The equation shown below is developed to calculate the upstream energy consumption and GHG emission (Wang et al. 2016).

$$UEE = \sum_{i=1}^{n} CE \cdot PE_i \cdot UEE_i \quad (1)$$

where, UEE = Upstream energy consumption (BTU/ton) or emission (g/ton); CE = Combustion energy (MMBTU/ton); and PE_i = Percent of the i_{th} type of energy in the energy matrix.

The life-cycle inventory data for the components of asphalt and concrete pavements and plant production can be found from the literature (Yang, 2014; Marceau, et al., 2007; Marceau, et al., 2006; Marceau, et al., 2003; Prowell et al., 2014; Argonne National Laboratory, 2007). Since GREET model does not provide emission factor for HDPE (High Density Polyethylene) material for underdrain pipes, the emission factors for polypropylene were used as an estimation (Worrell et al, 2000; GREET, 2013).

4.3 *LCA Comparisons between different pavement materials*

In order to reach the same performance level of storm water runoff quality control, instead of collecting the runoff and discharge into the nearby water bodies directly, the impermeable sidewalk pavements with BMPs are comparable with porous pavement. Although BMPs is considered in this study, due to the lack of detail data of the specific BMPs design, construction equipments and productivities, it is neglected in the LCA comparison part.

Table 5 presents the results of energy consumption and GHG emission of four alternatives for each structure layer of one-mile sidewalk pavement structure. The results cover combustion and upstream values associated with the material related activities of raw material acquisition and plant production. Regardless of maintenance activities, the results indicate that porous concrete design 1 requires more energy and releases more CO_2 eq. than conventional concrete pavement. It is noted that porous paving material that has less fine aggregate to achieve the higher rate of permeability usually requires more binding agent of cement (NJDEP, 2004). Moreover, small percentages of Portland cement and asphalt binder of the total paving material have the most critical contribution in the energy consumption and GHG emission in material acquisition and production phases. These reasons explain the difference of results between porous concrete and conventional concrete pavements. However, the use of slag cement as partial substitution of Portland cement considerably reduces energy consumption and GHG emission. For Porous Concrete design 2, 25% Portland cement is

Table 5. Energy Consumption (MJ) of one-mile sidewalk pavement structure.

Energy consumption (MJ/mile)		Porous concrete design 1	Porous concrete design 2	Porous HMA	Conventional concrete
1)	Surface	51609	39039	17974	36414
2)	Bedding	441	441	441	–
3)	Reservoir Layer	5469	5469	5469	–
4)	Subgrade	–	–	–	–
5)	Drainage Pipes	–	–	–	17137* (include feedstock energy)
Total		57519	44950	23885	53551

replaced by slag cement, significantly reducing the energy consumption comparing to porous concrete design 1.

The results also show that bedding and reservoir layer consist of aggregate has noticeably smaller impact on energy consumption and GHG emission, while drainage pipes, one of the plastic materials, has significant contribution in energy consumption of conventional concrete pavement. From Tables 5 and 6, it is obvious that porous asphalt pavement has the lowest energy consumption and GHG emission compared to the other two alternatives using cement concrete surface. The unit energy consumption is 1551 MJ/ton for cement but only 789 MJ/ton for asphalt binder.

Figure 1 presents the percentage distribution of energy consumption and CO_2 eq. emission based on the initial construction of one-mile sidewalk. The results cover plant production and raw material used in pavement construction including cement, asphalt, aggregate, slag, and underdrain pipes. For porous and conventional pavements, the material related processes have the greatest contribution to energy consumption and GHG emission. It is apparent that, compared to cement concrete production, manufacturing of porous asphalt pavement consumes more energy and release the greater amount of CO_2 eq. as well. The difference is mainly due to the amount of heating energy used to dry aggregate and to mix asphalt binder with aggregate at certain discharge temperature, usually around 160°C. With higher moisture content in aggregate, more energy is required to remove moisture from it before mixing. On the other hand, cement concrete production mixes cement, aggregate, and water without extra heating energy.

4.4 LCA Comparison with different life span ratios

The literature data show that the life span of porous asphalt and porous concrete design 1 is from 15 years to 40 years without major repair or replacement. However, the estimations cannot be specific without convincing evidence or laboratory testing. In order to avoid the arbitrary selection of life span of sidewalk pavement alternatives, sensitivity analysis of pavement life span was considered in this study assuming pavements need reconstructions after reaching the end of life during the study period of 40 years. Tables 5 and 6 show the environmental burdens of constructing one-mile porous asphalt sidewalk is smaller than the ones using porous concrete and conventional concrete. If the life span of porous asphalt is shorter than concrete pavements, the environmental benefits of using porous asphalt as sidewalk pavement will be offset by the need for more frequent maintenance of reconstruction actives.

Figure 2 (a) shows the breakeven point of pavement life ratio between porous asphalt and porous concrete is around 0.56 when the life of porous concrete is assumed to be 25 years. It means that the life span of porous asphalt is 12.5 yr and the porous concrete is 25 years. In this case, porous asphalt with three reconstructions has the same energy consumption with porous concrete with one reconstruction activity in the 40-year analysis period. When the life ratio is smaller than 0.5, porous asphalt will require more energy consumption than porous concrete; when the life ratio is bigger than 0.5, porous asphalt will have less energy consumption than porous concrete. On the other hand, Figure 2(b) shows that the breakeven point of life ratio between porous asphalt to porous concrete is 0.36 for CO_2 eq. emission.

Table 6. CO_2 eq. Emission (Kg) of one-mile sidewalk pavement structure.

CO_2 eq. Emission (Kg/mile)	Porous concrete design 1	Porous concrete design 2	Porous HMA	Conventional concrete
1) Surface	7614	5595	1354	4956
2) Bedding	20	20	20	–
3) Reservoir Layer	243	243	243	–
4) Subgrade	–	–	–	–
5) Drainage Pipes	–	–	–	191
Total	7876	5857	1617	5147

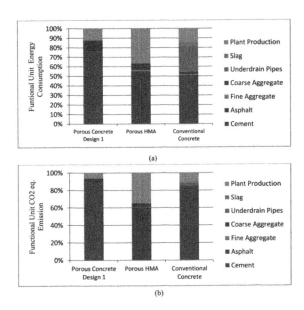

Figure 1. Percentage distributions of (a) energy consumption and (b) CO_2 eq. emission for porous concrete, porous HMA, and conventional concrete.

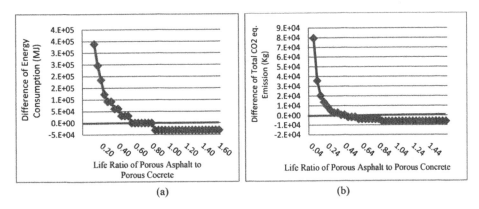

Figure 2. Breakeven points of pavement life ratios between porous asphalt and porous concrete for (a) energy consumption and (b) CO_2 eq. emission (the life of porous concrete is assumed to be 25 years).

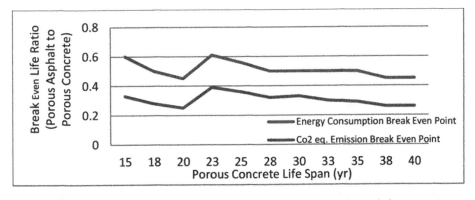

Figure 3. Breakeven point of life ration between porous asphalt and porous asphalt pavements.

Figure 3 shows the breakeven point of life ratios when the life span of porous concrete pavement varies from 15 to 50 years. It was found that the breakeven point of life ratio was kept relatively constant regardless of the selection of analysis period and the assumption of life span of porous concrete pavement.

5 LIFE CYCLE COST ANALYSIS

In order to be consistent with LCA results, the analysis period of LCCA conducted in this study is set to be 40 years. Three comparison alternatives are porous concrete design 1, porous asphalt, and conventional concrete sidewalk. In order to reach the same performance level of storm water runoff quality control, instead of collecting and discharging the runoff into the nearby water bodies directly, the impermeable sidewalk pavements with BMPs are comparable with porous pavement with reservoir layer. For porous pavements and BMPs, annual regular maintenances are required (EPA, 1999).

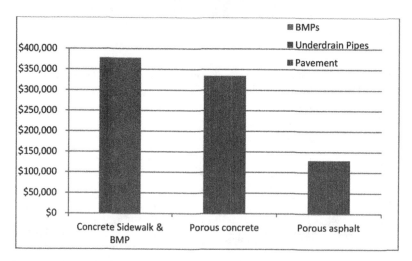

Figure 4. Initial construction cost ($/mile) of three sidewalk pavements.

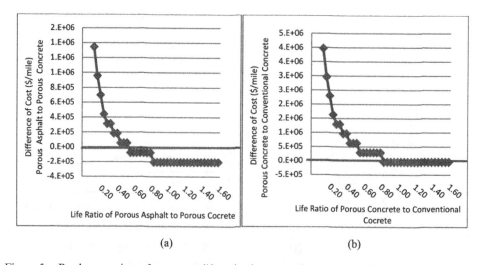

(a)   (b)

Figure 5. Breakeven points of pavement life ratios between (a) porous asphalt and porous concrete and (b) porous concrete to conventional concrete for cost ($/mile) in 40 yr analysis period, the life of porous concrete is assumed to be 25 years.

The cost data is primarily obtained from the recent three years biding report of New Jersey DOT (NJDOT, 2013, 2014, and 2015). The construction and maintenance cost of BMPs is from Preliminary Data Summary of Urban Storm Water Best Management Practices (EPA, 1999), which include construction cost, design and contingency cost, and annual maintenance costs. The report also covers the cost for annual maintenance activates of porous pavement. The discount rate is 3% which is the average annual federal inflation rate at 2016. In this study, the salvage value at the end of 40-year period is assumed to be zero due to uncertain life spans of the four alternatives.

Figure 4 shows the initial construction cost of three alternatives. Compared to asphalt pavement, conventional concrete pavements have the highest costs in the 40-year analysis period. It can be observed that underdrain pipes for conventional sidewalk pavement have significant contribution in the initial construction cost.

It is expected that the uncertainty of life spans of pavement alternatives will critically affect the result of life-cycle cost analysis. The life-cycle cost of regular maintenance of BMPs and porous pavement were included. Since the cost of initial construction of each alternative is already known, the breakeven points of pavement life ratio can be calculated using the method discussed above. As shown in Figure 5, for the life-cycle cost, the breakeven point of life ratio is 0.80 between porous concrete and conventional concrete, and 0.50 between porous asphalt and porous concrete.

6 CONCLUSIONS

In the LCA comparison of different paving material of the initial construction phase, the results illustrate that mix designs of the pavement surface course greatly influence the LCA comparison results. Porous concrete pavement with slag cement requires less energy than conventional concrete; porous asphalt pavement has the lowest energy consumption and GHG emission compared to the other alternatives using cement concrete surface. The largest environmental burdens are from the raw material and manufacturing processes of Portland cement and asphalt binder.

The sensitivity analysis of pavement performance indicates that pavement life ratio can significantly affect the comparison results between different pavement types due to the reconstruction activities for the one with shorter life span. It was also found that the breakeven life ratio was kept relatively constant regardless of the selection of analysis period and the assumption of life span of porous concrete pavement.

For life cycle cost analysis, pavement life ratio and maintenance frequency are critical factors in the results. The result also shows that underdrain pipes for conventional sidewalk pavement have significant contribution in the initial construction cost. Preliminary results from this study indicates that porous pavement is not always environmentally preferable and less expensive option when we use the life-cycle approach to quantify cost and environmental impact of energy consumption and GHG emission.

For a more complete life cycle assessment, the construction phase cannot be neglected. Detail information about sidewalk construction including equipment type and productivity should be collected and added to the life cycle inventory. It is also needed to consider the LCA of construction and maintenance of BMPs which is included in the life cycle cost analysis. At the same time, maintenance activity specifically for sidewalk pavement is not yet well established and their costs need to be considered in the LCCA. This study is in progress and future work will address these limitations.

REFERENCES

Argonne National Laboratory. (2007). GREET 2.7 vehicle cycle model. Argonne National Laboratory.
Basch et.al. (2012). *Roadmap for Pervious Pavement in NYC*. NYC DOT.
California Stormwater Quality Association. (2003). *California Stormwater BMP Handbook-New Development and Redevelopment*. California Stormwater Quality Association.

Chapter 3 The propylene Chain. (n.d.). Retrieved from http://www1.eere.energy.gov/manufacturing/resources/chemicals/pdfs/profile_chap3.pdf

City of Olympia, WA. (n.d.). *North Street Reconstruction Project Summary of Porous Concrete Sidewalk.* Retrieved from file:///E:/928/North_Street_Reconstruction_Project_Porous_Concrete_Summar.pdf

Dave Rogge, Elizabeth A. Hunt. (1999). *Development of Maintenance Practices for Oregon F-Mix.* Washington, D.C: Federal Highway Administration.

EPA. (1999). *Preliminary Data Summary of Urban Storm Water Best Management Practices.*

EPA. (2008). *Reducing Urban Heat Islands: Compendium of Strategies.* EPA.

Ernst Worrell, Dian Phylipsen, Dan Einstein, Nathan Martin. (2000). *Energy Use and Energy Intensity of the U.S. Chemical Industry.* University of California Berkeley.

FHWA. (2010). *Stormwater Best Management Practices in an Ultra-Urban Setting, Porous Pavement Fact Sheet.* FHWA.

Gerald Huber. (2000). *NCHRP Synthesis of Highway Practice 284: Performance Survey on Open-Graded Friction Course Mixes.* Transportation Research Board.

Hui Li. (2012). *Evaluation of Cool Pavement Strategies for Heat Island Mitigation.* Davis: Institute of Transportation Studies, University of California, Davis.

International Organization for Standardization. (2006). *ISO 14044, Environmental management—life cycle assessment—requirements and guidelines.* Geneva, Switzerland: International Organization for Standardization.

Lebens, M. (2012). *Porous Asphalt Pavement Performance in Cold Regions.* Maplewood: Minnesota Department of Transportation.

Marceau ML, Nisbet MA, Vangeem MG. (2007). Life cycle inventory of Portland cement concrete (No. SN2011). Skokie, Illinois: Portland Cement Association.

Marceau ML, VanGeem MG. (2003). Life cycle inventory of slag cement manufacturing process, CTL Project Number 312012. Skokie, IL: Construction Technology Laboratories.

Marceau ML, Nisbet MA, Vangeem MG. (2006). Life Cycle Inventory of Portland Cement Manufacture (No. SN2095b). Skokie, Illinois: Portland Cement Association.

Marty Wanielista, Manoj Chopra. (2007). Performance Assessment of a Pervious Concrete *Pavement Used as a shoulder for a interstate rest area parking lot.* Florida Department of Transportation.

Muench, S. T. (2010). Roadway construction sustainability impacts review of life-cycle assessments, Transportation Research Record, No. 2151. Transportation Research Board, 36–45.

NJ DOT. (2013). *New Jersey Department of Transportation Capital Contracts Bid Price History 2013.* New Jersey Department of Transportation.

NJ DOT. (2014). *New Jersey Department of Transportation Capital Contracts Bid Price History 2014.* New Jersey Department of Transportation.

NJ DOT. (2015). *New Jersey Department of Transportation Capital Contracts Bid Price History 2015.* New Jersey Department of Transportation. (2015). *Porous Asphalt Pavements.* Wisconsin asphalt pavement association.

Product Notes. (1999). Retrieved from http://www.ads-pipe.com/pdf/en/Product_Note_3.108_N-12_Specification_for_Leachate.pdf

Sandra A. Blick, Fred Kelly, Joseph J. Skupien. (2004). *New Jersey Stormwater Best Management Practices Manual 2004.* New Jersey Department of Environmental Protection.

Sirje Vares, Sakari Pulakka. (2015). *LCA and LCCA for conventional and permeable pavement walkways.* VTT Techincial Research Centre of Finland.

Street reconstruction with pervious pavement—North Gay Avenue, Portland, Oregon. (2005). Retrieved from The City of Portlad, Oregon: https://www.portlandoregon.gov/bes/article/77074

Ting Wang. (2010). *A Framework for Life-Cycle Cost Analyses and Environmental Life-Cycle Assessments for Fully Permeable Pavements.* Davis, California: Institute of Transportation Studies, University of California, Davis.

Tosomeen, C. (2008). *Pervious Concrete Bicycle Lanes—Roadway Stormwater Mitigation within the Right-of-Way.* City of Olympia, Washington.

Wang MQ. (1999). *GREET 1.5—transportation fuel-cycle model, volume 1: methodology, development, use, and results.* Argonne National Laboratory.

Yang RY. (2014). Development of a pavement life cycle assessment tool utilizing regional data and introducing an asphalt binder model, Master Thesis. University of Illinois at Urbana-Champaign.

LCCA for silent surfaces

F.G. Praticò
University Mediterranea of Reggio Calabria, Reggio Calabria, Italy

ABSTRACT: Traffic noise (and in more detail rolling noise) affects the health of thousands of citizens. Consequently a number of silent technologies (quiet surfaces) are available, with different weaknesses, noise performance, and expected life. This notwithstanding, addressing noise impacts from a Life Cycle Cost Analysis (LCCA) perspective is still an issue. Indeed, characterization factors, inventory data, underlying methodology are neither clear nor well established. Complexity increases when not only environmental costs but also agency and/or user costs are considered as a whole. This implies that the assessment of the "overall" appropriateness of a "silent technology" may emerge as uncertain. In the light of the above, the study described in this paper focuses on setting up a methodology for the synergistic consideration of agency, user and environmental/noise-related costs. The methodology set up was applied to a given case study. Results can benefit both researchers and practitioners.

1 INTRODUCTION

1.1 *Overview*

According to the standard ISO 15686-5, "Life Cycle Costing (LCC) is a valuable technique which is used for predicting and assessing the cost performance of constructed assets".

LCC can be defined as "the cost of an asset or its part throughout its life cycle, while fulfilling the performance requirements". LCC includes construction, operation, maintenance, end of life, environment (including energy and utilities) costs (ISO 15686-5). The following definitions apply: i) External Costs: costs associated with an asset which are not necessarily reflected in the transaction costs between provider and consumer (e.g. business staffing and productivity, user costs, etc). Collectively these elements are referred to as externalities. ii) Environmental cost impacts: …costs (or savings via rebates) to LCC depending on the effects.. on the environment. Examples could include cost premiums for the use of non-renewable resources or for green house gas emissions. Where these costs are external to the constructed asset they will form part of a Whole Life Cost (WLC) analysis. iii) WLC elements: typically the difference to LCC is that the elements of WLC include a wider range of externalities or non-construction costs, such as finance costs, business costs and income streams.

LCC (or LCCA, Life Cycle Cost Analysis) entails the assessment of the estimated service life (see BS ISO 15686-1). To this end, the standard ISO 15686-1 focuses on service and design life, especially for buildings. For environmental impacts and their inclusion in LCCA, the standard BS ISO 15686-6 clarifies main environment-related components. LCC algorithms build on Present Values of costs (PV). PV is the current worth of a future sum of money given a specified rate of return and inflation. Future cash flows are discounted, and the higher the discount rate, the lower the present value of the future cash flows.

Note that unlike the nominal rate, the real interest rate takes the inflation rate into account.

Key issues in LCCA include the following: i) the same end-of-life concept implies the introduction of a salvage (or residual, remaining service life) value. This notwithstanding, there are different methods to derive these values (ACPA, 2011; West et al, 2012). Furthermore, from an epistemic and mathematic standpoint, several issues emerge (Weeds, 2001; Praticò and Casciano, 2015); ii) The discount rate (interest and inflation rates) can be assesses, in the U.S., through the OMB circular A-94, Appendix C. Anyhow, it is extremely variable and results may depend on the selected options; iii) the estimate of user costs (UC, delay costs, crash costs, vehicle operating costs) is debated, uncertain, and difficult (even if software is available, see Caltrans, 2011). Furthermore, there are doubts about merging agency and user costs in LCCA analyses (current FHWA policy does not recommend this practice) and UC value is usually higher or comparable with respect to agency costs (AC, see Delwar and Papagiannakis, 2001); iii) functional, structural, premium properties of pavements have to be considered in terms of agency costs, including their variability of time for the given pavement type (macrotexture modeling,,,,,, a study on the relationship......; iv) the assessment of the environmental impacts is included in LCCA, but the synergistic consideration of environmental and "traditional" (i.e., UC, AC) impacts poses theoretical and practical issues (Praticò et al., 2010; Praticò et al., 2012). This consideration pertains also to the simultaneous assessment of different environmental impacts (e.g., noise and carbon footprint of a pavement); v) several authors and road agencies (Mallela and Sadasivam, 2011) observe that the monetized impacts of user costs include travel delay costs, vehicle operating costs, crash costs, emissions costs, impacts of nearby projects. At the same time, they classify under other, non-monetary impacts the following items: noise, business impacts, inconvenience to local community. This method implies that environmental impacts are split into two different classes: monetized and non-monetary. Although necessary from a practical standpoint, from a logical standpoint, this division is quite unsatisfactory because: i) the same emission costs were once non-monetary and are now monetized through a given methodology; ii) this brings to variable results, based on CO2e unitary cost.

For noise-related impacts, Cucurachi and Heijungs, 2014 provided characterization factors to allow for the quantification of noise impacts on human health in the LCA framework. Garrain et al, 2009 proposed a specific indicator as the best unit to measure the negative impacts of noise upon human health. Doka, G. (2009) provided practical figures to estimate road, rail and airplane noise damages for Life Cycle. Buckland and Muirhead, 2014 analyzed the expected reduction in property value for an increase of 1dB(A). They found reductions in the range 0.2–2.2%. In terms of monetary evaluation of road traffic noise, values which range from zero (when the 24-hour A-weighted equivalent outdoor sound level is about 50 dB) to about 1040 €/year/exposed person (for 72dB) were proposed (see Vanchieri, 2014).

Overall, despite the simplicity of the logical bases for the derivation of noise costs (the higher the distance from a given threshold, around 50–55dB, the higher the cost), the following specific criticalities may be highlighted: i) order of magnitude of the unitary costs (per dB, per person, per year); ii) linear versus non-linear dependency on dB; iii) variability over time of pavement noise performance (Licitra, 2014).

1.2 Objectives

Based on the above, the study described in this paper aims at:

i. Setting up a series of algorithms to carry out comprehensive LCCAs on transportation infrastructures;
ii. In more detail, considering the effect of noise-related performance of pavements.

2 METHODOLOGY

2.1 The template file

The methodology set up and implemented builds on the estimate of the present value of each agency, user, and environmental cost.

The analysis of costs over time is carried out in terms of Present Values (PV). Based on Fisher equation (Fisher, 1930), the following expression is used in PV estimates:

$$R = \left(\frac{1+INF}{1+INT}\right) \tag{1}$$

where INF is the inflation rate (e.g., 0.04) and INT is the nominal interest rate (e.g., 0.08). Note that based on 1-st order Taylor expansion, it is possible to demonstrate the relationship between the real interest rate (r) and R:

$$\frac{1}{1+r} \approx R = \left(\frac{1+INF}{1+INT}\right) \tag{2}$$

Importantly, for the above quantities, the Fisher equation is as follows:

$$INT \approx r + INF \tag{3}$$

For User Costs (UC), the following primary contributions can be taken into account: i) Delays (D) due to Rehabilitations (REH); ii) delays due to Resurfacing (RES); Vehicle Operating Costs (VOC). In the following two equations E is the expected life of rehabilitation and O is the expected life of successive resurfacing, Cs refer to costs.

$$PV_{UC\_D\_REH} = \left(1 + \frac{R^E}{1-R^E}\right) \cdot C_{UC\_D\_REH} \tag{4}$$

$$PV_{UC\_D\_RES} = \left(\frac{\sum_{i=1}^{i=INT\left(\frac{E}{O}-1\right)} R^{i \cdot O}}{1-R^E}\right) \cdot C_{UC\_D\_RES} \tag{5}$$

Note that i represents the index of summation, 1 is the lower bound of summation, and the upper bound of summation is given by the integer derived based on E and O. The index, i, is incremented by 1 for each successive term, stopping when i = N, where

$$N = INT\left(\frac{E}{O} - 1\right) \tag{6}$$

Vehicle Operating Costs (VOC) are given by:

$$PV_{UC\_VOC} = \left(\frac{R}{1-R}\right) \cdot C_{VOC} \tag{7}$$

It results:

$$PV_{UC} = PV_{UC\_D\_REH} + PV_{UC\_D\_RES} + PV_{UC\_VOC} \tag{8}$$

For Agency Costs (AC), the following primary parts are considered:

$$PV_{AC\_REH} = \left(1 + \frac{R^E}{1-R^E}\right) \cdot C_{REH} \tag{9}$$

$$PV_{AC\_RES} = \left(\frac{\sum_{i=1}^{i=INT\left(\frac{E}{O}-1\right)} R^{i \cdot O}}{1-R^E}\right) \cdot C_{RES} \tag{10}$$

From above, it derives:

$$PV_{AC} = PV_{AC\_REH} + PV_{AC\_RES} \tag{11}$$

For environmental costs (EX), the following main quantities are considered: i) CO2e-related impact in rehabilitation and resurfacing (which include production/quarrying-related, recycling-related, and landfill-related impacts); pollution-related impacts; iv) noise-related impact; v) positive impact from trees planting:

$$PV_{EX\_CO2e\_REH} = \left(1 + \frac{R^E}{1-R^E}\right) \cdot C_{EX\_CO2e\_REH} \tag{12}$$

$$PV_{EX\_CO2e\_RES} = \left(\frac{\sum_{i=1}^{i=NINT\left(\frac{E}{O}-1\right)} R^{i \cdot O}}{1-R^E}\right) \cdot C_{EX\_CO2e\_RES} \tag{13}$$

$$PV_{EX\_POLL} = \left(\frac{R}{1-R}\right) \cdot C_{EX\_POLL} \tag{14}$$

$$PV_{EX\_TREES} = \left(\frac{R}{1-R}\right) \cdot C_{EX\_TREES} \tag{15}$$

$$PV_{EX\_Noise} = \left(\frac{R}{1-R}\right) \cdot C_{EX\_Noise} \tag{16}$$

$$C_{EX\_Noise} = \left(\frac{\alpha}{(USL-L_{dn})^\beta} - \frac{\alpha}{USL^\beta}\right) \cdot L_{dn} \cdot (V \cdot L) \tag{17}$$

where α and β are coefficients to calibrate, L_{dn} is a sound pressure level, USL is the corresponding upper specification limit, V is the number of vehicles per year, L is the length of the road stretch.

It follows:

$$PV_{EX} = PV_{EX\_CO2e\_REH} + PV_{EX\_CO2e\_RES} + PV_{EX\_Noise} + PV_{EX\_POLL} + PV_{EX\_Trees} \tag{18}$$

For the environmental costs, note that the quantification of the CO2e emissions is a well-established practice (European Union emissions trading system, see Sijm et al, 2006). On the other hand, the fluctuation of the carbon price is a matter of fact and the price of carbon dioxide (€ per tonne) is highly variable. This may increase the variability and unreliability of the results in terms of EX. To this end, the following procedure is herein proposed (for each i-th project or solution among the k ones). The first step is to derive the indicator which refers to the critical relationship between "traditional" and environmental costs:

$$v = \min_{i=1,2,k} \frac{PV_{ACi} + PV_{UCi}}{PV_{EXi}} \tag{19}$$

The second step is to operate in this vector space of the "internalized" factors to get a linear magnification:

$$PV_{EXi}^* = v \cdot PV_{EXi} \tag{20}$$

In pursuit of the estimation of the overall gain for each solution, note that Agency Costs (AC), User Costs (UC) and Environmental costs (EX) contribute to the overall present value, PV given as:

$$PV = PV_{AC} + PV_{UC} + PV_{EX}^* \tag{21}$$

Based on the above the comparison between two competing projects/alternatives 1 and 2 can be carried out based on the lowest present value, i.e., based on gains, G:

$$G = PV_1 - PV_2 = \left(PV_{AC1} - PV_{AC2}\right) + \left(PV_{UC1} - PV_{UC2}\right) + \left(PV_{EX1}^* - PV_{EX2}^*\right) \tag{22}$$

Note that equations 1–4, 8, 9, 11, 12, 14–16, 18–22 build on (Fisher, 1930; Pratico' et al, 2011; Pratico', 2017). In contrast, equations (5–7) and 10, 13, 17 are new. Note that in the above algorithms other parts may be added, based on specific issues.

3 DESIGN OF EXPERIMENTS AND RESULTS

Tables 1 and 2 and Figures 1–5 illustrate main inputs and outputs. Table 1 focuses on the options considered. Three different friction courses were considered: dense-graded (DGFC, 30 mm), open-graded (PA, 50 mm), and recycled open graded (RPA, 50 mm, 55% of reclaimed porous asphalt). The underlying pavement structure was the same: binder course (BIC, 40 mm), base course (unbound, BA, 200 mm), and subgrade (MR = 310 MPa). A total surface area of 10000 m^2 was considered. Note that the structure of the equations above basically presents two terms: a terms which allows deriving the impact of costs over time (the one containing R) and a coefficient (e.g., $c_{UC\_D\_REH}$). Table 2 illustrates the coefficients derived (see equations above, Euro), which are the main inputs used in the case-study.

Figures below illustrate the main results. Figure 1 illustrates how the present value of agency costs (y-axis, M€) varies over pavement life (x-axis, years), for the three selected options (see table 1). Note that Porous Asphalts (PA) have costs which are higher than the ones of DGFCs and a durability which is lower. It turns out that present values for PAs are higher than the ones for DGFCs. In contrast, due to the lower initial costs, recycled porous asphalts yield a present value which is slightly lower that the one exhibited by PAs (see Figure 2).

By referring to Figure 2, note that the carbon footprint of PAs is higher than the one of DGFCs, due to the different durability. In contrast, when an appreciable percentage of recycling is considered (about 50%, RPA) there is a gain in terms of CO2e (a lower quantity of

Table 1. Factorial plan of experiments.

	DGFC		PA		RPA	
Pavement structure	DGFC	30	PA	50	RPA	50
	BIC	40	BIC	40	BIC	40
	BA	200	BA	200	BA	200

Table 2. Main inputs and outputs.

Coefficients (rounded)	DGFC	PA	RPA
UC_D_REH	59000	59000	59000
UC_D_RES	29000	29000	29000
UC_VOC	70000	70000	70000
AC_REH	124000	175000	128000
AC_RES	55000	106000	78000
EX_CO2e_REH	1147000	1855000	1476000
EX_CO2e_RES	491000	708000	329000
EX_Noise	557000	392000	392000
EX-POLL	61000	61000	61000
Main outputs		PVAC	
		PVUC	
		PVEX	

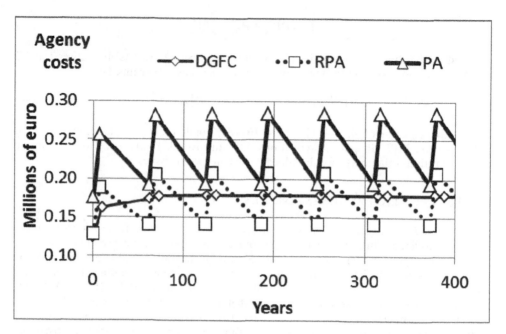

Figure 1. Present value of agency costs (y-axis, M€) over pavement life (x-axis, years).

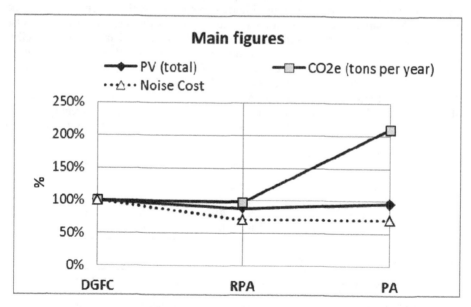

Figure 2. Total present value, noise impact, carbon footprint ad a percentage of the reference option, for the three options (DGFC, RPA, PA).

bitumen and aggregates is required). This gain balances the lower durability of PAs/RPAs with respect to DGFCs. It turns out that the RPA achieve the best result in terms of CO2e (see also Figure 2).

For the noise, note that under the hypothesis of having a similar sound pressure level for PAs and RPAs, the financial impact of noise is higher when DGFCs are used instead of porous asphalts (PAs or RPAs).

Figure 3 summarizes the three classes of costs (AC, UC, EX) under the three abovementioned hypotheses (DGFC, RPA, PA). Note that the present value of environmental costs

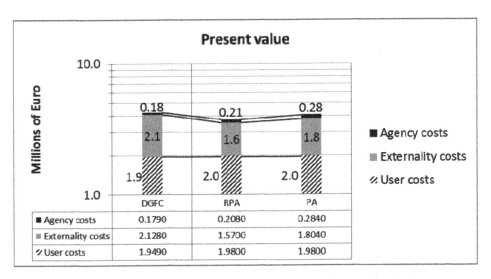

Figure 3. Present value of agency, user, environmental costs (millions of euro) for the three options (DGFC, RPA, PA).

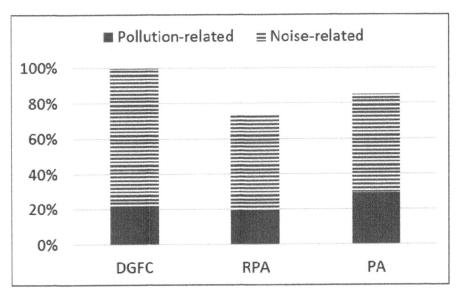

Figure 4. Noise-related versus pollution-related impacts (percentages by the total environmental costs for DGFC).

includes operation—and rehabilitation-related impacts. It takes into account pollution, noise, and carbon footprint. It results higher when DGFCs are used (Figure 3). Positive outcomes derive from recycling (e.g., RPAs) and planting trees (herein not considered). Note that user costs have a magnitude which is about ten times the one of agency costs. This is quite consistent with literature (Delwar and Papagiannakis, 2001). Finally, note that PA agency costs are higher the ones of RPA (due to material cost and durability). In turn, RPA agency costs are higher than DGFC agency costs, due to their lower durability.

Noise relative impact (with respect to the overall environmental impact of DGFCs) is depicted in Figure 4. Note that noise-related impact ranges from 56% to 79%, while pollution-related impact accounts for 18–29%.

Figure 5 focuses not only on present values but also on carbon footprint and noise cost, which are sensitive environmental targets. Gains are used, i.e., differences of PVs, with respect to PA (equation 22). It may be observed that RPA and DGFC perform better than PA when agency costs or carbon foot print are considered. In contrast, by referring to the overall Environmental impact (EX) and to the overall PV, RPA perform better than the remaining ones, while DGFC perform worse than the remaining ones.

By referring to waste balance (Figure 6), note that this is affected by hot mix asphalt durability (the higher the durability is the lower waste production is) and by recycling percentage

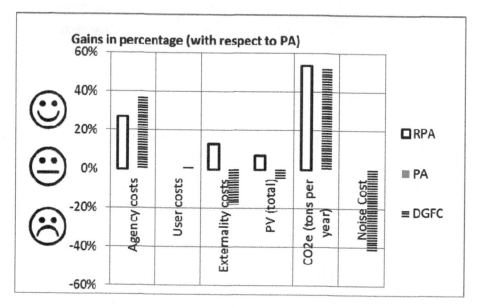

Figure 5. Gains in percentage (y-axis), with respect to Porous Asphalt (PA), for the for the three options (DGFC, RPA, PA), in terms of costs (AC, UC, EX, total), carbon footprint, and noise.

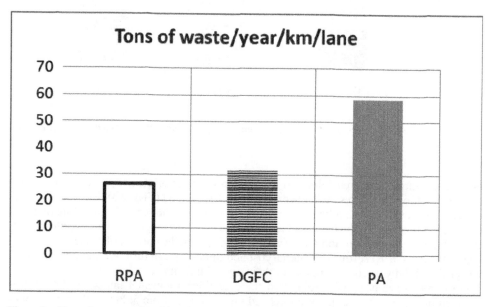

Figure 6. Tons of waste produced (y-axis ton/year/lane) for the for the three options (DGFC, RPA, PA).

(the lower the recycled percentage is the higher waste production is). It is noted that porous asphalt recycling implies an appreciable reduction of waste (Reclaimed Asphalt Pavement, RAP) production (per year and per lane). In more detail, RPA exhibits a RAP production that is lower than the ones of DGFCs or PA, despite the differences in expected life.

4 CONCLUSIONS AND SUMMARY

Life cycle costing is a complex procedure which handles simple items such as costs of construction and delay. Issues arise in terms of holistic approaches, which, in turn, are needed to get sound conclusions. A set of algorithms was set up to deal synergistically with a large number of impacts and an analysis period theoretically unlimited. A case study was considered. Based on the results above the following conclusions may be drawn: i) The impact of recycling is outstanding and can make the difference; ii) Durability is a key-factor which plays an important role in favour of traditional solutions (DGFCs). The unsatisfactory durability of porous asphalt has to be balanced through supplementary characteristics. High recycling percentages and low carbon footprint can be the right answer; iii) Densely-populated areas experience high levels of noise pollution, as do areas in which the land is not used in a smart and sustainable way. This impact has to be assessed singularly and objective thresholds must be in place. This concept is herein addressed by considering a non-linear cost curve with a vertical asymptote. On the other hand, noise and carbon footprint are different and concomitant aspects of the environmental problem. This concept is addressed by summing the concerned PVs. In turn, from a holistic standpoint, environmental and traditional PVs are features of the same problem. This concept is addressed by a standardization procedure. Under the above conditions, with respect to the overall environmental impact of a traditional, dense-graded solution, noise impact ranges from 56% to 79%, while pollution-related impact accounts for 18–29%.

Future research will address the application of the above algorithms to high-durable, high-cost solutions. Result of this study are supposed to benefit both researchers and practitioners.

REFERENCES

American Concrete Pavement Association (ACPA). 2011. Life-Cycle Cost Analysis: A Tool for Better Pavement Investment and Engineering Decisions. Engineering Bulletin EB011. American Concrete Pavement Association, Rosemont, IL.

Buckland, T. and Muirhead, M. (2014). QUESTIM, CEDR transnational road research programme (2014). Cost/benefit analysis and Life Cycle Costing of noise mitigation measures and methodology for implementation into pavement management systems, retrieved in September, 2016, from http://www.questim.org/questim.org/system/files/CEDR%20QUESTIM%20 WP5%20D5.pdf.

California Department of Transportation (Caltrans). 2011. Construction Analysis for Pavement Rehabilitation Strategies (CA4PRS): Caltrans "Rapid Rehab" Software. California Department of Transportation, Sacramento, CA.

Circular A-94 Appendix C, Revised November 2015, OMB Circular No. A-94, Discount Rates For Cost-Effectiveness, Lease Purchase, And Related Analyses, retrieved in September, 2016, from: https://www.whitehouse.gov/omb/circulars_a094/a94_appx-c.

Cucurachi S, Heijungs R. (2014). Characterisation factors for life cycle impact assessment of sound emissions, Sci Total Environ. 2014 Jan 15;468-469:280-91. doi: 10.1016/j.scitotenv.2013.07.080. Epub 2013 Sep 10.

Delwar, M. and Papagiannakis, A.T., (2001). Relative Importance of User and Agency Costs in Pavement LCCA, Fifth International Conference on Managing Pavements, August 11–14, 2001, Seattle, Washington, Conference Proceedings.

Doka, G. (2009). Estimates of road, rail and airplane noise damages for Life Cycle. Retrieved in September, 2015, from http://www.doka.ch/DokaNoiseFactorsForLCAv2.pdf.

Fisher, I. (1930). The Theory of interest. Philadelphia: Porcupine Press. ISBN 0-87991-864-0.

Garraín, D., Franco, V., Vidal, R., Casanova, S. (2009). The noise impact category in Life Cycle Assessment, January 2009, Selected Proceedings from the 12th International Congress on Project Engineering, Editors: AEIPRO.

ISO 15686-1:2011, Buildings and constructed assets—Service life planning—Part 1: General principles and framework.
ISO 15686-5:2008, Buildings and constructed assets—Service-life planning—Part 5: Life-cycle costing.
Licitra, G., Teti, L., Cerchiai, M., A modified Close Proximity method to evaluate the time trends of road pavements acoustical performances, Applied Acoustics, Volume 76, 2014, Pages 169–179.
Mallela, J., Sadasivam, S., FHWA-HOP-12-005, Work Zone Road User Costs—Concepts and Applications, December 2011.
Praticò F.G., Ammendola R., Moro A., Factors affecting the environmental impact of pavement wear, Transportation Research Part D: Transport and Environment, Volume 15, Issue 3, May 2010, Pages 127–133, ISSN 1361-9209, DOI: 10.1016/j.trd.2009.12.002. Elsevier Science Ltd.
Praticò F.G. & Vaiana R., Improving infrastructure sustainability in suburban and urban areas: is Porous asphalt the right answer? And how? Urban Transport: Urban Transport and the Environment in the 21st Century, Edited by: C. A. Brebbia, J. W. S. Longhurst, WIT Press, ISBN: 9781845645809, WIT Transactions on Ecology and the Environment, 2012, pp. 673–684.
Praticò F.G., Casciano A. (2015) Variability of HMA characteristics and its influence on pay adjustment, Journal of Civil Engineering and Management 21 (1), 119–130. DOI: 10.3846/13923730.2013.802713.
Praticò, F.G., Vaiana, R., A study on the relationship between mean texture depth and mean profile depth of asphalt pavements Construction and Building Materials, Volume 101, 30 December 2015, Pages 72–79.
Praticò, F.G., Vaiana, R., Iuele, T., Macrotexture modeling and experimental validation for pavement surface treatments, Construction and Building Materials 95 (2015) 658–666.
Sijm, J., Neuhoff, K., and Chen, Y., CO2 cost pass through and windfall profits in the power sector, Climate Policy, Volume 6, Issue 1, 2006.
Vanchieri, C. (2014). Cost-Benefit Analysis Noise Barriers and Quieter Pavements, Eric W. Wood, George C. Maling, Jr., and William W. Lang Editors, Copyright © 2014, Institute of Noise Control Engineering of the USA, Inc., ISBN: 987-0-9899431-1-6, Library of Congress Control Number: 2014937983.
Weed, R. M. (2001). "Derivation of equation for cost of premature pavement failure." Transp. Res. Rec., 1761(1), 93–96.
West, R., N. Tran, M. Musselman, J. Skolnick, and M. Brooks. 2012. A Review of the Alabama Department of Transportation's Policies and Procedures for LCCA for Pavement Type Selection. NCAT Report 13-06. National Center for Asphalt Technology, Auburn, AL.

Life cycle assessment and benchmarking of end of life treatments of flexible pavements in California

A. Saboori, J.T. Harvey, A.A. Butt & D. Jones
University of California Pavement Research Center, Davis, CA, USA

ABSTRACT: Pavement life cycle stages include materials, construction, maintenance and rehabilitation, use, and End-of-Life (EOL). The least studied stage in pavement Life Cycle Assessment (LCA) studies has been the EOL stage. Currently the options for EOL are recycling (in-place and in-plant) and landfilling. Recycling has always been closely linked to environmental stewardship and this, added to the scarcity of virgin materials resources in parts of California, has resulted in Caltrans aggressively pursuing recycling in their pavement projects. However, there are limited and unreliable data for quantifying the environmental impacts of EOL treatments. This paper provides benchmarking of alternative EOL treatments in California by developing locally-representative models for material production and simulating state of the practice construction activities in the state through field investigations and communication with contractors and experts.

1 INTRODUCTION

Sustainable transport infrastructure is a major area of focus for Caltrans as stated in its mission and has resulted in Caltrans pursuing innovative materials and construction processes in their projects for many years. As pavements reach their End-of-Life (EOL), there are multiple options available amongst which recycling is becoming more popular due to the general perception of recycling as a more sustainable alternative than using new materials. Moreover, virgin aggregate sources have also become scarce in many parts of California. For a better understanding of performance of different alternatives in terms of sustainability, objective quantification of their environmental impacts throughout their life cycle is needed.

Life Cycle Assessment (LCA) can be considered as a suitable tool for such purposes, and can answer a wide range of questions regarding pavements (Harvey et al., 2015). This paper uses LCA methodology for benchmarking the environmental impacts of the current methods in practice for flexible pavements at their EOL in California. The methods considered for recycling are in place recycling (cold in-place and full depth reclamation, with different additives and asphalt wearing courses on top) versus conventional methods such as asphalt overlay and mill-and-fill.

2 LIFE CYCLE ASSESSMENT FRAMEWORK

2.1 *Goal and scope definition*

The goal of this LCA study is to quantify the environmental impacts of the current EOL treatments in use for flexible pavements through a benchmark study. The intended application is to provide an estimate of how different alternatives perform in terms of the environmental impacts during material production, transportation, and construction stages. As this range does not cover the full life cycle of the alternatives, no comparison is made between treatments. Therefore, an attributional LCA, and not a comparative LCA, was conducted on a matrix of alternative treatments for EOL of flexible pavements in California. Table 1 shows the EOL treatments considered in this study.

The intended audience is Caltrans, local government transportation agencies in California, and other agencies for which the results may be applicable. The location of use is limited to the state of California and other environments using similar practices and materials, and the functional unit of the study is one ln-km of pavement. The physical boundary only includes the traveled lanes and not the shoulders. As the system boundary only includes the material production and construction stages, the analysis period is selected to be one year. For the functional unit of one ln-km, the LCI of each of the EOL treatment includes the following life cycle stages:

- Material production
- Transportation of the materials to the site
- Construction activities

The material production inventories are cradle-to-gate for all the materials used in the construction. This means that the LCI includes the energy consumption and emissions of all the processes during: raw material acquisition from the ground, transport to the plant, and further processing of raw materials in the plant until they are ready to be shipped at the gate. The models represent the conditions, technologies, and practices used in local plants and construction processes to the extent possible. For each of the construction materials, models were developed in GaBi™ (2015) and energy sources in the model were calibrated to represent the local conditions in terms of electricity grid mix and fuel type used in plant, which is natural gas in California. The grid mix was taken from the California Energy Almanac website, the table for year 2012 is reproduced here as Table 2.

For all the treatments, an 80 km transport distance was assumed for transportation of the materials to the site, which is typical in the state. It was assumed that heavy trucks (24 tonnes gross vehicle weight) are used.

The combustion of fuel in construction equipment plus the electricity and other energy sources used on site contribute to the impacts in the construction stage. To capture the energy

Table 1. The EOL treatments considered in this study.

#	Item
1	CIR (10 cm milled + Mech. Stab.) w. 2.5 cm of HMA OL
2	CIR (10 cm milled + Mech. Stab.) w. Chip Seal
3	FDR (25 cm milled + no stabilization) w. 6 cm RHMA OL
4	FDR (25 cm milled + 4% AE + 1% PC) w. 6 cm RHMA OL
5	FDR (25 cm milled + 3% FA + 1% PC) w. 6 cm RHMA OL
6	FDR (25 cm milled + 2% PC) w. 6 cm RHMA OL
7	FDR (25 cm milled + 4% PC) w. 6 cm RHMA OL
8	FDR (25 cm milled + 6% PC) w. 6 cm RHMA OL
9	HMA Overlay (7.5 cm)
10	HMA Mill & Fill (10 cm)

Mech. Stab.: mechanical stabilization
CIR: cold in-place recycling
FDR: full-depth reclamation
AE: asphalt emulsion
FA: foamed asphalt
OL: overlay
PC: portland cement
RHMA: rubberized hot mix asphalt (gap-graded)
Mill & Fill: milling of a portion of existing asphalt followed by overlay of thickness shown

Table 2. Electricity grid mix in California in year 2012 (Energy Almanac website).

Fuel type	Percent in California power mix
Coal	7.50%
Large Hydro	8.30%
Natural Gas	43.40%
Nuclear	9.00%
Oil	0.00%
Other	0.00%
Renewables	15.40%
Biomass	2.30%
Geothermal	4.40%
Small Hydro	1.50%
Solar	0.90%
Wind	6.30%
Unspecified	16.40%

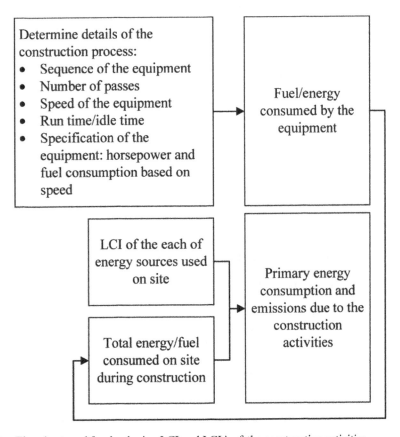

Figure 1. Flowchart used for developing LCI and LCIA of the construction activities.

consumption and emissions of the construction activities, the construction process was closely simulated for each of the in-place recycling techniques and the conventional rehabilitation methods. Figure 1 shows the flowchart for developing the LCI models for construction activities.

Table 3. Impact categories and inventories reported in this study.

Impact category/inventory	Abbreviation	Unit
Global Warming Potential	GWP	kg of $CO_{2\text{-}e}$
Photochemical Ozone Creation (Smog) Potential	POCP	kg of $O_{3\text{-}e}$
Particulate Matter less than 2.5 μm	PM2.5	kg
Primary Energy Demand from Renewable & Non-Renewable Resources, fuel (net calorific value)	PED (total)	MJ
Primary Energy Demand from Non-Renewable Resources, fuel (net calorific value)	PED	MJ
Primary Energy Demand from Non-Renewable Resources, non-fuel (feedstock) (net calorific value)	FE	MJ

Mix design and construction processes for each of the treatments were taken from Caltrans' Maintenance Technical Advisory Guide (MTAG) (Caltrans 2007) and through field investigations and inquiries from personal observation of the authors and local contractors and experts. The mix designs for the wearing surface were taken from Wang et al. (2012) and are typical of California practice.

The whole database developed under this study and previous LCA studies performed at the University of California Pavement Research Center (UCPRC) went through critical review, and was recently verified by a 3rd party committee. The details of all the assumptions and final results are available in the documentation of the UCPRC LCI database (Saboori et al., in prep.).

This study follows the US EPA TRACI 2.1 methodology (Bare 2012) for impact assessment. Main areas of concern are Primary Energy Demand (PED), Global Warming Potential (GWP), and air quality (ground-level ozone creation and particulate matter: PM2.5). Table 3 shows the LCI and LCIA items that are reported in this study.

3 RESULTS AND DISCUSSION

In this section the total impacts for each of the treatments are shown and discussed. Total impacts are the summation of the impacts for each indicator across the three stages discussed in the previous section: material production, transport to the site, and construction. Table 4 shows the summary of the total impacts for each of the EOL treatments. Table 5 shows the percentage share of each of the life cycle stages on the total impacts.

The results indicate that the material production stage is dominant in all of the impact categories for all the EOL treatments material except for POCP of CIR treatments where the construction stage is the main source of the impacts. Transportation has the lowest share of the impacts in all categories for all the in-place recycling treatments but this is not the case for the conventional rehabilitation methods in which construction and transportation have close shares in most categories.

Between the treatments, the GWP for one ln-km ranged between 7.65e3 and 2.12e5 kg of $CO_{2\text{-}e}$. The contribution of the material production stage to GWP ranged between 48% and 95%.

Photochemical ozone creation potential, indicator of smog formation, was the only category in which material production stage was not the dominant source of emission across all the treatments. This indicator varied between 2.23e3 to 1.60e4 kg of $O_{3\text{-}e}$ and the material production share of the total emissions in this category ranged between 27% and 81%. The construction stage POCP were higher than those of the material production stage only for the CIR treatments, with values of 62% and 70% for CIR with HMA overlay and CIR with chip seal overlay, respectively.

Table 4. Summary of the total impacts for each of the EOL treatments for 1 ln-km.

Surface treatment	GWP [kg CO_{2-e}]	POCP [kg O_{3-e}]	PM2.5 [kg]	PED (total), Fuel [MJ]	PED (non-ren), Fuel [MJ]	PED (non-ren), Non-Fuel [MJ]
CIR (10 cm milled + Mech. Stab.) w. 2.5 cm of HMA OL	1.64E+04	3.15E+03	1.09E+01	6.26E+05	6.20E+05	3.57E+05
CIR (10 cm milled + Mech. Stab.) w. Chip Seal	7.65E+03	2.23E+03	5.83E+00	4.01E+05	3.97E+05	1.62E+07
FDR (25 cm milled + no stabilization) w. 6 cm RHMA OL	4.20E+04	6.20E+03	2.74E+01	2.01E+06	1.98E+06	9.64E+07
FDR (25 cm milled + 4% AE + 1% PC) w. 6 cm RHMA OL	1.16E+05	1.52E+04	7.57E+01	6.67E+06	6.59E+06	3.11E+08
FDR (25 cm milled + 3% FA + 1% PC) w. 6 cm RHMA OL	1.02E+05	1.33E+04	6.60E+01	5.45E+06	5.40E+06	2.57E+08
FDR (25 cm milled + 2% PC) w. 6 cm RHMA OL	9.87E+04	9.48E+03	4.96E+01	2.27E+06	2.23E+06	9.64E+07
FDR (25 cm milled + 4% PC) w. 6 cm RHMA OL	1.55E+05	1.28E+04	7.19E+01	2.54E+06	2.48E+06	9.64E+07
FDR (25 cm milled + 6% PC) w. 6 cm RHMA OL	2.12E+05	1.60E+04	9.41E+01	2.81E+06	2.73E+06	9.64E+07
HMA Overlay (7.5 cm)	3.80E+04	4.48E+03	2.39E+01	1.72E+06	1.71E+06	3.57E+05
HMA Mill & Fill (10 cm)	5.13E+04	6.24E+03	3.23E+01	2.31E+06	2.28E+06	3.57E+05

Table 5. Share of each of the life cycle stages of the total impacts for each of the EOL treatments.

Item	Life cycle stage	GWP [kg CO_{2-e}]	POCP [kg O_{3-e}]	PM2.5 [kg]	PED (total), Fuel [MJ]	PED (non-ren), Fuel [MJ]
CIR (10 cm milled + Mech. Stab.) w. 2.5 cm of HMA OL	Material	64%	31%	64%	87%	87%
	Transport	8%	7%	4%	3%	3%
	Construction	27%	62%	32%	10%	10%
	Total	100%	100%	100%	100%	100%
CIR (10 cm milled + Mech. Stab.) w. Chip Seal	Material	48%	27%	50%	86%	86%
	Transport	6%	3%	3%	2%	2%
	Construction	46%	70%	48%	12%	12%
	Total	100%	100%	100%	100%	100%
FDR (25 cm milled + no stabilization) w. 6 cm RHMA OL	Material	79%	53%	81%	94%	94%
	Transport	8%	9%	4%	2%	2%
	Construction	13%	39%	16%	4%	4%
	Total	100%	100%	100%	100%	100%
FDR (25 cm milled + 4% AE + 1% PC) w. 6 cm RHMA OL	Material	92%	80%	93%	98%	98%
	Transport	3%	4%	2%	1%	1%
	Construction	5%	16%	6%	1%	1%
	Total	100%	100%	100%	100%	100%

(*Continued*)

Table 5. (Continued).

Item	Life cycle stage	GWP [kg CO$_{2-e}$]	POCP [kg O$_{3-e}$]	PM2.5 [kg]	PED (total), Fuel [MJ]	PED (non-ren), Fuel [MJ]
FDR (25 cm milled + 3% FA + 1% PC) w. 6 cm RHMA OL	Material	91%	77%	92%	98%	98%
	Transport	4%	5%	2%	1%	1%
	Construction	5%	18%	6%	1%	1%
	Total	100%	100%	100%	100%	100%
FDR (25 cm milled + 2% PC) w. 6 cm RHMA OL	Material	91%	69%	89%	94%	94%
	Transport	4%	6%	2%	2%	2%
	Construction	6%	25%	9%	3%	3%
	Total	100%	100%	100%	100%	100%
FDR (25 cm milled + 4% PC) w. 6 cm RHMA OL	Material	94%	76%	92%	95%	95%
	Transport	2%	5%	2%	2%	2%
	Construction	4%	19%	6%	3%	3%
	Total	100%	100%	100%	100%	100%
FDR (25 cm milled + 6% PC) w. 6 cm RHMA OL	Material	95%	81%	94%	95%	95%
	Transport	2%	4%	1%	2%	2%
	Construction	3%	15%	5%	3%	3%
	Total	100%	100%	100%	100%	100%
HMA Overlay (7.5 cm)	Material	84%	65%	88%	95%	95%
	Transport	11%	15%	6%	3%	3%
	Construction	6%	21%	7%	2%	2%
	Total	100%	100%	100%	100%	100%
HMA Mill & Fill (10 cm)	Material	83%	62%	86%	95%	94%
	Transport	11%	14%	5%	3%	3%
	Construction	7%	24%	8%	2%	2%
	Total	100%	100%	100%	100%	100%

Table 6. Increase in total impacts as the transport distance is increased from 80 to 160 km.

Surface treatment	GWP [kg CO$_2$e]	POCP [kg O$_3$e]	PM2.5 [kg]	PED (total), Fuel [MJ]	PED (non-ren), Fuel [MJ]
CIR (10 cm milled + Mech. Stab.) w. 2.5 cm of HMA OL	8%	7%	4%	3%	3%
CIR (10 cm milled + Mech. Stab.) w. Chip Seal	6%	3%	3%	2%	2%
FDR (25 cm milled + no stabilization) w. 6 cm RHMA OL	8%	9%	4%	2%	2%
FDR (25 cm milled + 4% AE + 1% PC) w. 6 cm RHMA OL	3%	1%	0%	1%	1%
FDR (25 cm milled + 3% FA + 1% PC) w. 6 cm RHMA OL	4%	1%	0%	2%	2%
FDR (25 cm milled + 2% PC) w. 6 cm RHMA OL	4%	6%	2%	2%	2%
FDR (25 cm milled + 4% PC) w. 6 cm RHMA OL	3%	5%	2%	2%	2%
FDR (25 cm milled + 6% PC) w. 6 cm RHMA OL	2%	4%	1%	2%	2%
HMA Overlay (7.5 cm)	11%	15%	6%	3%	3%
HMA Mill & Fill (10 cm)	11%	14%	5%	3%	3%

PM2.5 emissions for all the treatments was mainly due to the material production stage, ranging from 5.83 to 94.1 kg for one ln-km of treatment and a range of 50% to 94% share of the total. For the CIR treatments, construction activities had the largest share among all treatments, between 32% and 48%.

Total primary energy demand for all treatments ranged between 3.97e5 to 4.77e6 MJ for one ln-km of pavement and was mainly due to the material production stage, with a share of 87% to 95% across all the treatments.

3.1 *Sensitivity analysis on the transportation distance*

A sensitivity analysis was conducted on travel distance to see how it impacts the final results. The model was run under 80 and 160 km for the transport distance and the percent increases in each of the impact categories were calculated for each of the EOL treatments. As Table 6 shows, the increase in emissions are higher for the conventional methods where more materials are transported to the site, therefore the total impacts of those treatments are more affected by the transportation distance. The increase for the POCP is highest, 15%, for the HMA overlay.

4 CONCLUSIONS AND FUTURE WORKS

This study was conducted to benchmark the environmental impacts of some of the common EOL treatments used in California for flexible pavements at their end of service life. The system boundary consists of material production, transport to the site, and construction activities and does not include the other life cycle stages such as use stage, maintenance and rehabilitation in the future, and traffic delays during construction. Ten treatments were

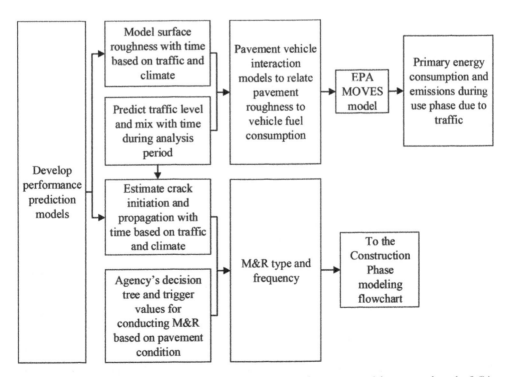

Figure 2. Flowchart of implications of developing the performance models to complete the LCA model for doing comparative LCA between EOL treatments.

considered consisting of eight in-place recycling (CIR and FDR with different stabilization methods and wearing courses on top) and two conventional treatments (HMA overlay and HMA mill-and-fill).

There are large variabilities between treatments in each of the impact categories studied in this research but as discussed earlier due to limited scope of the study and the fact that the system boundary of the study does not include all life cycle stages, no comparison can be made between treatments until the full life cycle is assessed, including performance in the use stage and subsequent EOL.

The results show that for all treatments and all impact categories that material production is the main contributing factor, the only exception being POCP of CIR treatments in which construction had a larger share of the total impacts. The findings of this study provides an estimate of the environmental impacts of alternative EOL treatments for flexible pavements in California and the percentage share of each of the life cycle stages considered.

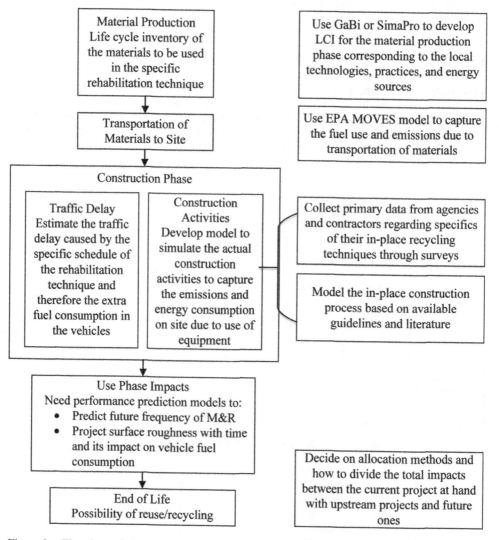

Figure 3. Flowchart of the approach to be used in the comparative LCA study between the EOL treatments.

From a pavement management perspective, it is ideal to have an LCA tool that can supplement the life cycle cost analysis in comparing possible alternatives for pavements at their EOL. The findings of this study cannot provide such information as the system boundary defined in the goal and scope definition is not inclusive of the whole life cycle of the treatments, and therefore comparing treatments is not possible. There are unsolved questions regarding:

- How does pavement surface roughness, which directly affects vehicle fuel consumption, change with time under each treatment?
- How does cracking initiation and propagation, that determines future maintenance and rehabilitation frequency and the section service life, differ between alternatives?

What is needed for modeling the use stage, future maintenance and rehabilitation frequencies, and service life of each treatment, is performance models. At this point the effect of recycling treatments on the performance of the section is not fully understood. The next stage of this study will focus on developing the performance models for the common EOL treatments in California so that fair comparison can be made between alternatives. Figure 2 shows how these prediction models will be used to provide the required information for completing the LCA model to conduct a comparative LCA study between EOL treatments. The LCA study to be done in the next step of this project will follow the flowchart presented in Figure 3.

The final step will be to combine the LCA and life cycle costs for the alternative treatments to provide recommendations to Caltrans and local governments.

ACKNOWLEDGEMENTS

The research presented in this paper was requested and sponsored by the California Department of Transportation (Caltrans). Caltrans sponsorship of that work is gratefully acknowledged. Special thanks are for Nick Burmas, Joe Holland, and Deepak Maskey for their support for and contributions to this research project. The contents of this paper reflect the views of the authors and do not necessarily reflect the official views or policies of the State of California, or the Federal Highway Administration.

REFERENCES

Bare, J. 2012. Tool for the Reduction and Assessment of Chemical and Other Environmental Impacts (TRACI). TRACI Version 2.1 User's Guide. EPA/600/R-12/554 2012. Environmental Protection Agency, Cincinnati, OH. Web Link.
Caltrans Maintenance Technical Advisory Guide (MTAG) 2007. California Department of Transportation, Sacramento, CA.
GaBi Version 6.3. 2015. Life Cycle Assessment Software. PE International, San Francisco, CA.
Harvey, J., A. Kendall, and A. Saboori. 2015. White Paper: Reduction of Life Cycle GHG Emissions from Road Construction and Maintenance and the Role of Life Cycle Assessment. National Center for Sustainable Transportation, University of California, Davis, CA.
Saboori, A. in prep. Documentation of the UCPRC Life Cycle Inventory (LCI) Used in CARB/Caltrans LBNL Heat Island Project and other Caltrans' LCA Studies. University of California Pavement Research Center, Davis, CA.
Wang, T., L. In-Sung, J.T. Harvey, A. Kendall, E.B. Lee, and C. Kim. 2012., UCPRC Life Cycle Assessment Methodology and Initial Case Studies on Energy Consumption and GHG Emissions for Pavement Preservation Treatments with Different Rolling Resistance. 2012, UCD-ITS-RR-12-36, Institute of Transportation Studies, Davis, CA.

Life cycle assessment of pavements under a changing climate

O. Valle
Sustainability Fellow, University of New Hampshire, USA

Y. Qiao, E. Dave & W. Mo
Department of Civil and Environmental Engineering, University of New Hampshire, New Hampshire, USA

ABSTRACT: Each of pavements' life cycle phases such as material extraction, transportation, construction, operation, rehabilitation, and end-of-life emits greenhouse gases. Such emissions can be influenced by climate change via changes in the rates of pavement deterioration and thus intensity and frequency of maintenance and rehabilitation. However, climate change has not been given full considerations in previous pavement Life Cycle Assessments (LCA). This research introduced a methodology to integrate the effects of climate change within pavement LCA. A case study was performed to calculate the life cycle Global Warming Potential (GWP) and costs of several typical interstate pavements due to climate change, using downscaled climate data obtained from the Coupled-Model Intercomparison Project Phase 5 (CMIP5). Pavement performance (roughness) was determined using the Pavement ME system. Rehabilitaiton alternatives were also assessed according to predefined pavement roughness triggers. Whereby, roughness triggers were alternatives assessed using LCA. Use of Reclaimed Asphalt Pavement (RAP) was evaluated as an alternative of local virgin materials. SimaPro and PaLATE were used to convert material and energy consumptions into GWP values and equations from HDM4 model were used for determining operational GWP impacts and costs. Asphalt mix production costs were used on the basis of values obtained from local asphalt contractor. For each of the case study scenario, LCA was conducted using historic climatic data, as the current state of practice in field of pavement LCA, as well as future climatic projections. The results of this research demonstrate the importance of considering future climate change in pavement LCAs. This study also presents a generalizable framework for climate change informed pavement LCAs.

1 INTRODUCTION

Over the last decade, Life Cycle Assessment (LCA) has been increasingly applied to quantify the cradle-to-grave environmental impacts of pavements (For example, Santero et al. 2011, Harvey et al. 2014, Huang and Parry 2014). The life cycle of a pavement typically consists of five phases, including materials extraction and production, construction, use, maintenance and end-of-life disposal (FHWA 2016). There is a dynamic interaction between climate and pavements. On one hand, pavements' life cycle generates a large amount of Greenhouse Gas (GHG) emissions, which calls for a comprehensive understanding of emissions associated with each phase to guide future pavement design and management (Santero 2009). On the other hand, climate change may accelerate pavement deterioration (Mills et al. 2009) and increase its life cycle GHG emissions, energy use, and costs (Qiao et al. 2015). Current pavement LCA methodology typically assumes static climate, which may not be suitable for long-term planning into the future. Hence, it is important to incorporate climate change into pavement LCA in order to improve our understandings on the coupled effects and to provide more reliable results.

Climate change refers to changes in climate stressors in the future, such as increases in temperature, precipitation, and extreme weather e.g. hurricane and flooding (Meyer et al. 2013). For flexible pavements, climate stressors such as temperature, precipitation, wind speed, solar radiation, and groundwater table can be influential to pavement performance and thus changes in these stressors may result in changes of pavement performance and service lives. Previous research found that temperature is the most influential climate stressors (Qiao et al, 2013) and therefore downscaled future temperature was obtained to represent future climate and applied as an input for pavement responses.

The overall goal of this research is to develop a method to incorporate climate change in pavement LCA, using a segment of Interstate-95 (I-95) located in southern New Hampshire as a case study.

2 METHODOLOGY

This study started with identifying a typical road section on I-95 and collecting its structural data. Pavement deterioration was modeled using pavement ME (AASHTO, 2016). In order to assess the effects of climate change on road performance, the default climate data embedded in Pavement ME (hourly data of air temperature, rainfall, humidity, percent of sunshine and groundwater table) were modified based on downscaled daily temperature and precipitation (for the period of 2020–2040) obtained from the Coupled Model Intercomparison Project Phase 5 (CMIP5) using RCP 4.5 scenario. International Roughness Index (IRI) was adopted as performance criteria of the road section, which can trigger maintenance at certain levels. It is also related to user fuel consumptions (NCHRP 1985, Ockwell 1999, Greenwood and Christopher 2003). The life cycle road performance was incorporated in an attributional LCA to calculate and compare the Global Warming Potential (GWP) and energy consumptions of different pavement structures (see Table 1), with different maintenance regimes over a design life of 20 years. The LCA was conducted using SimaPro (for the production, transportation machinery operation and maintenance phases) and PaLATE (for the construction phase) software. A methodology flowchart is shown in Figure 1. Particularly, two scenarios were considered for road M & R, including a "do nothing" and a mill & fill regime. The mill & fill is applied by milling top 3-inch asphalt layers and fill with a new asphalt layer, which is typically applied in New Hampshire. As a responsive strategy, the triggers were set at IRI of 120, 140, 160, 180, and 200 inch/mile respectively for comparison purposes. The analysis period was assumed to be 60 years to show the average effects of maintenance cycles to the LCA results. In the end, the life cycle GHG emissions and costs of the investigated road segment were estimated.

Table 1. Pavement structures.

Structures	All Virgin mixture (V)	Virgin mixture with 40% percent RAP (V/R)	All RAP (R)
Standard	AC: 6 inch GB: 28 inch SB: 8 inch SG: clay	AC: 6 inch GB: 28 inch SB: 8 inch SG: clay	AC: 6 inch GB: 28 inch SB: 8 inch SG: clay
Medium strength	AC: 9 inch GB: 18 inch SB: 8 inch SG: clay	AC: 9 inch GB: 18 inch SB: 8 inch SG: clay	AC: 9 inch GB: 18 inch SB: 8 inch SG: clay
Deep strength	AC: 12 inch GB: 12 inch SB: 8 inch SG: clay	AC: 12 inch GB: 12 inch SB: 8 inch SG: clay	AC: 12 inch GB: 12 inch SB: 8 inch SG: clay
Full depth	AC: 16 inch SB: 8 inch SG: clay	AC: 16 inch SB: 8 inch, sand SG: clay	AC: 16 inch SB: 8 inch SG: clay

Note: AC: asphalt concrete, GB: granular base, SB: subbase, SG: subgrade (1 inch = 25.4 mm).

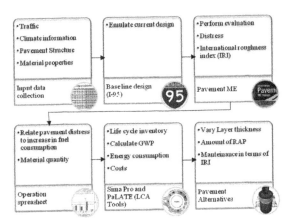

Figure 1. Methodology flowchart.

Table 2. Traffic inputs.

Input	Value	Source
AADT	88,000	NHDOT Traffic Counts
Percent trucks	10%	NHDOT Traffic Counts
Operational speed	70 mph (112 km/h)	Default
Percent trucks in design lane	95%	Default

2.1 Pavement sections

An eighteen-mile roadway of I-95 was studied. Four typical pavement structures were identified and applied in the case study, including a standard (baseline), medium strength, deep strength, and full depth structures (see in Table 1). For each structure, three types of asphalt mixtures were investigated: virgin asphalt mixture, virgin mixture with 40% RAP, and 40% RAP for all asphalt layers (see in Table 1). RAP is an alternative to virgin aggregates in Hot-Mix Asphalt (HMA) production and in the construction of base or subbase (Yang 2014).

Material properties inputs for the granular base, sand subbase, and clay subgrade were all assumed to be the default values available in Pavement ME Design. The asphalt material properties including density, air void, binder contents, and dynamic modulus were taken from lab measurements. These properties are used to differentiate long-term performance between virgin binder and RAP. The dynamic modulus inputs are at a higher detailed calibration level and thus likely leads to more accurate performance prediction. It should however be noted that the distress predictions here are made using national calibrations and not regional or local calibrations.

Traffic information was shown in Table 2. Traffic growth was assumed to be 0% as to derive the differences in emissions and costs caused by climate change alone.

2.2 Climate

Climate inputs of Pavement ME are hourly data including temperature, precipitation, wind speed, percent sunshine, and ground water level. As the CMIP5 can only provide daily climatic data, an hourly temperature generator was developed. The generator assumed that the daily minimum temperature (t_{min}) occurs at the sunrise and the maximum temperature (t_{max}) occurs at 2 p.m. in the afternoon. This method was initially presented by De Wit et al. (1978) and was obtained from the subroutine WAVE in ROOTSIMU V4.0 by Hoogenboom and Huck (1986). This method requires t_{min} of the next day and divides the day into two segments, from sunrise to 2 p.m. and from 2 p.m. to sunrise of the next day. The intervening temperatures are calculated from the following equations:

$$\text{for } 0{:}00 < h < \text{rise and } 14{:}00 < h < 24{:}00, \ T(h) = t_{ave} + \text{amp}(\cos(\pi \times h')/(10 + \text{rise})) \quad (1)$$

$$\text{for rise} < h < 14{:}00, \ T(h) = t_{ave} - \text{amp}(\cos(\pi(h' - \text{rise})/(14 - \text{rise}))) \quad (2)$$

where

$$h' = h + 10 \quad \text{if } h < \text{rise} \quad (3)$$
$$h' = 14 \quad \text{if } h > 14 \quad (4)$$

where rise = time of sunrise in hours; $T(h)$ = temperature at any hour; h = time in hours, $h' = h + 10$ if $h <$ rise, $h' = 14{:}00$ if $h > 14{:}00$; $t_{ave} = (t_{min} + t_{max})/2$; amp = $(t_{max} - t_{min})/2$. Daily precipitation was assumed to have occurred on a random hour of each rainy day. Other climatic factors were kept unchanged. For wind speed and sunshine percentage, it is usually considered that they do not dominate pavement performance e.g. IRI and thus their effects are negligible (Qiao et al. 2013). IRI is the most sensitive to temperature and thus it is important to include its impact. Changes in groundwater level may also be influential for IRI in some cases, especially for thin asphalt (Qiao et al. 2013). However, groundwater projections are not available. Because of the relatively thick asphalt layers in this case study, the impacts of groundwater change are not considered.

2.3 Life Cycle Assessment

Material production includes production of asphalt, gravel, and sand materials. Using the cross section of the road, the total amount of materials needed for each structure was estimated. The transportation stage was quantified considering the use of conventional dumping trucks of ten cubic yards. Construction process considered emissions from asphalt paving, rolling, grading, and compaction of unbound materials, and machinery operations. Rehabilitation includes asphalt mill & fill. Gasoline and diesel consumptions were considered in the road use phase. PaLATE is an Excel-based tool which performs LCA based upon user inputs of detailed road design, material type, machinery information. We use default user input values embedded in PaLATE to estimate constructional impacts of the target road

Table 3. Impact Inputs (1 ton = 907 kg, 1 mile = 1.6 km, 1 yd³ = 0.84 m², 1 gal = 3.8 L, and 1 lb = 0.46 kg).

Impact Input	Units	Value	Source
Production			
Asphalt Concrete	MJ/ton	641	SimaPro
Asphalt Concrete	kg CO_2 eq/ton	84.7	SimaPro
Gravel	MJ/ton	265	SimaPro
Gravel	kg CO_2 eq/ton	14.10	SimaPro
Sand	MJ/ton	61.8	SimaPro
Sand	kg CO_2 eq/ton	4.25	SimaPro
Transportation			
Dump Truck Transportation	MJ/ton*mile	5.134	SimaPro
Dump Truck Transportation	kg CO_2 eq/ton*mile	0.321	SimaPro
Construction			
Asphalt Paving (Productivity)	ton/hr	10	PaLATE
Asphalt Rolling—Tandem (Productivity)	ton/hr	395	PaLATE
Unbound Material Compaction (Productivity)	ton/hr	1832	PaLATE
Construction Machine Operation	MJ/hr	10816	SimaPro
Construction Machine Operation	kg CO_2 eq/hr	72	SimaPro
Maintenance			
Asphalt Milling	MJ/yd³	6.23	SimaPro
Asphalt Milling	kg CO_2 eq/yd³	0.409	SimaPro
Operation			
Gasoline	MJ/gal	130	EPA
Gasoline	lb CO_2/gal	19.64	EPA
Diesel	MJ/gal	137	EPA
Diesel	lb CO_2/gal	22.38	EPA

section. We also use SimaPro and information collected from the EPA to estimate impacts of use and maintenances phases. Table 3 provides a list of required materials and equipment in this study and their associated costs and impacts. Five vehicle groups were used to classify the traffic, including car, vans, SUV, light truck and articulated truck. IRI values at each year were used to determine the fuel consumption for each class of vehicle at a certain year.

3 RESULTS AND DISCUSSIONS

Using the methodology laid out previously, results for each test scenario in the project scope were assessed and global warming potentials for each alternative within the scenario were compared for both historical climate conditions as well as future climate projections.

3.1 *Scenario 1: Comparison of different pavement structures and levels of recycling*

As previously mentioned, this scenario was used to understand the impacts and costs associated with varying the pavement structure types and the pavement material compositions. Four amounts of Hot Mix Asphalt (HMA) were used in the various pavement structures tested. In terms of recycled content, three different material combinations were tested.

The impacts in terms of GWP of the twelve alternatives are shown in Figure 2. All the values are normalized to have a fair comparison between the different alternatives. Normalization is very important since each combination has different life spans at which it reached terminal serviceability level. The structures are separated into three groups, each with four results. In each group, the leftmost bar represents the "Standard" pavement structure, and the thickness of HMA used in the pavement layer increases in each subsequent bar to the right. The leftmost group shows the results analyzed using no recycled content (V), the middle shows the virgin/recycled mix (V/R), and the rightmost shows the HMA with 40% recycled asphalt material (R). Each bar is split into the portion of impacts or costs associated with either the construction (solid fill pattern) or operation (cross-hatched fill pattern) life cycle phase. As it can be seen in the plots, the construction phase GWP between the 12 cases vary only slightly when compared to the operational phase impacts. The small change is primarily coming from use of different structures (affecting quantities of various material production as well as construction process) and to some extent due to use of 40% RAP in asphalt layers for some cases.

Analysis results using future climate projections (cross hatched fill) as opposed to historical climate data (solid fill) are compared in Figure 3. For the first eight structures, less GWP is realized for the alternatives with Virgin materials and Virgin +RAP mixtures when using the historical climate scenario. An opposite trend is noticed for the four remaining structures containing 40% RAP in asphalt mixtures, for these cases the GWP dropped when using future climate projections.

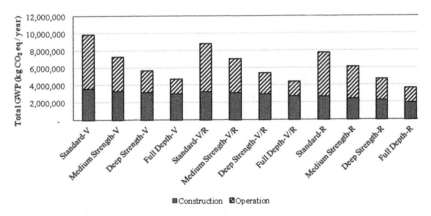

Figure 2. Global Warming Potential (Scenario 1).

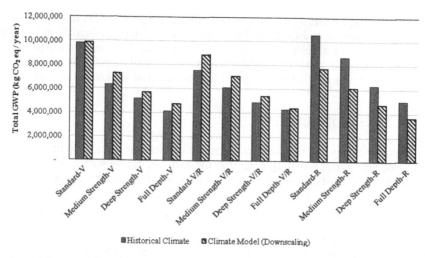

Figure 3. GWP comparison for different pavement structure and asphalt mix recycling alternatives with use of historical climate data and future climate projections.

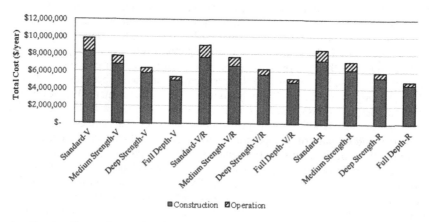

Figure 4. Construction and operational costs for Scenario-1 alternatives using historic climate data.

These findings indicate the importance of using appropriate climate information when conducting a comprehensive pavement LCA. As seen in this case study, use of future climate projections (range of 5 to 18°C of daily average air temperatures) alter the ranking of pavement structure and mix combinations as compared with use of historical climate data (range of 2 to 15°C of daily average air temperatures), which is the current status quo for pavement analysis. For standard pavement cross-section, when using historical climate data, the use of asphalt mixtures with 40% in all lifts (indicated as Standard-R) shows a higher GWP than alternative of using 40% RAP in only lower asphalt lifts (indicated as Standard-V/R). However when using future climate projections, the GWP of these two alternatives reverse. The counter intuitive trends of the plot can be explained by the physical characteristic difference that were observed between different alternatives. With future climate projections, there was lower roughness in roadways with RAP in mixture as the climate trend is in warming direction. Thus, the total GWP from lower vehicle emissions (due to lower IRI) lead to lowering of GWP in case of using RAP mixes in all asphalt layers.

Information in Figure 4 gives the costs associated with each test scenario that was analyzed. The construction cost is considered in terms of activities related with the transportation agency working on the roadway, while the operation costs are directly connected to drivers of the roadway in terms of fuel consumption. The comparative construction and

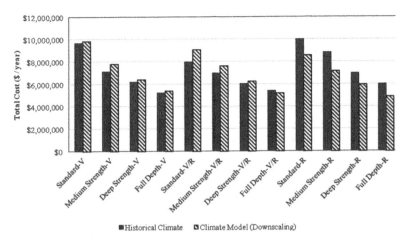

Figure 5. Construction and operational cost comparisons for Scenario-1 using historic and future climate projections.

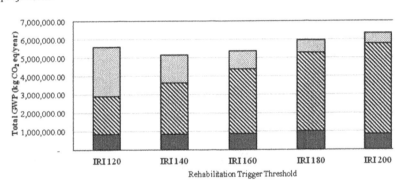

Figure 6. Global Warming Potential for different pavement roughness thresholds to trigger mill and overlay rehabilitation.

operational costs of all alternatives for both scenarios (historic climate and future climate projections) are shown in Figure 5. The first 7 cases (structures with Virgin materials and V+R) show lower accumulated cost for the scenario using historical climate data. The Full Depth V+R structure is the only one of this group that has a slightly higher value on the scenario of historical climate data. When comparing the structures with RAP in the four cases, it presents a considerable lower cost in the scenario of climate change. This is associated with better performance of such pavement options in terms of IRI.

3.2 *Scenario 2: Pavement roughness threshold for overlay rehabilitation*

As discussed in the introductory section of this paper, the second scenario assessed in this work is to make comparisons between different rehabilitation roughness thresholds at which overlay rehabilitation is triggered. Five distinct alternatives were assessed in terms of the IRI at which pavement rehabilitation is undertaken through replacement of top 3 inches of asphalt surface via milling and overlay. Each of these alternatives were simulated for three full rehabilitation cycles. Furthermore, the simulation continued after the third rehabilitation cycle until the terminal IRI of 200 inch/mile (3.2 m/km) is reached. This was necessary to ensure that each alternative was fairly assessed and there was no bias in results due to shorter analysis periods.

For each of the alternatives, annual GWP was calculated and plotted in Figure 6 using the historical climate data. As with previous scenario, the GWP is presented in terms of initial construction (solid fill pattern) and operation (cross-hatched fill pattern). The impact of rehabilitation activity is also shown on the plot (dotted fill pattern). It has to be noted that there is an optimal point in term of GWP at around IRI of 140 inch/mile (2.2 m/km). At this point, there seems to be a balance between the operational emissions from rough pavements and the emissions associated with construction and rehabilitation processes.

A primary focus of this research is to use consequential LCA approach in conjunction with the future climate projections to build additional reliability in LCA results and to demonstrate further need for inclusion of climate change data in current LCA applications to pavement engineering. Total GWP in terms of annual CO_2 equivalent for the five rehabilitation IRI trigger levels using historic climate data (solid fill pattern) and future climate projections (cross-hatched fill pattern) are presented in Figure 7. These GWP values include emissions from initial construction, operation and rehabilitation activities. The results once again demonstrate that the use of future climate projections substantially change the assessment of alternatives. For example, the difference in GWP for using 140 inch/mile (2.2 m/km) versus 160 inch/mile (2.5 m/km) as trigger IRI for mill and overlay is much smaller (4%) when using future climate projections than the one from use of historic climatic data (16%). Since present analysis did not account for factor such as construction related congestion and delays, the use of

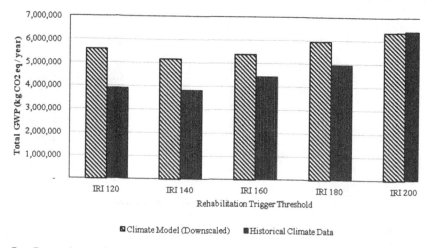

Figure 7. Comparisons of Global Warming Potential for different mill and overlay pavement roughness thresholds for analysis conducted using historical climate and future climate projections (100 in/mi = 1.58 m/km).

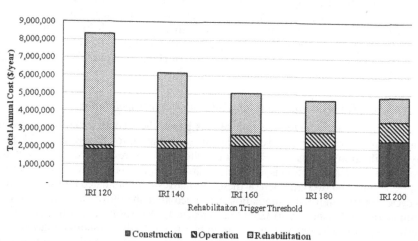

Figure 8. Total costs for different mill and overlay pavement roughness thresholds.

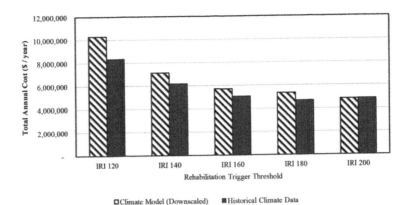

Figure 9. Comparison of total annual costs for different mill and overlay pavement roughness thresholds for analysis conducted using historic climate data and future climate projections.

IRI of 160 inch/mile (2.5 m/km) might be adopted by highway agency as sacrifice of 4% GWP increase when using future climate projections. In fact, the choice of the maintenance trigger is also dependent on the local traffic volume, which can influence congestion and delays (Wang, et al. 2014). However, these influences are out of the scope of this study.

As with previous scenario, the total equivalent annual cost (initial construction, vehicle operation in terms of fuel consumption and rehabilitation costs) were also determined for each alternative. The breakdown and total cost for each alternative is shown in Figure 8. As with GWP, the costs also show an optimality condition for the 180 inch/mile IRI threshold for mill and overlay activity.

Finally, total cost comparisons between analyses using historic climatic data (solid fill pattern) and future climatic projections (cross-hatched fill pattern) are presented in Figure 9. Life cycle costs from use of downscaled climatic data from future projections again show substantially different results as compared to use of historic climate data.

4 SUMMARY, CONCLUSIONS AND FUTURE RESEARCH

This paper presents results of a consequential LCA conducted on a segment of interstate highway in the northeastern United States. The uniqueness of this research is to combine future climatic data in pavement performance evaluation to increase robustness of the LCA findings. Downscaled climatic projections from CMIP5 were utilized in this research. Two scenarios were assessed for making comparative analysis: evaluation of different pavement structure and asphalt mixture recycling amounts, and use of different pavement roughness trigger values for mill and overlay decision process. As with other research, it was found that the operational phase of roadways has a substantial impact on the GWP and life cycle costs, however it was clearly seen in results that the optimal alternative from LCA change substantially when using downscaled future climate projections as opposed to historic climate data. Major conclusions form this research can be summarized as following:

- LCA can provide a good decision process to compare different design and planning alternatives, such as, selection of different materials and structures or rehabilitation decision processes. Use of pavement performance model is critical in such process to ensure that reliable operational phase quantification is made.
- Results presented herein clearly showed that LCA findings change drastically with use of future climate information as opposed to historic climate data.
- Framework presented in this paper provides demonstration of using climate change aware LCA that can be easily implemented by public highway agencies to provide design and operational guidance for roadways.

A number of areas for future research were also identified during the present research effort. Major necessary efforts include:

- Present work utilized Pavement ME as primary tool for pavement performance curve development. While Pavement ME is quite comprehensive, it requires a detailed calibration process to make it reliable for a local region. An alternative would be to use pavement performance curves from pavement or asset management systems that are relevant to a specific region for the materials in that region.
- With changing climate, it is necessary to use reliable future climate projections in LCA process. In reality, the emissions and GWP from analysis like the ones presented in this paper have a certain and quantifiable effect on future climate. Thus, this work can be extended to also take the future climate impact of the comparison alternatives into consideration.
- Effect of climate change on the equipment and vehicle efficiencies need to be accounted to improve reliability of the GWP calculations.

ACKNOWLEDGEMENTS

A portion of this study was funded by The Sustainability Institute of the University of New Hampshire and the University of New Hampshire Center for Infrastructure Resiliency to Climate (UCIRC). The authors would like to acknowledge financial support from these entities. The contents of this paper reflect the views and opinions of the authors who are responsible for the facts and accuracy of the data presented herein and do not necessarily reflect the official views or policies of any agency or institute.

REFERENCES

AASHTO (2016). AASHTOWare Pavement ME Design. *American Association of State Highway and Transportation Officials.* Washington D.C.

FHWA (2016). *Pavement Life Cycle Assessment Framework.* Federal Highway Administration. Report FHWA-HIF-16-014, Washington D.C. USA.

Greenwood, I.D. and R.B. Christopher (2003). HDM-4 Fuel Consumption Modelling.

Harvey, J., T. Wang and J. Lea (2014). Application of LCA Results to Network-Level Highway Pavement Management. *Climate Change, Energy, Sustainability and Pavements,* Springer: 41–73.

Huang, Y. and T. Parry (2014). Pavement Life Cycle Assessment. *Climate Change, Energy, Sustainability and Pavements,* Springer Berlin Heidelberg.

Meyer, M., Flood, M., Dorney, C., Leonard, K., Hyman, R. & Smith, J. 2013. Synthesis of Information on Projections of Climate Change in Regional Climates and Recommendation of Analysis Regions. National Cooperative Highway Research Project.

Mills, B.N., Tighe, S.L., Andrey, J., Smith, J.T., & Huen, K. (2009). *Climate change implications for flexible pavement design and performance in southern Canada.* Journal of Transportation Engineering, 135(10), 773–782.

NCHRP (1985). Life-Cycle Costs Analysis of Pavements. Washington D.C.

Ockwell, A. (1999). Pavement management: Development of a Life Cycle Costing Technique. B. o. T. a. C. Economics.

Qiao, Y., Dawson, A.R., Parry, T., Flintsch, G. (2015). *Evaluating the effects of climate change on road maintenance intervention strategies and Life-Cycle Costs.* Transportation Research Part D: Transport and Environment (41). 492–503.

Qiao, Y., Flintsch, G., Dawson, A., Parry, T. (2013). *Examining effects of climatic factors on flexible pavement performance and service life.* Transportation Research Record (2349).100–107.

Reicosku, D.C., Winkelman, L.J., Baker, J.M., Baker, D.G. (1989) *Accuracy of hourly air temperatures calculated from daily minima and maxima.* Agricultural and Forest Meteorology (46). 193–209.

Santero, N.J. (2009). *Global Warming Potential of Pavements.* Environ. Res. Lett. 4(2009), 034011.

Santero, N.J., E. Masanet and A. Horvath (2011). "Life-cycle assessment of pavements. Part I: Critical review." *Resources, Conservation and Recycling* 55(9–10): 801–809.

Wang T., Harvey, J., Kendall, A. (2013). Reducing greenhouse gas emissions through strategic management of highway pavement roughness. Environmental Research Letters, 9(3), 1–10.

Yang, R., Ozer, H., Kang, S., Al-Qadi, I.L. (2014). Environmental Impacts of Producing Asphalt Mixtures with Varying Degrees of Recycled Asphalt Materials. *International Symposium on Pavement LCA.* Davis, California.

Capitalizing green pavement: A method and valuation

Xiaoyu Liu, David Choy, Qingbin Cui & Charles W. Schwartz
Department of Civil and Environmental Engineering, University of Maryland, College Park, MD, USA

ABSTRACT: The use of recycling techniques in pavement construction offers significant benefits of resource conservation and emission reduction. While the environmental benefits are generally considered as business expenses, a new trend is to turn sustainable construction into profit. By generating financial revenue via carbon trading mechanism, asphalt producers and road builders can compensate their sustainable efforts and adoption of recycling techniques for reducing Greenhouse Gas (GHG) emissions. This carbon trading mechanism rewards emission reduction projects with carbon offsets that can be sold at a certain price in the market. This paper develops a framework for pavement recycling projects to verify GHG reductions and register for carbon offsets. Unlike most standards today that use a project-by-project method to quantify carbon offsets, a performance standard is presented based on the life cycle environmental impact analysis to facilitate credit assessment, generation, and verification. The results of thirteen pavement project cases demonstrate that, even with a 20% market share, cold recycling techniques will be able to achieve an emission reduction of 2.8 million tons of GHGs every year. If these GHG reductions can be quantified into carbon offsets for trading, cold recycling techniques will open up more than 20 million revenue sources beyond direct product sales.

1 INTRODUCTION

Asphalt is the most widely used pavement material in the world. In the United States, more than 92% of the 4 million kilometers (km) of paved roads and highways are surfaced with asphalt; in Europe, more than 90% of the total 5.2 million km are surfaced with asphalt. Canada has about 415,000 km of paved roads, of which about 90% are surfaced with asphalt (Mangum, 2006). The U.S. Environmental Protection Agency (EPA) reported that asphalt pavement emits 0.48 metric ton (t) of GHG equivalent per thousand dollar, which is approximately three times greater than that of power and communication lines (Truitt, 2009). In addition to GHG, asphalt pavement also emits 25~34t of air pollutants including Sulfur Dioxide (SO_2), Nitrogen Oxide (NO_x), Carbon Monoxide (CO), Volatile Organic Compounds (VOC), and volatile Hazardous Air Pollutant (HAP) organic compounds (EPA, 2009). Reclaimed Asphalt Pavement (RAP) has been increasingly used in road construction to reduce air pollutants. More than 90 million t of RAP are produced every year, which accounts for approximately 80% of asphalt pavement excavated in the United States (FHWA, 1993). New Jersey, in particular, has doubled the use of RAP since 2001 (Copeland, 2011). Increased use of RAP has significantly reduced GHG emissions during asphalt production, which has been demonstrated by the Fairfield consulting study (Fairfield, 2008).

As one example, Foam Stabilized Base (FSB) is manufactured using 100% RAP in combination with a small amount of hot bitumen blended together with 1~2% potable water. The use of FSB reduces GHGs because it eliminates the use of energy stocks to heat aggregates and the need to quarry and transport virgin aggregates. Previous research showed the total GHG emissions of in-plant and in-situ FSB are 78.4 kg and 23.4 kg CO_2 equivalent (CO_2e) per ton of FSB. This represents approximately 42% and 83% reductions in carbon emissions compared to the most conservative estimate of emissions from Hot Mix Asphalt (HMA) (Liu et al., 2016). This result shows a great business opportunity for FSB producers to label their green products and potentially generate carbon offsets from FSB production and construction.

The challenge lies in the technical barriers to convert emission reductions to carbon offsets. The emission reductions from recycled pavements must be examined and verified by specific registries based on their standards and requirements for benchmark and additionality. First, existing carbon registries verify that emission offsets are real, additional, permanent, and verifiable on a project basis. There is a technical barrier to adapting emission reductions of recycled pavements, which are based on material processing and construction, to the existing project-based carbon credit framework. Second, existing registries require the demonstration of additionality by comparing to business-as-usual condition for the same emission source. There is a technical barrier to establish an acceptable business-as-usual benchmark for recycled pavements. Although HMA emissions seem an obvious and fair comparator, there is a need to establish an industry-wide benchmark that considers the differences among individual HMA plants to comply with carbon registry requirements.

The objective of this paper is to develop a framework for pavement recycling projects to verify GHG reductions and register for carbon offsets. Unlike most standards today that use a project-by-project method to quantify carbon offsets, a performance standard is presented based on the life cycle environmental impact analysis to facilitate credit assessment, generation, and verification. Carbon offsets are estimated based on the comparison of emissions from recycled pavements against those from HMA pavements that serve as performance benchmarks in the standard. The performance benchmarks are determined from the emission distribution of a group of hot mix facilities and placement projects. An illustrative example is presented to show the application of the performance standard in quantifying carbon offsets by using pavement recycling techniques. The results can be valuable to public agencies that are increasingly mandated to establish GHG emission standards used for attaining the low-carbon pavement practices, and asphalt producers who want to get recognition for their efforts to reduce GHG emissions.

2 CARBON TRADING FOR PAVEMENTS

2.1 Overview

Reducing GHGs from pavement projects has not been mandated in any states. Voluntary emission reductions efforts are rewarded through generating carbon offsets from carbon registries. Some pavement project proponents have joined carbon registries in order to document early actions taken to reduce their GHGs in advance of future mandatory regulations. By registering their emissions, these project proponents are likely to receive offsets for their early actions in future regulatory programs. Registering carbon emission reductions helps provide guidance, infrastructure, and quality standards for early carbon reductions, and encourage broad adoption of low-carbon practices with significant economic and environmental benefits.

The federal government first started a voluntary carbon reporting program in 1992, known as the "1605(b) program" of the Energy Policy Act, under the administration of the Department of Energy. This program was followed by the American Carbon Registry (ACR), the first non-profit nationwide private voluntary carbon registry. The ACR focuses on carbon emission reductions resulted from forestry and agriculture projects. It develops its own registering standards and protocols and provides an online registry system for registered projects to track and trade their verified emission reductions branded as Emission Reduction Tons (ERTs). At the state level, New Hampshire, Wisconsin and California have developed their own carbon reporting systems to support state-wide carbon emission reduction initiatives. These states encourage local facilities owners and project developers to voluntarily report emission reductions. A few states also collaborated together to establish regional carbon emissions registries. For example, 10 Northeastern and Mid-Atlantic states created the Eastern Climate Registry in 2003 to support voluntary carbon reduction programs and mandatory carbon markets in these states. The goal of the registry was to standardize the best practices in data reporting, verification, tracking, and management and therefore to establish a set of common quantification protocols that could be used throughout the region.

As two major carbon registries currently in the U.S., Verified Carbon Standard (VCS) and the Climate Registry started in 2008. These two registries standardized the best practices around the country and established a set of common reporting protocols that corporations, governments and nonprofit organizations can use and track across the states boundaries. Their common standards were based on previous registries, such as the California Climate Action Registry and the Eastern Climate Registry, and therefore were consistent with standards and protocols defined in ISO14064, PAC2050, and other standards accepted by the United Nations Framework Convention on Climate Change (UNFCCC). Proponents of the new registries view it as a critical first step to successfully implement emerging regulatory measures to reduce GHG emissions.

VCS is recognized as the world's most widely used voluntary carbon reduction program and has registered more than 1000 projects since 2008. As it features projects coming in all shapes and size, recycled pavement projects can be registered in the VCS. This provides the potential to create an alternative revenue stream from carbon offset trading in addition to tracking and managing their environmental liabilities. After award of verification from the VCS, project proponents could obtain 1) carbon offsets that can be traded in compliance and voluntary markets; 2) a verification-attached emission reduction logo, which usually has different levels showing the degree of their achievements; 3) a publicly acknowledged emission reduction dataset, which has a potential value of improving corporate reputation.

2.2 *Carbon offset standards*

Project-based standards have been widely used for years in the VCS. The standards produce carbon offsets based on the comparison of project emissions against a counterfactual baseline that represents a level of emissions that would occur in the absence of the project (Fischer, 2005). The standards have faced substantial criticism in the last few years, and there is evidence that a significant number of offsets come from projects that would have been undertaken anyway (Millard-Ball, 2013). These non-additional offsets, when traded to the regulated entities, implicitly expand the emission caps in compliance schemes and result in failing to achieve the desired emission targets. This means that the counterfactual baseline may be systematically biased. The bias is particularly prominent in evaluating project-based reduction against a counterfactual baseline that cannot be reasonably predicted. As the certifying agency is limited in its ability to propose such a counterfactual baseline, it must consign this task to the individual project proponents. This leaves great uncertainty regarding the integrity of baseline determination.

As an alternative approach, performance standards address this weakness in that they no longer rely on evaluating individual projects but use a pre-defined baseline to streamline the process of determining additionality (VCS, 2012). In this way, performance standards can establish an emissions threshold for a class of project activities. Individual projects that meet or exceed the threshold automatically qualify as additional projects, obviating the need for each project to determine additionality in its own right. Performance standards can be used by project developers, industry associations or governments to deliver reductions swiftly and affordably across multiple projects, industries or sectors. By lowering costs and helping speed project approval, this ensures industries and governments can curb GHG emissions at the pace and scale required to address climate change. The first comprehensive framework for performance standards were released by the VCS in 2012, which has been used in building and agriculture projects but has not been used for pavement projects.

2.3 *Performance benchmark*

Performance standards use benchmarks to both determine additionality and establish crediting baselines. A benchmark threshold is established at the outset, and all performance that meets or exceeds the threshold is considered additional, provided other qualifying criteria are met as well. A performance benchmark can also serve as the baseline for crediting emission reductions and removals. For pavement projects, the benchmark is determined based on emission distribution of HMA projects because more than 90% of pavements in the U.S. are constructed using HMA (NAPA, 2006).

HMA production throughout the country is being done in the same way other than difference in additives, such as crumb rubber, polymers, antistripping agents etc., even though the polymers are added their percent weight by mix is less than 2% (Mundt et al, 2009). This can be understood as the process of manufacturing HMA is same throughout the country irrespective of the mix designs. GHG emission performance of HMA plants depends on their production variables including percentage of RAP used as aggregate in HMA, type of fuel used for plant combustion, and aggregate hauling distance. The current distributions of HMA production performance are summarizes as follows. The average percentage of RAP ranges from 13% to 19% according to the studies of NAPA (2013) and Federal Highway Administration (2011). Typical fuel types include natural gas, oil and propane. EPA (2000) reported that natural gas fuel is used to produce 70% to 90% of the HMA. The remainder of the HMA is produced using oil, propane, waste oil, or other fuels (EPA, 2000). Aggregate hauling distance is typically less than 40 miles when projects are using local aggregates and larger than 40 miles when projects are importing aggregates from other places.

Sixteen HMA producers and ten placement projects are surveyed to determine performance benchmarks. Performance benchmarks are represented by GHG emission intensities (CO_2 equivalent per metric ton HMA, CO_2e/t) from the sampling projects, which are the sum of emissions from raw material production, the hot mix facility and the placement process. Each producer reported the consumption of raw material and energy, and material delivery distance on a quarterly basis. The estimation of GHG emission intensities is detailed in our previous work (Cui, 2014). Due to the significant impact of project type and transport distance on GHG emissions, performance benchmarks are stratified on project types and one-way distances between the HMA plant and job site. Stratum 1 is for patching projects with hauling distance less than 40 miles, while Stratum 2 is for is for patching projects with hauling distance larger than 40 miles, finally, Stratum 3 is for roadway projects. The performance benchmarks for all three strata are summarized in Table 1.

The average emission intensity (μ) of surveyed HMA producers is 134.8 $kgCO_2e/t$ HMA and the standard deviation (σ) is 15.5 $kgCO_2e/t$ for Stratum 1, a represented in Figure 1. The average emission intensity (μ) of surveyed HMA producers is 170.3 $kgCO_2e/t$ HMA and the standard deviation (σ) is 33.6 $kgCO_2e/t$ for Stratum 2. The average emission intensity (μ) of surveyed HMA producers is 121.9 $kgCO_2e/t$ HMA and the standard deviation (σ) is 19.8 $kgCO_2e/t$ for Stratum 3. After stratification, each stratum has a performance benchmark. According to UNFCCC (2006), the performance benchmark is defined as a threshold that surpasses the 80th percentile of existing HMA producers. Given the HMA emission approximates a normal distribution, the performance benchmark 121.9 $kgCO_2e/t$ HMA (equals to $\mu - 0.84\sigma$) for Stratum 1 (patching projects with hauling distance less than 40 miles), which is illustrated in Figure 1.

Projects that emit less than the predetermined benchmark are determined to have additionality. Mathematically, the additionality is determined using the project emission intensity (derived from section 3.2) minus the performance benchmark. The project can be determined additional if the figure is less than 0; otherwise the project is not additional.

Table 1. Performance benchmark for patching projects and roadway projects $kgCO_2e/t$.

Stratum	Project type	Hauling distance	Average emission intensity (μ)	Standard deviation (σ)	Performance benchmark
1	Parking lot	≤ 40 miles	134.8	15.5	121.9
2	Parking lot	> 40 miles	170.3	33.6	142.4
3	Roadway	Undefined	121.9	19.8	105.5

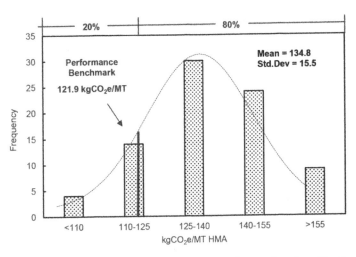

Figure 1. Illustration of performance benchmark for hauling distance less than 40 miles.

Performance benchmarks change over time. The changing trend is decided in the following way: use of recycled raw materials saves significant GHG by eliminating the emissions from mining, processing and transporting crushed stone and bitumen binder. According to NAPA (2012), when the use of RAP increases by 1 t, 10 kg emission can be avoided accordingly. As such, if the percentage of RAP increases by 1%, 0.1 kg emission can be avoided for producing 1t HMA. According to NAPA (2013), the use of RAP in HMA is expected to increase by 1% every year. Therefore, performance benchmarks decreases by 0.1 $kgCO_2e/t$ annually.

3 RECYCLED PAVEMENT EMISSION REDUCTIONS

3.1 *Types of recycled pavements*

Five types of recycling methods are commonly used for asphalt pavements: (*i*) hot recycling, (*ii*) hot in-place recycling, (*iii*) cold in-place recycling, (*iv*) cold central plant recycling, and (*v*) full depth reclamation.

- Hot recycling is the process in which RAP materials are combined with new materials, sometimes along with a recycling agent, to produce HMA mixtures. Both batch and drum type hot mix plants are used to produce recycled mix. The RAP materials can be obtained by milling or ripping and crushing operation. The mix placement and compaction equipment and procedures are the same as for regular HMA.
- Hot In-place Recycling (HIR) consists of a method in which the existing pavement is heated and softened, and then scarified/milled to a specified depth. New HMA and recycling agent may be added to the scarified RAP materials during the recycling process. HIR can be performed either as a single pass or as a multiple pass operation. In single pass operation, the scarified in-place material can be combined with new material if needed. In multiple pass operation, the restored RAP material is re-compacted first, and a new wearing surface is applied later. The depth of treatment varies between ¾ in to 2 in. The advantages of HIR are that surface cracks can be eliminated, ruts and shoves and bumps can be corrected, aged asphalt is rejuvenated, aggregate gradation and asphalt content can be modified, traffic interruption is minimal, and hauling costs are minimized.
- Cold In-Place Recycling (CIR) involves reuse of the existing pavement material without the application of heat. Except for any recycling agent, no transportation of materials is usually required, and aggregate can be added, therefore hauling cost is very low. The process includes pulverizing the existing pavement, sizing of the RAP, application of recycling agent, placement, and compaction. The processed material is deposited in a windrow from

the mixing device, where it is picked up, placed, and compacted with conventional hot mix asphalt laydown and rolling equipment. The depth of treatment is typically from 3 to 4 in. The advantages of CIR include significant structural treatment of most pavement distress, improvement of ride quality, minimum hauling and air quality problems, and capability of pavement widening.
- Cold Central Plant Recycling (CCPR) is similar to both a traditional asphalt plant and a CIR process. With CCPR, material removed from an existing asphalt pavement is transported to a central location—either on the project site or an existing asphalt plant. The removed material can be crushed and screened to make a uniform product or and be simply screened prior to feeding the cold plant. The cold plant, like the CIR process, uses either asphalt emulsion or foamed asphalt as a binding agent. Once the material and the binding agent are mixed, it can be discharged into a truck and taken to a project for paving. An advantage of the CCPR process is the ability to use existing materials at an asphalt plant. In many urban areas, contractors have excess quantities of RAP. By processing the RAP on hand, asphalt contractors can use the cold plant to produce a new asphalt base mix that can be used in various ways including new construction.
- Full Depth Reclamation (FDR) is a recycling method where all of the asphalt pavement section and a predetermined amount of underlying base material is treated to produce a stabilized base course. It is basically a cold mix recycling process in which different types of additives are added to obtain an improved base. The four main steps in this process are pulverization, introduction of additive, compaction, and application of a surface or a wearing course. If the in-place material is not sufficient to provide the desired depth of the treated base, new materials may be imported and included in the processing. This method of recycling is normally performed to a depth of 4 to 12 in.

GHG emission reductions from recycled asphalt pavements versus conventional HMA pavements are as follows: (i) less or no virgin aggregates are required, eliminating the energy and resources needed for excavating machines and trucking, (ii) less liquid asphalt/bitumen is required, reducing embodied material emissions, and (iii) aggregates in cold recycling process do not have to be heated, which reduces the use of energy for heating. In most applications, but especially in rural areas, the GHG emissions from trucking are significantly reduced. This is because in-place recycling process are done on the project site.

3.2 *Carbon offset quantification*

Emission intensity of a pavement project (EI) represents the quantity of GHG emitted from producing and placing 1 metric ton of pavement. It is the summation of material emission intensity (EI_M), to-plant delivery emissions intensity (EI_{PD}), in-plant production emission intensity (EI_P), to-site delivery emissions intensity (EI_{SD}) and on-site installation emission intensity (EI_I). EI should be calculated as follows:

$$EI = EI_M + EI_{PD} + EI_{SD} + EI_P + EI_I \qquad (1)$$

The material EI (EI_M) should be calculated as follows:

$$EI_M = \frac{EF_M \times W_M}{\text{Project amount}} \qquad (2)$$

where:
EI_M = Emission intensity of raw material production (kgCO$_2$e/t)
EF_M = Material emission factor (kgCO$_2$e/kg)
W_M = Material weight (kg)
Project amount = Amount of pavement product manufactured (t)

The to-plant delivery EI (EI_{PD}) and to-site delivery (EI_{SD}) should be calculated as follows:

$$EI_{PD} = \frac{Distance_P \times EF_T}{Project\ amount} \tag{3}$$

$$EI_{SD} = \frac{Distance_S \times EF_T}{Project\ amount} \tag{4}$$

where:
EI_{PD} = Emission intensity of to-plant delivery (kgCO$_2$e/t)
EI_{SD} = Emission intensity of to-site delivery (kgCO$_2$e/t)
$Distance_P$ = Distance to plant (mile)
$Distance_S$ = Distance to site (mile)
EF_T = Truck emission factor (kgCO$_2$e/mile)
$Project\ amount$ = Amount of pavement product manufactured (t)

The In-plant production EI (EI_P) should be calculated as follows:

$$EI_P = EI_D + EI_E \tag{5}$$

$$EI_D = \frac{EF_{EQ} \times HR_{EQ}}{Project\ amount} \tag{6}$$

$$EI_E = \frac{EF_{EL} \times C_{EL}}{Project\ amount} \tag{7}$$

where:
EI_P = Emission intensity of in-plant production (kgCO$_2$e/t)
EI_D = Emission intensity of diesel-consuming activities (kgCO$_2$e/t)
EI_E = Emission intensity of electricity-consuming activities (kgCO$_2$e/t)
EF_{EQ} = Equipment emission factor (kgCO$_2$e/hour)
EF_{EL} = Electricity emission factor (kgCO$_2$e/kWh)
HR_{EQ} = Equipment operation hours (hour)
C_{EL} = Electricity consumption (kWh)
$Project\ amount$ = Amount of pavement product manufactured (t)

The on-site installation EI (EI_I) should be calculated as follows:

$$EI_I = \frac{EF_{EQ} \times HR_{EQ}}{Project\ amount} \tag{8}$$

where:
EI_I = Emission intensity of pavement installation (kgCO$_2$e/t)
EF_{EQ} = Equipment emission factor (kgCO$_2$e/hour)
HR_{EQ} = Equipment operation hours (hour)
$Project\ amount$ = Amount of pavement installed (t)

Therefore, the net emission reductions for a pavement recycling project (ER) should be the emission intensity differences adjusted by the weight differences (φ)[1]. The reductions should be calculated according to Equation 9.

$$ER = (PB - \varphi EI) \cdot Project\ amount / 1{,}000 \tag{9}$$

---

[1] Here is an example showing how to determine weight difference (φ). On average, various Departments of Transportation are considering a structural layer coefficient of 0.32 for FSB (Schwartz and Khosravifar, 2013). The structural layer coefficient for a 19 mm HMA base mix is 0.40. Accordingly, substituting FSB and asphalt emulsions for HMA on a project would, on average, require the FSB layer to be approximately 25% thicker than the HMA layer. The densities of FSB and HMA are 130 lb/cu.ft and 160 lb/cu.ft. After factoring in these density differences, the use of FSB should be 2% more than HMA base by weight for the same length of paved road. The weight difference (φ) is therefore 1.02 for FSB project.

where:

ER = Net emission reduction of a pavement recycling project (tCO_2e)
PB = Performance benchmark, referring to Table 1 (kgCO_2e/t)
Φ = Adjustment factor for weight difference
EI = Emission intensity of a pavement recycling project (kgCO_2e/t)
Project amount = Amount of pavement product manufactured (t)

4 CASE STUDY

A case study of FSB projects is performed to illustrate the use of performance standards in project crediting. In this case, FSB is manufactured with CCPR process. The amount of carbon offsets from FSB projects are quantified and the gains from offset trading are analyzed to demonstrate economic viability of performing pavement recycling projects.

Production data relevant to CCPR was collected onsite at the Global Emissionairy, LLC. at Forestville, MD from March 2012 to May 2013. The company suspended the production from October 2012 to February 2013 due to cold weather. During the production periods, a total of 1,337t of FSB was produced using 95.6% RAP in combination with 2.3% hot bitumen blended together with 1% potable water and 1% Portland cement. Transportation distances for the shipment of input materials were collected from production records. Material production and transportation emission were calculated using Equations 2 to 7.

The placement procedures for a CCPR overlay are the same as for HMA. Similar to the estimation for HMA placement, the equipment operation information for CCPR was gathered from thirteen patching projects conducted in Maryland and Virginia. Primary equipment used for placing FSB includes milling machines, backhoes, loaders, sweeper, paver, rollers and trucks. The operation information for equipment was obtained from contractor's daily reports and truck driver reports. Equipment emissions were calculated using Equation 8.

Figure 2 shows the distribution of GHG emissions from FSB production and placement using CCPR. The GHG emission intensity of CCPR is 76.8 kgCO_2e per metric ton of FSB pavement for the monitoring period. The top four emission sources are paving (20.2 kg CO_2e/t, 26.2%), milling (13.7 kg CO_2e/t, 17.8%), bitumen (11.7 kgCO_2e/t, 15.2%), and cement (9.1 kg CO_2e/t, 11.8%). Nearly half of emissions (54%) are generated at the job site and the remaining half (46%) are generated off site.

Comparing against the performance benchmark of 121.9 kgCO_2e/t, performing FSB projects is eligible to generate 43.5 kgCO_2e/t of carbon offsets (Fig. 3). Within the three-month

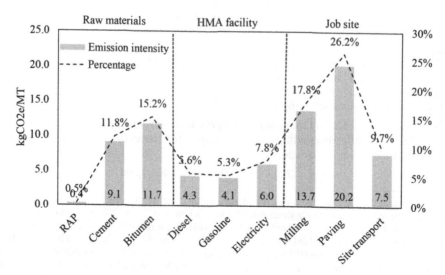

Figure 2. Distribution of GHG emissions from FSB pavement using CCPR.

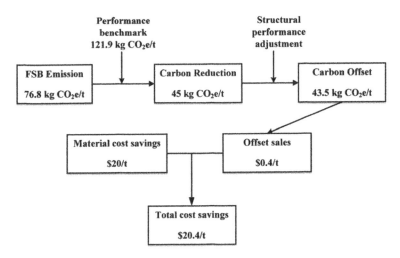

Figure 3. Cost savings with the use of FSB pavement.

production periods, a total of 58.2 t of carbon offsets can be generated with the production of 1,337 t of FSB. The revenues from offset sales can amount to $582 assuming the offset price is $10/t CO_2e. In addition to the offset revenues, the use of FSB can save material costs by approximately $20/t. When cost savings from materials are accounted for, the total cost savings from replacing HMA with FSB are approximately $20.4/t. The National Asphalt Pavement Association (NAPA) estimated HMA production is 317 million t in 2014 (NAPA, 2015). Even with a 20% market share, FSB producers will be able to obtain 2.8 million t of carbon offsets in a year. If all of these offsets can be traded in the market, FSB producers will open up more than 20 million revenue sources beyond direct FSB sales.

5 CONCLUSION

Performance benchmarks have been developed to serve as baseline emissions for quantifying GHG removals by using recycled asphalt materials. The baseline configuration for determining the benchmarks is the production of conventional HMA. A carbon offset quantification method has been established to measure the GHG emissions from producing and placing HMA. Sampling surveys of typical HMA facilities and placement projects form a data pool to support the determination of conservative performance benchmarks that represent different geographic areas, pavement structures, and production techniques. An illustrative example is used to demonstrate that significant financial rewards can be granted by adopting recycled asphalt pavements, with the aid of the proposed benchmark under an increasingly expanded carbon trading market. Primary findings include:

- The performance benchmarks are determined as 121.9 kgCO_2e/t HMA for patching projects with hauling distance less than 40 miles, 142.4 kgCO_2e/t HMA for patching projects with hauling distance larger than 40 miles, and 105.5 kgCO_2e/t HMA for roadway projects. These values represent emission levels that 80% of existing HMA producers are unable to reach and reasonably avoid the occurrence of free-riders. The boundary covers the GHG emissions from raw material production, hot mix facility and placement process.
- Rapidly evolving carbon trading markets increase the competitive advantages of "green" producers by generating financial revenues via offset sales. For example, performing FSB projects is eligible to generate 43.5 kgCO_2e/t of carbon offsets. When cost savings from materials are accounted for, the total cost savings from replacing HMA with FSB are approximately $20.4/t. Considering 317 million t of HMA production per year, even with a 20% market share, FSB producers will be able to obtain 2.8 million t of carbon offsets in

a year. If all of these offsets can be traded in the market, FSB producers will open up more than 20 million revenue source beyond direct FSB sales.

REFERENCES

Copeland, A., Reclaimed Asphalt Pavement in Asphalt Mixtures: State of the Practice, FHWA-HRT-11-021, Turner-Fairbank Highway Research Center, Federal Highway Administration, McLean, VA, 2011.

Cui, Q., Methodology for pavement application using foam stabilized base, Verified Carbon Standard, Washington, D.C., 2014.

EPA, Hot mix asphalt plants emission assessment report, EPA 454/R-00-019, Office of Air and Radiation, United States Environmental Protection Agency (EPA), Washington, D.C., 2009.

EPA. Hot mix asphalt plants emission assessment report. United States Environmental Protection Agency (EPA), Washington, D.C., 2000.

Fairfield, Australia's Future Infrastructure Requirements, Infrastructure Australia Discussion Paper 1, Fairfield City, 2008.

FHWA, A Study of the Use of Recycled Paving Materials: A Report to Congress, FHWA-RD-93-417, Federal Highway Administration (FHWA), Washington, D.C., 1993.

Fischer, C., Project-based mechanism for emissions reductions: balancing trade-offs with baselines, Energy Policy 33, 1808, 2005.

Liu, X., Cui, Q., Schwartz, C. Introduction of mechanistic-empirical pavement design into pavement carbon footprint analysis. International Journal of Pavement Engineering, 2016. DOI: 10.1080/10298436.2016.1205748

Mangum, M., Asphalt Paving Sector Presentation, Health Effects of Occupational Exposure to Emissions from Asphalt/Bitumen Symposium, Dresden, 2006.

Millard-Ball, A., The trouble with voluntary emissions trading: Uncertainty and adverse selection in sectoral crediting programs, Journal of Environmental Economics and Management 65, 40, 2013.

Mundt D.J., Marano K.M., Nunes A.P., Adams R.C. A review of changes in composition of hot mix asphalt in the United States, Journal of Occupation Environmental Hygene, 2009.

NAPA. Annual asphalt pavement industry survey on recycled materials and warm-mix asphalt usage: 2009–2012. National Asphalt Pavement Association, Lanham, MD, 2013.

NAPA. Annual asphalt pavement industry survey on recycled materials and warm-mix asphalt usage: 2011–2014. National Asphalt Pavement Association, Lanham, MD, 2015.

NAPA. National Asphalt Pavement Association Comments to Midwest Regional Planning Organization: Interim White Paper on Candidate Control Measures to Reduce Emissions from Hot Mix Asphalt Plants, National Asphalt Pavement Association, Lanham, MD, 2006.

Truitt, P., Potential for Reducing Greenhouse Gas Emissions in the Construction Sector, Environmental Protection Agency (EPA), McLean, VA, 2009.

UNFCCC. Report of the Conference of the Parties serving as the meeting of the Parties to the Kyoto Protocol. United Nations Framework Convention on Climate Change, Born, 2006.

VCS, Scaling up: VCS Standardized Methods. Verified Carbon Standard, Washington D.C., 2012.

A methodology for sustainable mechanistic-empirical pavement design

N. Soliman & M. Hassan
Louisiana State University, Baton Rouge, Louisiana, USA

ABSTRACT: Roads cause significant impacts on the environment, such as contributing to emissions responsible for climate change. For this reason, pavement design methodology should integrate environmental as well as economical impacts. The objective of this study was to incorporate environmental impacts as part of the Mechanistic-Empirical (ME) pavement design procedure using Environmental Product Declarations (EPDs). This was accomplished through integration of environmental and economic impacts into the current ME pavement design framework. To facilitate the use of EPDs, a windows-based tool with a graphical user interface, was developed. The simple interface tool enables the user to filter and query the environmental impacts based on criteria such as compressive strength, product name, and product mix number. The final selection criteria can be adjusted by the user, based on feedback from the stakeholders. Two design case studies are presented to demonstrate the use of the developed tool.

1 INTRODUCTION

The Mechanistic-Empirical Pavement Design Method (MEPDG), referred to as Pavement ME, has been implemented by a number of state agencies as a tool for pavement design and performance prediction. This approach calculates pavement responses such as stresses, strains, and pavement deflections under different climatic conditions. The procedure then accumulates pavement damage over the design period and relates calculated damage over time to pavement distresses and smoothness, based on performance models (What Is Mechanistic-Empirical Design 2012).

The MEPDG does not incorporate the environmental impacts into the design framework. Design analyses performed do not quantify the environmental impacts of pavements such as Global Warming Potential (GWP), Ozone Depletion Potential (ODP), Acidification Potential (AP), Eutrophication Potential (EP), and Photochemical Ozone Creation Potentials (POCP). If the MEPDG does not address current global trends by targeting environmentally-responsible design, it will be of little value in the long-term. Therefore, an incorporation of environmental impacts into the overall design, as well as in the transportation stage, is becoming of primary importance. One of the tools to quantify environmental impact is Life-Cycle Assessment (LCA). However, it is time-consuming and requires extensive amount of data (Inyim et al. 2016).

Another method to quantify environmental impacts is Environmental Product Declaration (EPD), which represents an internationally standardized method (ISO 14025 standard—ISO, 2006c). EPDs are documents, which communicate information about the environmental performance of a product based on LCA. EPDs also provide quantifiable environmental data, based on pre-determined parameters (ISO 2006c). A recent literature review showed that the use and disclosure of environmental information through EPDs is gaining acceptance (Fazil et al. 2016). The intended audience for those EPDs includes consumers and professional buyers involved in the decision-making process (Fazil et al. 2016). However, it should be noted that while LCA may be used to quantify the environmental impact in any stage of the lifecycle of the product (e.g., cradle to grave), EPDs only cover a cradle to gate analysis.

2 OBJECTIVE AND SCOPE

The objective of this study was to integrate environmental and economic design criteria as part of the mechanistic empirical Pavement ME design framework. The developed framework was used to evaluate two case studies in Texas. The first case study was SH-121 consisting of a Continuously-Reinforced Concrete Pavement (CRCP) design. It compared conventional vs. internally cured concrete. The second case study was on Interstate I-45 consisting of a Jointed-Plain Concrete Pavement (JPCP) design. It also compared conventional vs. internally Cured Concrete (ICC). The environmental impact considered the following phases: raw materials extraction, transportation to the manufacturing location, and manufacturing, which was obtained from EPDs. The transportation impact from the manufacturing to construction phase was evaluated using LCA. The final design selection was evaluated using a weighted average criterion considering both economic and environmental criteria.

3 BACKGROUND

3.1 *Tools for evaluating environmental impact*

3.1.1 *Life-Cycle Assessment (LCA)*

LCA presents a method that can be used to evaluate the net environmental impacts of products and services across a set of environmental matrix inclusive of all interactions with human and natural systems (Fazil et al. 2016). It consists of four main steps (Strazza et al. 2016):

a. Goal and scope definition: Defines the goal for conducting a life-cycle assessment for a given product.
b. Inventory analysis: Documents the resulting emissions, materials, and energy used in the atmosphere, land, and water.
c. Impact analysis: The effects of resource use and emissions are grouped and quantified into a number of impact categories, which can be weighted for importance.
d. Interpretation: The results are interpreted for evaluation towards reducing the environmental impacts of a product.

Many studies have used LCA to quantify the environmental impacts of pavement (Santero et al. 2010). However, these studies used different functional units, different data sources, and evaluated different environmental impact categories. This renders those LCA studies incomparable. Therefore, a more consistent and comparable approach is needed to evaluate the environmental impacts.

3.1.2 *Environmental Product Declarations*

An emerging method for quantifying the environmental impacts of a product employs Environmental Product Declarations. EPDs are defined as independently verified and registered documents that communicate transparent and comparable information regarding the life-cycle assessment of a product (What Is Mechanistic-Empirical Design 2012). EPDs are the summary of the data collected in LCA. These are verified by a third party to guarantee transparency based on ISO 14025. However, the phases in EPDs only cover cradle to gate analysis: raw materials acquisition, manufacturing, and transportation from the manufacturing place to the use phase. Hence, the use phase is not included in the analysis. The environmental impacts covered include (a) Global Warming Potential (GWP), (b) Ozone Depletion Potential (ODP), (c) Acidification Potential (AP), (d) Eutrophication Potential (EP), and (e) Photochemical Ozone Creation Potential (POCP).

3.1.3 *Product Category Rule*

Product Category Rules (PCRs) are defined in ISO 14025 as a set of specific rules, requirements, and guidelines for developing EPDs for one or more products that can fulfill equivalent functions. PCRs determine what and how information should be gathered for an environmental product declaration. This is important for facilitating the comparison of EPDs. The product category rule considered in this study focused on concrete products and thereby enables the development of EPDs related to this product from cradle to gate. The considered PCR outlines both mandatory and optional impact categories that may be included. The PCR was developed for use in North America and in Canada, as well as other countries according to the following standards: ASTM C94, ASTM C90, and CSA A231.1/A23.2, UNSPSC code 30111500. In the considered product category rule, the transportation phase to construction site is optional. However, the following factors in the transportation phase were considered: (a) fuel and truck type, (b) average miles per gallon (or liters per kilometers) of gasoline or diesel, (c) total annual distance traveled for each type of truck, and (d) the impact from the truck backhaul (return trip of the truck to the plant) (ISO 2006).

4 RESEARCH METHODOLOGY

Environmental impacts, associated with a specific product, were queried from EPDs, and were then integrated into the overall design framework of Pavement ME. The developed algorithms analyze the results from the output file as follows: (a) The results are analyzed to assess whether the design passed or failed technical criteria; (b) If the design is technically acceptable, the user selects the number of material alternatives to be evaluated. Selection can be based on strength requirements such as the required compressive strength for concrete materials and transportation radius. Once the required compressive strength value is selected, the available products are displayed along with their respective environmental impacts. Based on the user's selection, the tool provides the products that match the required compressive strength along with the associated environmental impacts (GWP, AP, EP, ODP, and POCP). After selecting feasible products, the transportation distance from the manufacturer location to the project location is calculated and the environmental impacts are evaluated. The environmental impacts from EPD are then added to the environmental impacts of transportation to get the total environmental impacts. Values are then normalized and summed to get one final score for the environmental impact.

The design alternatives are then evaluated for cost analysis. The cost analysis uses the net present value method for evaluating different design alternatives. This includes factors such as initial cost, maintenance and rehabilitation costs, and salvage value. Finally, depending on stakeholder decision criteria, each component is assigned a weight for environmental and cost performances. For example, weights can follow the weights recommended by Building for Environmental and Economic Sustainability (BEES), which were assigned by the EPA Science Advisory Board. The final design selection is based on a weighted average value between these two factors, considering that these products have already satisfied the technical design criteria. The default weight value is 0.5 for both the economic and environmental impacts.

This process was incorporated on a windows-based design tool for ease of use. As illustrated in Figure 1, the software allows the analysis of multiple designs and layers. The design layer thickness is inputed as well as the project zip code. As shown in Figure 1(b), weights for environmental impacts are inputed as well as the weights for economic and environmental criteria. As shown in Figure 1(c), the vehicle type as well as the type of fuel used is selected to evaluate the environmental impacts of transportation. Furthermore, the material cost is inputed and is discounted to the net present value. A final report is displayed showing the results of the environmental and economic impacts, Figure 1(d).

4.1 *Calculating the environmental impacts of transportation*

The environmental impact of the transportation phase is an important factor, which may affect the results of the analysis. The proposed framework adopted the following equation to estimate the emissions associated with the transportation phase (Heather and Lester 2016):

$$E_{of\ substance} = 2 \times (VKT \times FC \times substance\ content \times Factor) \qquad (1)$$

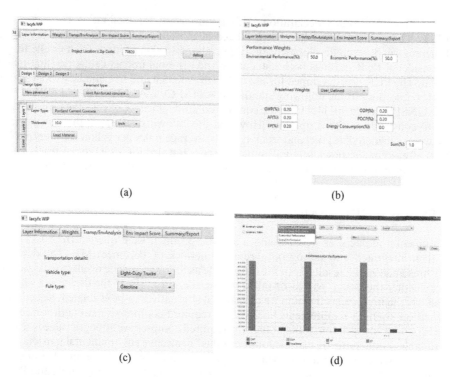

Figure 1. Layout of the developed computer tool.

where $E_{of\ substance}$ is the emissions of the required substance in grams; VKT is the vehicle kilometers traveled (input by the user); and FC is the fuel consumption (L/100 km) (Heather and Lester 2016). FC can be assumed 12.8 for light trucks (Heather and Lester 2016). It is noted that the emission calculated by Equation (1) is multiplied by a factor of 2 to account for the backhaul distance, as per the PCR requirements. Substance content represents the grams of the material per liter of fuel recommended by EPA and other agencies (Wilde et al. 1999) as follows: For carbon emissions (global warming potential): 2,421 grams-carbon / gallon for gasoline (Carbon Dioxide Emissions Resulting from Gasoline and Diesel 2005); and for sulfur emissions (acidification potential): 30 ppm for gasoline. The Factor in Equation (1) accounts for converting the molecular mass of carbon to carbon dioxide to account for global warming potential and to convert the molecular mass of sulfur to sulfur dioxide for acidification potential. The conversion factors used were (44/12) for global warming potential and (64/32) for acidification potential.

4.2 *Evaluating life cycle cost analysis*

Life-cycle cost of the different design alternatives are evaluated based on the net present worth method (life cycle cost analysis in pavement designs 1998). This method considers the initial cost, rehabilitation cost, and salvage value.

5 DEMONSTRATION OF THE DEVELOPED TOOL IN CASE STUDIES

5.1 *Case study #1*

5.1.1 *Project description*

The selected project was located in SH 121 west of I-75 and east of the Dallas North Tollway. (Rao and Darter 2003) falling in The Dallas-Fort Worth weather station. The pavement is expected to serve moderate traffic volume with an Average Annual Daily Traffic (AADT) of 23,400 with a linear traffic growth of 4%.

5.1.2 *Design inputs*

The design analysis period was assumed to be 30 years with a CRCP, Initial International Roughness Index (IRI) limit of 63 and a terminal IRI of 160 with a reliability level of 90%. The terminal thresholds for transverse cracking, longitudinal cracking, and corner cracking represented 10% of the slabs cracked (Rao and Darter 2003).

5.1.3 *Initial and alternative designs*

The original vs. the alternative trial designs are illustrated in Figure 2. The alternative design has thinner concrete thickness, consisting of Internally Cured Concrete (ICC), leading to lower environmental impacts and lower cost.

5.1.4 *Concrete products description*

The concrete mix designs that were used in this analysis were 5,500 psi for ICC and 6,000 psi for conventional concrete. The 5,200 psi was extrapolated to 5,500 psi, to match the value in EPD's, and the 6,000 psi was used as listed.

5.1.5 *Technical analysis*

Designs were re-analyzed to optimize the design thicknesses. This was performed by decreasing the thickness of the CRCP design by increments of ½ inch. The optimization process led to thicknesses of 10 in. for ICC at a 97% reliability. The performances of both designs were identical. In fact, when the performance of those designs was plotted on the same chart, the predictions coincided for a period of 30 years (Rao and Darter 2003).

5.1.6 *Environmental performance*

The environmental impacts for the mix design alternatives are presented in Figure 3, based on the EPDs and the developed tool; the environmental impacts for the five categories (i.e., GWP, ODP, AP, EP, POCP) are presented. As shown in Figure 3, the ICC alternative was more environmentally-friendly in terms of GWP and POCP, while all the other impacts were negligible.

5.1.7 *Environmental impacts of transportation*

The environmental impacts of the transportation phase are illustrated in Table 1, indicating that the ICC alternative contributed less emissions compared to the conventional one, since it was transported from a shorter distance.

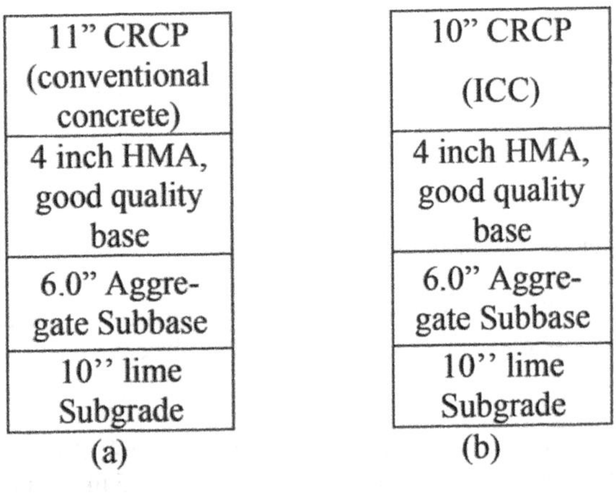

11" CRCP (conventional concrete)	10" CRCP (ICC)
4 inch HMA, good quality base	4 inch HMA, good quality base
6.0" Aggregate Subbase	6.0" Aggregate Subbase
10" lime Subgrade	10" lime Subgrade
(a)	(b)

Figure 2. (a) initial design vs. (b) alternative design.

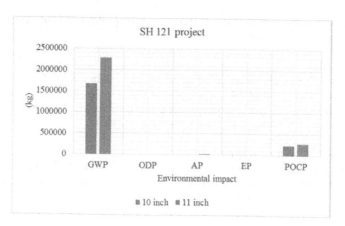

Figure 3. Environmental impact for conventional vs. ICC alternative.

Table 1. Environmental impact of transportation.

Project	Thickness (inch)	Distance (km)	Global warming potential (kg)	Acidification potential (kg)
SH121	10	5	3.001	0.0768
	11	6	3.60	0.09216

5.1.8 *Economic impact analysis*

Values were calculated per directional mile length of a roadway consisting of two lanes and a shouldee. The analysis period was selected to incorporate at least one rehabilitation activity (LCCA 1998). Therefore, the analysis period was selected at 60 years. Materials unit cost was obtained from Dallas-Fort Worth area. Results are presented in Table 2 for the conventional vs. the optimized designs. As shown in Table 2, the initial cost was lower for the ICC.

5.1.9 *Final design selection*

As illustrated in Table 3, the final design criteria were selected based on a weighted average between the economic and the environmental impacts, resulting in a total score of 0.46 for the ICC design and 0.54 for the conventional design. These results indicate lower environmental impacts and lower costs for the ICC design.

5.2 Case study #2

5.2.1 *Project description*

The selected project is located on I-45 east of the UP Intermodal Terminal falling in the Dallas-Fort Worth weather station. It is a connector that connects Dallas and Houston. This highway has a two-way Annual Average Daily Truck Traffic (AADTT) of 7,200 trucks.

5.2.2 *Design inputs*

The design analysis period was assumed to be 30 years with a Joint-Plain Concrete Pavement (JPCP), initial International Roughness Index (IRI) limit of 63 and a terminal IRI of 160 with a reliability level of 90%. The terminal thresholds for transverse cracking, longitudinal cracking, and corner cracking represented 10% of the slabs cracked (Rao and Darter 2003).

5.2.3 *Alternative designs*

The original and alternative designs are illustrated in Figure 4. Both designs are the same, except that the original design had a concrete thickness of 11.5 in. and the alternative design consist of internally cured concrete of 10.5 in.

Table 2. LCCA Comparison for SH 121 Project. Conventional concrete and ICC alternative.

Alternative (1) - Conventional concrete		Alternative 2 - ICC	
Description (year)	Total NPV ($)	Description (year)	Total NPV ($)
Initial cost	3,727,390	Initial cost	3,281,530
Full-depth repair (15)	4108	Full-depth repair (15)	4,108
Diamond Grind Existing Surface (25)	59,626	Diamond Grind Existing Surface (25)	59,626
Full-depth pavement repairs (25)	382	Full-depth pavement repairs (25)	459
Full-depth pavement repair (42)	1,156	Full-depth pavement repairs (40)	245
Full-depth pavement repairs (50)	18,066	Full-depth pavement repairs (60)	4,685
Place asphalt tack coat (9 sy/gal)	961	Place asphalt tack coat (9 sy/gal) (60)	715
2.0-in HMAC binder	36,690	2.0-in HMAC binder (60)	27,301
2.0-in HMAC surface	36,690	2.0-in HMAC surface (60)	27,301
Salvage value	−57,754	Salvage value	−75,002
Total (NPV)	3,827,315	Total (NPV)	3,330,968

*Note the analysis is performed for the top layers only, since all the other layers are similar. Design costs, overheads, etc. are assumed zero

Table 3. Summary of Environmental and Economic analysis.

Project	Design #	Thickness (in.)	Env. impact	weight	Sum (Env.)	Env. impact (0.5)	Ec. impact	Ec. impact (0.5)	Total weight
SH121	1	10	GWP	0.2	0.44	0.22	0.48	0.24	0.46
			ODP	0.2					
			AP	0.2					
			EP	0.2					
			POCP	0.2					
	2	11	GWP	0.2	0.56	0.28	0.52	0.26	0.54
			ODP	0.2					
			AP	0.2					
			EP	0.2					
			POCP	0.2					

Env. = Environmental; Ec. = Economic.

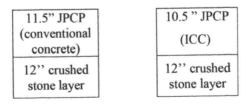

Figure 4. Design and design alternative—I-45 project.

5.2.4 Concrete product description

The concrete mix designs that were used in this analysis for ICC were 5,120 psi, extrapolated to 5500 psi to match the product database, and 6,000 psi for conventional concrete.

5.2.5 Technical analysis

The performances of both designs were identical over the entire analysis period, when plotted on the same chart, at a reliability level of 97% (Rao and Darter 2003).

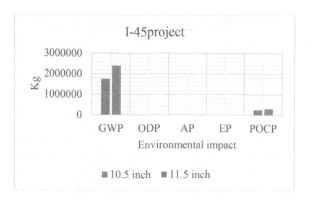

Figure 5. Environmental impact for different alternatives—I-45 project.

Table 4. Transportation environmental impact data for different alternatives.

Project	Thickness (in.)	Distance (km)	Global warming potential (kg)	Acidification potential (kg)
I-45	10.5	10	6.00	0.1536
	11.5	12	7.24	0.18432

Table 5. LCCA comparison for I-45 optimized design alternatives with conventional concrete and ICC. The analysis period is 60 years.

Alternative (1) - Conventional concrete		Alternative 2 - ICC	
Description (year)	Total NPV ($)	Description (year)	Total NVP ($)
Initial cost	3,896,816	Initial cost	3,445,606
Full-depth pavement repairs (15)	2,862	Full-depth pavement repairs (15)	2,862
Full-depth pavement repairs (25)	4,259	Full-depth pavement repairs (25)	4,259
Diamond Grind Existing JPCP (40)	38,271	Diamond Grind Existing JPCP (40)	38,271
Full-depth pavement repairs (40)	64,241	Full-depth pavement repairs (40)	72,442
Full-depth pavement repairs (50)	5,085	Full-depth pavement repairs (50)	5,085
Full-depth pavement repairs (60)	2,838	Full-depth pavement repairs (60)	5,676
Place asphalt tack coat (60)	715	Place asphalt tack coat (60)	715
2.0-in HMAC binder (60)	27,301	2.0-in HMAC binder (60)	27,301
2.0-in HMAC surface (60)	27,301	2.0-in HMAC surface (60)	27,301
Total (NPV)	4,069,689		3,629,518

*Note the analysis is performed for the top layers only, since all the other layers are similar. Design cost is assumed zero.

5.2.6 *Environmental performance*

As illustrated in Figure 5, both alternatives had similar environmental impact, except for GWP and POCP, where the conventional concrete alternative had a higher impact due to greater thickness.

5.2.7 *Environmental impact of transportation*

The environmental impact of transportation is illustrated in Table 4. The conventional concrete resulted in more emissions, since it was transported for longer distances.

5.2.8 *Economic impact analysis*

The cost-analysis was performed for a 60-year analysis period. The use of ICC resulted in a thickness reduction leading to cost savings. However, in the long-term, the maintenance and rehabilitation costs were higher by 6.7%. However, the overall net present value decreased by 1.2%. Results are illustrated in Table 5. The maintenance and rehabilitation cost varied.

Table 6. Summary of Environmental and Economic analyses.

Project	Design #	Thickness (inch)	Env. impact	weight	Sum (Env.)	Env. impact (0.5)	Ec. impact	Ec. impact (0.5)	Total weight
I-45	1	10.5	GWP	0.2	0.45	0.23	0.49	0.24	0.47
			ODP	0.2					
			AP	0.2					
			EP	0.2					
			POCP	0.2					
	2	11.5	GWP	0.2	0.55	0.27	0.51	0.26	0.53
			ODP	0.2					
			AP	0.2					
			EP	0.2					
			POCP	0.2					

Env. = Environmental; Ec. = Economic.

The Control CRCP will have maintenance in years 15, 25, 42 and a rehabilitation at year 50. The ICC will have maintenance at years 15, 25, 40 and a rehabilitation at year 60, offsetting the salvage value.

5.2.9 *Final design selection*

Detailed environmental and economic results are illustrated in Table 6. The final weighted average criteria was 0.47 for the ICC design vs. 0.53 for the initial design indicating lower emissions and costs for the ICC design.

6 CONCLUSIONS AND RECOMMENDATION

This paper developed a tool incorporating the environmental and economical impacts into the Mechanistic—Emperical Pavement Design (MEPDG). The environmental impact was included by incorporating Environmental Product Declarations (EPDs) to evaluate the environmental impacts from cradle to gate. This method evaluates the environmental impact in a comparable way, thereby solving the limitations of LCA. The use of the developed tool enabled the evaluation of alternative materials in terms of performance, environmental impacts, and economical impacts. Alternative materials such as internally cured concrete, based on this analysis, proved to have a better performance when compared to conventional concrete. This is due to its lower thickness and saving in intial cost as well as its better performance saving maintenance and rehabiltiation costs on the long term.

ACKNOWLEDGMENTS

The authors would like to thank the following companies for providing information facilitating the completion of this study: GCC Concrete Company, National Ready Mixed Concrete Association (NRMCA), CEMEX, and Portland Cement Association (PCA).

REFERENCES

Average Carbon Dioxide Emissions Resulting from Gasoline and Diesel Fuel. (2005, February). Retrieved from http://www.carbonsolutions.com/Resources/Average_Carbon_Dioxide_Emissions_Resulting_from_Gasoline_and_Diesel_Fuel.pdf.

BEES online. (2009, March). Retrieved from http://www.nist.gov/el/economics/BEESSoftware.cfm/

Fazil N. et al. Effective Environmental Policy toward Reducing Greenhouse Gas Emissions produced from Transportation. Accessed November 8, 2016.

Heather L., Lester B. Lifecycle assessment of Automobile/Fuel options, Mellon University, Pittsburgh, www.cmu.edu/gdi/docs/lca-of-automobile.pdf. March 1st, 2016.

ICC Evaluation Service, LLC (ICC-ES). (n.d.). Retrieved April 15, 2016, from http://www.icc-es.org/

International Organization for Standardization, Environment Management-Life Cycle Assessment—principles anf Framework. ISO 14040:2006 (E). 2006.

Inyim, P., Pereyra, J., Bienvenu, M., & Mostafavi, A. (2016). Environmental assessment of pavement infrastructure: A systematic review. *Journal Of Environmental Management, 176*128–138. doi:10.1016/j.jenvman.2016.03.042

Life-Cycle Cost Analysis in Pavement Design. (1998, September). Retrieved April 15, 2016, from http://www.wsdot.wa.gov/NR/rdonlyres/7A7CC34A-6336-4223-9F4A-22336DD26BC8/0/LCCA_TB.pdf

Loren Carl Mars. Environmental impact analysis: Gas vs. diesel in light-duty highway applications in the U.S. http://webpages.charter.net/lmarz/Diesel.pdf. Accessed March 27th, 2016.

Mack, J. (2011, September). Understanding and using the mechanistic emperical pavement design.

Rao, C., & Darter, M. (2003, September). Evaluation of internally cured concrete for paving applications. Retrieved august 1, 2016, from http://www.escsi.org/uploadedfiles/technical_docs/internal_curing/eval of Icc for paving apps report.pdf

Santero, N., Masanet, E., and Horvath, A., Life-Cycle Assessment of Pavements: A Critical Review of Existing Literature and Research, SN3119a, Portland Cement Association, Skokie, Illinois, USA, 2010, 81 pages.

Santeroa, N., Masanet, E., & Horvath, A. (2011). Life-cycle assessment of pavements. Part I: Critical review. Retrieved June 3, 2016.

Strazza, C., Del Borghi, A., Magrassi, F., & Gallo, M. (2016). Using environmental product declaration as source of data for life cycle assessment: a case study. Journal Of Cleaner Production, 112(Part 1), 333–342. doi:10.1016/j.jclepro.2015.07.058

Swei, O., Noshadravan1, Gregory, & Kirchain. (n.d.). Uncertainty management in comparative life-cycle assessment of pavements. Retrieved August 1, 2016, from https://cshub.mit.edu/sites/default/files/documents/uncertinaty management in comparative life-cycle assessment of pavements.pdf

The International EPD system. www. environdec.com/en/What-is-an-EPD. Accessed March. 8, 2016.

U.S Fuels: Diesel and Gasoline. February 2105. www.ransportpolicy.net/index.php?title=US:_Fuels:_Diesel_and_Gasoline#Diesel. Accessed March 27th, 2016.

What Is Mechanistic-Empirical Design? – The MEPDG and You. (2012.). Retrieved April 17, 2016, from http://www.pavementinteractive.org/2012/10/02/what-is-mechanistic-empirical-design-the-mepdg-and-you/

Wilde, W., Waalkes, S., & Harrison, R. (1999, September). Life cycle cost analysis of Portland cement concrete pavements. Retrieved April 15, 2016, from http://www.utexas.edu/research/ctr/pdf_reports/1739_1.pdf

Energy consumption and Greenhouse Gas Emissions of high RAP central plant hot recycling technology using Life Cycle Assessment: Case study

Yong Lu, Hao Wu, Aihua Liu, Wuyang Ding & Haoran Zhu
JSTI Group Co. Ltd., Nanjing, China
National Engineering Laboratory for Advanced Road Materials, Nanjing, China

ABSTRACT: Traditional LCA of central plant hot recycling shows that, the higher RAP added, the less energy consumed. This conclusion is mainly based on the calculation of reduced raw material. However, when the content of Reclaimed Asphalt Pavement mixture (RAP) reaches some level, it is needed to take measures such as using additive or improved construction method for maintain the performance of asphalt pavement. For the situation of high RAP central plant hot recycling technology has become more and more widely used, in order to evaluate its effect of energy saving and emission reduction, this paper investigate and compare the life cycle energy consumption and greenhouse gas emissions with the method of LCA, with considering the matters of RAP addition, asphalt-recycling agent, warm agent and transport distance. The study results indicate that, it is not appropriate say that the higher RAP added, the less energy used. Raw material transportation distance is the key factor of energy consumption and greenhouse gas emissions, while asphalt-recycling agent is the least factor. Besides, with theoretical calculation of different warm mix method, the result show that mechanical foaming is more energy saving than the organic additive warm agent, while being used in high RAP central plant hot recycling.

Keywords: High RAP central plant hot recycling; LCA; Warm agent; Transport distance

1 INTRODUCTION

Against the background of accelerating global warming due to the effect of greenhouse gases, efforts to reduce greenhouse gas emissions are being made worldwide. Among public works, road construction, which involves numerous processes, leads to greenhouse gas emission. Construction industry, in particular, is one of the high energy consumption industries in which the problem of a lack of efficient energy conservation in construction projects has been identified. Owing to the increase in project scales in recent times, the construction stage of public works requires mechanized construction and involves a significant amount of energy consumption and emission of environmental pollutants. Efficient use of fuel and material resources, reduction in greenhouse gas emissions and control of environmental impacts have become important to the construction industry, including pavement engineering. Life Cycle Assessment (LCA) is one important way of estimating the scale and environmental impacts of resource use and emissions to the environment (1,2). In several productive fields, recycling techniques represent one of the most promising strategies to achieve economic and environmental sustainability goals. Also in the pavement industry, recycling has gained increasing importance for the production of new asphalt mixtures.

The results of this Life Cycle Assessment (LCA) contribute information on the relative environmental impacts of reclaimed asphalt pavement methods and choice to regulatory and market decision makers in China.

In fact, it allows the optimization of the materials in terms of costs and natural resources saving (3, de la Roche et al. 2013). Many laboratory studies concerning the evaluation of performance of recycled mixtures with high RAP content have been carried out (4 Aurangzeb et al. 2012; 5, Frigio et al. 2014; 6, Stimilli et al. 2013). However, it is widely recognized that mixtures prepared in reclaimed asphalt pavement could perform excellent natural resources saving and is good to achieve economic and environmental sustainability goals when large amounts of RAP are incorporated in the new mixture. In fact, in the case of high content RAP mixtures the production process requires particular sequences to improve the performance of mixtures. The warm mix agent or warm mix method has to be used at the stage of the production process to avoid the excessive smoke during large amount RAP healing process, in addition of being benefit for the main properties of a flexible pavement (compact ability, stiffness, cracking aptitude and fatigue) (7). Moreover, recycling agent is utilized in reclaimed asphalt mixtures to improve such properties. Meanwhile, different transportation distance has influence on energy consumption and greenhouse gas emission. In this sense, many aspects related to the reclaimed asphalt pavement with large amount of RAP, can alter the energy consumption and greenhouse gas emission. So that combination of different factors with reclaimed asphalt pavement is necessary to assess by LCA methods for widely demonstrate the actual use of recycling technology with large amount of RAP in the new mixtures.

In this paper, the research focuses on life cycle assessment on recycling technology. Due to LCA concept, milling, production and transportation of materials, mixing and paving, etc. such production process are divided to calculate energy consumption and greenhouse gas emission in every stage. As well as, many factors including different content of RAP such as 10 wt.%, 30 wt.%, 50 wt.% in whole mixture (10 wt.% is defined as that the asphalt mixture has 10% content RAP at weight in the whole mixture), milling and repaving, in particular, based on 50 wt.%RAP, three warm mix methods and three content of recycling agent are considered to compare the energy consumption and greenhouse gas emission with combination of different factors.

2 METHODOLOGY

2.1 *Indicators*

In order to evaluate the environmental benefits of high RAP central plant hot recycling, depletion of minerals (resource saving and recycling), depletion of fossil fuels (energy consumption), Global Warming Potential (GHG emission) were selected as indicators to evaluate the environmental benefits of asphalt pavement construction and maintenance. The energy consumption data during the construction were directly investigated, and were converted to the unified energy unit according to the conversion coefficient of average low calorific value. The GHG emission including Carbon Dioxide (CO_2), Methane (CH_4), Nitrous Oxide (N_2O) was calculated through the energy consumed. They were converted to equivalent CO_2 emission using Global Warming Potential (GWP); the conversion coefficients of CH_4 and N_2O were 25 and 298 at a time horizon of 100 years, respectively (8). Other impact category indicators, for example, acidification, photo oxidant formation and so on, were not considered at this initial stage, because sufficient inventory data is still unavailable in China till now. Only the energy consumption and GHG emission are insufficient to indicate the advantage of materials recycling and the value of natural resources. The materials recycled were calculated for evaluation of the technology.

2.2 *Functional unit*

For comparison of different asphalt mixtures, the energy consumption and GHG emission of per ton mixture is adopted as the functional unit. For comparison of different pavement maintenance techniques, per unit area pavement is used.

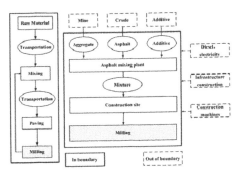

Figure 1. System boundary of LCA for reclaimed asphalt pavement.

Table 1. Energy consumption and GHG emission of some raw materials.

Raw materials		Energy consumption MJ/t	GHG emission kg/t
RAP		3.08E+1	2.09E+0
Virgin asphalt		2830	1.89E+02
Warm mix agent	Organic additives	54.63	5.02 E+0
	Mechanical foaming	/	/
Recycle agent		/	/
Aggregates		31.82	2.40E+00

2.3 System boundary

As the case study mainly focuses on the influence with RAP content, Recycling Agent content, warm mix technology and transportation distance to energy consumption and CO_2 emission, meanwhile there is not adequate data and calculation models for service life of pavement at present, only the processes directly related to pavement construction are analyzed in this paper, including the production and transportation of asphalt, aggregate and other raw materials, mixing and paving. While the indirectly related processes as infrastructure construction, production of machines and service life of pavement are excluded. Meanwhile, some raw materials with small addition which have little influence on the final results are negligible. According to the cut-off rule suggested by ISO 14040, the constituent parts with a proportion to the total results less than 1% by mass will be ignored, but the total proportion ignored should not exceed 5%. Figure 1 shows the system boundary analyzed in this research.

3 LIFE CYCLE IMPACT ASSESSMENT RESULTS

3.1 Production of raw materials

The raw materials of asphalt pavement include aggregate, asphalt, recycle agent and warm mix agent. Recycle agent has the same ingredient with asphalt, so the energy consumption and GHG emission is regarded as the same as asphalt. The energy consumption and GHG emission of aggregate refers to Chinese Life Cycle Database (CLCD) (9). Global Warming Potential (GWP) was characterized using factors reported by the Intergovernmental Panel on Climate Change in 2007(10–12). The database of European Bitumen Association provides the basic LCI data for virgin asphalt (8). Table 1 shows energy consumption and GHG emission of some raw materials.

Table 2. Energy consumption and GHG emission of equipment during production process of mixture.

Production process	Types	Energy consumption MJ/t	GHG emission kg/t
Mixing	Xi'an Construction Machine J3000	2.56E+2	1.98E+1
Transportation	30t	5.438E-1	4.030E-2
Paving	Vogele1800	1.651E+1	1.223E+0
Compaction	Steel roller	3.785E+0	2.805E-1
	Tire roller	5.441E+0	4.033E-1
Others	/	8.445E+0	6.259E-1

"Others" means energy consumption and GHG emission of loader and screening machine, because of low values, in this paper, their values are the average by survey data.
kg/t is the unit of GHG emission which means during the production process 1 ton asphalt mixture would produce mass of GHG emission in weight.

Referring to CLCD, the type of equipment is selected as the standard transportation mode. The energy consumption and GHG emission are shown in Table 2.

3.2 Mixture production

Typical pavement asphalt mixtures were chosen for analysis, such as Sup-20 (dense-graded HMA using Superpave method) with base asphalt by asphalt-aggregate ratio of 4.8% (asphalt content of 4.58%). Compared with new asphalt pavement, the construction process adds the milling process. In the mixing process, the Reclaimed Asphalt Pavement (RAP) is mixed with new asphalt and aggregate. The production of raw materials includes the mill and transport of RAP to the plant. The heating temperature of new aggregate is increased for heating RAP to the mixing temperature. Therefore, the fuel consumption of the plant is about 1 kg·t-1 higher than the typical new HMA based on collected data. The hot in-plant recycling mixture is typically used for the 8 cm thick bottom course in China, and the asphalt-aggregate ratios of both RAP and virgin asphalt mixtures are designed as 4.8%.

The paper focused on hot in-plant recycling mixture with different amount of RAP, such as the weight of 10%, 30% and 50%. Besides, warm technology as organic additives and machine foaming, different content of recycle agent and milling and repaving and so on, different factors among hot in-plant recycling technology are considered to comparing the energy consumption and GHG emission.

The production of asphalt mixture usually uses batch asphalt mixing plant in China. Nearly twenty experienced asphalt pavement construction companies were investigated for the accurate and reliable energy data. Average energy consumption is taken as a representative value in this research.

Currently, installation of auxiliary building asphalt mixture recycling equipment is mainly technology. Under this background, the energy consumption and GHG emission is assessment. Due to survey, the average fuel oil amount is approximately 6.0 kg/t, within the moisture content of RAP is 3–5%, heating to the temperature 900°C to 1100°C. During the hot in-plant recycling production, the heating temperature of new aggregate is higher than conventional asphalt mixture, so the energy consumption is higher than conventional batching plant with about 1 kg/t.

3.3 Paving and compaction

The speed of asphalt paver should be adapted to the production of mixing plant, thickness and width of pavement. The paving speed of reclaimed asphalt mixture is usually lower than other asphalt mixture, and typical construction speeds of 2.5 m·min^{-1} and 3 m·min^{-1} were selected for SUP pavements. The energy consumption of paving and compaction was calculated according to the power of machines and production efficiency. The average result is taken as the representative value.

3.4 *Summarization of data*

To describe clearly the combination of different factors in hot in-plant recycling mixture with high content RAP, Table 3 listed every process in construction and the content of raw materials.

Based on the analysis and calculation results, the energy consumption and GHG emission of reclaimed asphalt pavement are shown in Table 4.

The results show that, hot in-plant recycle technology with different factors has different energy consumption and GHG emission. In this paper, aimed at different factors, the LCA assessment method is used to analysis hot in-plant recycle with high RAP. The factors are followed.

- RAP content, different contents of RAP, such as 0% (in other words, milling and repaving), 30%, 40%, 50%, four kinds of RAP contents are considered for analysis.
- Recycle agents, when the RAP content is 50 wt.%, different contents of recycle agents are utilized.
- Warm mix technology, to ensure the performance and compaction of hot in-plant recycling mixture with high RAP, warm mix technology is usually method. Meanwhile, there are many kinds of warm mix technology, and during construction course, it usually no heating reduction method to make for the compaction.
- Distance, when the transportation distance of raw materials is different, it is obvious that the energy consumption and GHG emission is variant.

Table 3. Different factors combination for LCA assessment.

Construction process		Milling and repaving	RAP content (In weight)			RAP content, 50%, warm mix recycling	
			10%	30%	50%	Organic additive	Mechanical foaming
Temperature reduction		/	/	/	/	30°C or 0	30°C or 0
Raw materials production	Virgin asphalt	4.8%					
	Aggregate	/	/	/	/	/	/
	Recycling agent	/	3%, 5%, 7%, 9% weight in old asphalt			/	/
	Warm mix agent	/	/	/	/	3% weight in asphalt	Water is 1.5% weight in asphalt.
	RAP (80 km)	/	10%	30%	50%	50%	50%
Raw materials transportation distance		50 km, 100 km, 150 km, 200 km, 250 km					
Mixture mixing							
Mixture transportation		80 km					
Pavement construction		Paving Compacting Others					

Table 4. Energy consumption and GHG emission of different maintenance measures.

Construction process		Energy and emission	Milling and repaving	RAP content (In weight)			RAP content, 50%, warm mix recycling	
				10%	30%	50%	Organic additive	Mechanical foaming
Raw materials production	Virgin asphalt	Energy consumption (MJ·t⁻¹)	136.46	122.81	95.52	68.23	68.23	68.23
		GHG (kg·t⁻¹)	9.11	8.20	6.38	4.56	4.56	4.56
	Aggregate	Energy consumption (MJ·t⁻¹)	30.29	27.26	21.20	15.14	15.14	15.14
		GHG (kg·t⁻¹)	2.28	2.06	1.60	1.14	1.14	1.14
	Recycle agent content 3%	Energy consumption (MJ·t⁻¹)	/	0.36	1.09	1.82	1.82	1.82
		GHG (kg·t⁻¹)	/	0.02	0.07	0.12	0.12	0.12
	5%	Energy consumption (MJ·t-1)	/	0.61	1.82	3.03	3.03	3.03
		GHG (kg·t⁻¹)	/	0.04	0.12	0.20	0.20	0.20
	7%	Energy consumption (MJ·t⁻¹)	/	0.85	2.55	4.24	4.24	4.24
		GHG (kg·t⁻¹)	/	0.06	0.17	0.28	0.28	0.28
	9%	Energy consumption (MJ·t⁻¹)	/	1.09	3.27	5.45	5.45	5.45
		GHG (kg·t⁻¹)	/	0.07	0.22	0.36	0.36	0.36
	Warm mix agent	Energy consumption (MJ·t⁻¹)	/	/	/	/	39.51	/
		GHG (kg·t⁻¹)	/	/	/	/	3.63	/
	RAP (80 km)	Energy consumption (MJ·t⁻¹)	/	11.73	35.18	58.63	58.63	58.63
		GHG (kg·t⁻¹)	/	0.80	2.39	3.98	3.98	3.98
Raw materials transportation distance	50	Energy consumption (MJ·t⁻¹)	27.19	24.47	19.03	13.61	13.63	13.61
		GHG (kg·t⁻¹)	2.02	1.81	1.41	1.01	1.01	1.01
	100	Energy consumption (MJ·t⁻¹)	54.38	48.94	38.07	27.22	27.26	27.22
		GHG (kg·t⁻¹)	4.03	3.63	2.82	2.02	2.02	2.02
	150	Energy consumption (MJ·t⁻¹)	81.57	73.81	58.28	42.75	42.75	42.75
		GHG (kg·t⁻¹)	6.05	5.47	4.32	3.17	3.18	3.17
	200	Energy consumption (MJ·t⁻¹)	108.76	97.88	76.13	54.45	54.53	54.45
		GHG (kg·t⁻¹)	8.06	7.25	5.65	4.04	4.04	4.04
	250	Energy consumption (MJ·t⁻¹)	135.95	122.36	95.17	68.06	68.16	68.06
		GHG (kg·t⁻¹)	10.08	9.07	7.06	5.04	5.04	5.04
Mixture mixing (no heating reduction with warm mix technology)		Energy consumption (MJ·t⁻¹)	256.00	256.00	256.00	256.00	256.00	256.00
		GHG (kg·t⁻¹)	19.80	19.80	19.80	19.80	19.80	19.80
Mixture mixing (heating reduction with warm mix technology)		Energy consumption (MJ·t⁻¹)	256.00	256.00	256.00	256.00	204.80	204.80
		GHG (kg·t⁻¹)	19.80	19.80	19.80	19.80	15.84	15.84
Mixture transportation distance (80 km)		Energy consumption (MJ·t⁻¹)	43.50	43.50	43.50	43.50	43.50	43.50
		GHG (kg·t⁻¹)	3.22	3.22	3.22	3.22	3.22	3.22
Pavement construction	Paving	Energy consumption (MJ·t⁻¹)	16.51	16.51	16.51	16.51	16.51	1.22
		GHG (kg·t⁻¹)	1.22	1.22	1.22	1.22	1.22	1.22
	Compacting	Energy consumption (MJ·t⁻¹)	9.23	9.23	9.23	9.23	9.23	0.68
		GHG (kg·t⁻¹)	0.68	0.68	0.68	0.68	0.68	0.68
	Others	Energy consumption (MJ·t⁻¹)	8.45	8.45	8.45	8.45	8.45	0.63
		GHG (kg·t⁻¹)	0.63	0.63	0.63	0.63	0.63	0.63

4 DISCUSSION

4.1 *The effect of RAP content*

The main difference among milling and repaving, hot in-plant recycle mixture with 10%, 30%, 50% amount RAP in the whole mixture is the difference of the content of RAP. RAP is waste mixture recycling. When the other situations are the same, with the more RAP used, the energy consumption and GHG emission is less. Based on the date in section 3.2 and Table 2, the energy consumption and GHG emission of every construction process and total production can be calculated. The result is shown in Figure 2.

Results show that, based on some assumptions, with the increasing of the RAP content, the energy consumption and GHG emission is reduction. Actually, when the RAP content is increased, to ensure the reclaimed asphalt pavement performance, many improvement actions are used inevitably. These actions such as warm mix technology and recycle agent would have some energy consumption and CO_2 emission. Thus these factors will be considered in the paper.

4.2 *The effect of recycle agent*

Under the above date base, different content of recycle agent, such as 3%, 5%, 7%, 9% accounted of old asphalt mass are taken into consideration to be added in reclaimed asphalt mixture. Then, the energy consumption and CO_2 emission are calculated and be shown in Table 5.

Results showed that, the energy consumption and GHG emission of hot in-plant asphalt mixture with different content RAP is nearly the same while the amount of recycle agent is changed. Maybe it is because, the energy consumption and CO_2 equivalent value has no primary date. Considering the chemistry composite is similar with asphalt, so the energy consumption and CO_2 equivalent value of recycle agent is referred to asphalt. Besides the mass of recycle agent is a little account for asphalt, consequently, the mass of recycle agent could be ignored while compared with the mass of asphalt. That is to say, from the point of recycle agent, aimed to hot in-plant recycling technology with high RAP, the change of recycle agent has no harmful influence to energy consumption and GHG emission.

4.3 *The effect of warm mix technology*

Based on the section 3.2 assumptions, the energy consumption and GHG emission are calculated combined with two kinds of warm mix technology such as organic additives and mechanical foaming. So as to consider the actual construction operation, two kinds of situations including heating reduction (the temperature is reduced by 30°C with 20% energy saving) and no heating reduction. The calculation result is shown in Figure 3.

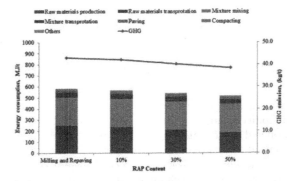

Figure 2. Energy consumption and GHG emission of different content of RAP.

Table 5. Energy consumption and GHG emission of different content of recycle agent.

Construction process		Energy and emission	Milling and repaving	10%	30%	50%
Raw materials (Including recycle agent)	3%	Energy consumption (MJ·t^{-1})	166.74	162.16	152.99	143.82
		GHG (kg·t^{-1})	11.40	11.08	10.44	9.80
	5%	Energy consumption (MJ·t^{-1})	166.74	162.40	153.72	145.03
		GHG (kg·t^{-1})	11.40	11.09	10.49	9.88
	7%	Energy consumption (MJ·t^{-1})	166.74	162.64	154.44	146.24
		GHG (kg·t^{-1})	11.40	11.11	10.54	9.96
	9%	Energy consumption (MJ·t^{-1})	166.74	162.89	155.17	147.46
		GHG (kg·t^{-1})	11.40	11.13	10.58	10.04
Raw materials transportation (150 km)		Energy consumption (MJ·t^{-1})	81.57	73.81	58.28	40.79
		GHG (kg·t^{-1})	6.05	5.47	4.32	3.03
Mixture mixing		Energy consumption (MJ·t^{-1})	256.00	256.00	256.00	256.00
		GHG (kg·t^{-1})	19.80	19.80	19.80	19.80
Mixture transportation (80 km)		Energy consumption (MJ·t^{-1})	43.50	43.50	43.50	43.50
		GHG (kg·t^{-1})	3.22	3.22	3.22	3.22
Construction	Paving	Energy consumption (MJ·t^{-1})	16.51	16.51	16.51	16.51
		GHG (kg·t^{-1})	1.22	1.22	1.22	1.22
	Compacting	Energy consumption (MJ·t^{-1})	9.23	9.23	9.23	9.23
		GHG (kg·t^{-1})	0.68	0.68	0.68	0.68
	Others	Energy consumption (MJ·t^{-1})	8.45	8.45	8.45	8.45
		GHG (kg·t^{-1})	0.63	0.63	0.63	0.63
Total	3%	Energy consumption (MJ·t^{-1})	582.00	569.66	544.96	518.30
		GHG (kg·t^{-1})	43.00	42.10	40.31	38.39
	5%	Energy consumption (MJ·t^{-1})	582.00	569.50	544.50	519.50
		GHG (kg·t^{-1})	43.00	42.09	40.28	38.47
	7%	Energy consumption (MJ·t^{-1})	582.00	569.74	545.23	520.71
		GHG (kg·t^{-1})	43.00	42.11	40.33	38.55
	9%	Energy consumption (MJ·t^{-1})	582.00	569.98	545.95	521.93
		GHG (kg·t^{-1})	43.00	42.12	40.38	38.63

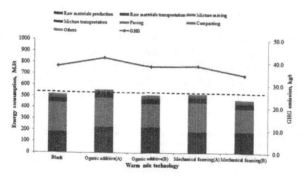

Figure 3. Energy consumption and GHG emission of different warm mix technology.

Results showed that, the energy consumption and GHG emission is increased to ensure the construction workability of hot in-plant recycling with high RAP, that is to say, no heating reduction is considered while warm mix technology is used. Furthermore, warm mix technology with organic additive has more energy consumption and GHG emission. Yet the energy change of mechanical foaming is less which can be ignored. However, when heating reduction is considered, the warm mix technology with mechanical foaming can saving more energy consumption and GHG emission.

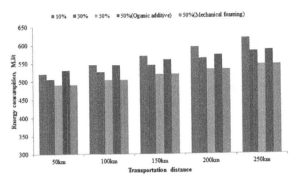

Figure 4. Energy consumption of different transportation distance.

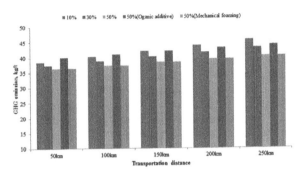

Figure 5. Energy consumption and GHG emission of different warm mix technology.

4.4 *The effect of transportation distance*

So as to analysis in influence of transportation distance to energy consumption and GHG emission, five distances are taken into consideration including 50 km, 100 km, 150 km, 200 km and 250 km under the situation of different RAP content and warm mix technology. The calculation result is shown in Figure 4 and Figure 5.

From Table 4, Figure 4 and Figure 5, we can see that, the energy consumption and GHG emission has obviously change with the transportation for five kinds of asphalt mixture, including reclaimed asphalt with different content RAP, organic additive and mechanical foaming used in reclaimed asphalt with 50% RAP. Furthermore, the change trend is different with different transportation distance.

When the transportation distance is 50 km, the reclaimed asphalt mixture with the amount of 50% RAP in the whole mixture that used organic additive has the most energy consumption and GHG emission. And with the increasing of transportation distance, the gap is gradually decreased. Above-mentioned asphalt mixture has less energy consumption and GHG emission than the reclaimed asphalt mixture with the amount of 10% RAP in the whole mixture at the distance of 150 km. Even the same with reclaimed asphalt mixture with the amount of 30% RAP in the whole mixture at the distance of 250 km. that is to say, When the transportation distance is 150 km, the reclaimed asphalt mixture with amount of 50% RAP in the whole mixture used organic additive has less energy consumption and GHG emission than reclaimed asphalt mixture with amount of 10% RAP in the whole mixture, but more energy consumption and GHG emission than reclaimed asphalt mixture with amount of 30% RAP in the whole mixture. And when the transportation distance is father, the reclaimed asphalt mixture with amount of 50% RAP in the whole mixture began to appear advantage of low RAP recycle asphalt mixture.

5 CONCLUSIONS

This research investigated the energy consumption and GHG emission of high RAP central plant hot recycle technology under different factors including RAP content, recycle agent content, warm mix technology and the raw materials transportation. Such effect factors are actually problems during construction production process. By this study, the preliminary findings show many different results, instead that generally it is think that with the increasing of RAP content, it would appear more and more energy consumption and GHG emission. The following conclusions were drawn:

1. The transportation distance is more significant influence factor than warm mix technology for energy consumption and GHG emission. Warm mix technology which take more energy consumption appear less with the more transportation.
2. Without warm mix technology, with the increasing of RAP content, the energy consumption and GHG emission are more and more. In fact, during the construction production process of high RAP central plant hot recycling technology, warm mix technology is commonly method to improve the compaction and performance of reclaimed asphalt mixture.
3. The energy consumption and GHG emission effect of recycle agent which added into high RAP central plant hot recycling asphalt mixture could be ignored, yet the content change of recycle agent make no difference to energy consumption and GHG emission.
4. To ensure construction workability and compaction of high RAP central plant hot recycling asphalt mixture, warm mix technology is utilized and no heating reduction during every construction process. In this situation, when the transportation distance is less than 150 km, the reclaimed asphalt mixture with amount of 50% RAP in the whole mixture has no advantage over the mixture with low amount of RAP.

ACKNOWLEDGEMENTS

The research activities described in this paper was sponsored in part by the Research Project of Jiangsu DOT under Grant No. 2015T01. This support is greatly acknowledged. Data analysis and opinions are those of the authors and do not necessarily reflect those of the sponsoring agency.

REFERENCES

[1] Wang, T., J. Harvey and A. Kendall. Reducing greenhouse gas emissions through strategic management of highway pavement roughness. Environmental Research Letters [J], Vol. 9, No. 3, 2014, pp. 1–10.
[2] http://www.ce.berkeley.edu/~horvath/palate.html.
[3] Aurangzeb, Q., Al-Qadi, I.L., Abuawad, I.M., Pine, W.J. & Trepanier, J.S. 2012. Achieving Desired Volumetrics and Performance for High Rap Mixtures. Trans Res Rec 2294:34–42.
[4] de la Roche, C., Van de Ven, M., Planche, J.P., Van den Bergh, W., Grenfell, J., Gabet, T. et al. 2013. Hot Recycling of Bituminous Mixtures. In M.N. Partl, H.U. Bahia, F. Canestrari, C. de laRoche, H. Di Benedetto, H. Piber & D. Sybilski (eds) State-of-the-Art-Reports 9: Advances in Inter Laboratory Testing and Evaluation of Bituminous Materials:361–428. Springer.
[5] Frigio F., Pasquini, E. & Canestrari, F. 2014. Laboratory Study to Evaluate the Influence of Reclaimed Asphalt Content on Performance of Recycled Porous Asphalt. JOTE 43(6).
[6] Stimilli A., Ferrotti G., Graziani A. & Canestrari F. 2013. Performance evaluation of cold recycled mixture containing high percentage of reclaimed asphalt. RMPD, 14:149–61.
[7] Leon, J.G., Jensen, P.H. Environmental aspects of warm mix asphalt produced with chemical additives. 5th Eurasphalt & Eurobitume Congress, Istanbul. 2012.
[8] IPCC Guidelines for National Greenhouse Gas Inventories, Volume 2 Energy, 2006, Page 1.21–1.24.
[9] Chinese Life Cycle Database (CLCD), Sichuan University & IKE.
[10] Temren, Z., Sonmez, I. A study on energy consumption and carbon footprint of asphalt and concrete mixtures [C]. 5th Eurasphalt & Eurobitume Congress, Istanbul, 2012.
[11] http://www.arra.org/.
[12] https://www.fhwa.dot.gov/.

Implementation of life cycle thinking in planning and procurement at the Swedish Transport Administration

S. Toller
Swedish Transport Administration, Stockholm, Sweden

M. Larsson
KTH Royal Institute of Technology, Stockholm, Sweden

ABSTRACT: According to transport policy objectives, limiting the energy use and climate impact of the transport system, including infrastructure, is an important task for the Swedish Transport Administration (STA). Choices that affect climate performance of transport infrastructure are made at different stages in the planning process. The STA has developed the Klimatkalkyl climate calculation model for an efficient, consistent life cycle calculation of infrastructure greenhouse gas emissions and energy use. As a result of the model's implementation in planning and procurement, a life cycle perspective is now being used on a regular basis for environmental procurement claims, decision support and monitoring purposes.

1 INTRODUCTION

According to transport policy objectives, limiting the energy use and climate impact of the transport system is an important task for the Swedish Transport Administration (STA). This task includes limiting the climate impact of both traffic and infrastructure.

The transport system affects the climate through emissions from traffic as well as emissions from the construction, operation and maintenance of infrastructure. The STA is responsible for long-term planning for the domestic transport systems of all types of traffic, as well as the construction, operation and maintenance of public roads and railways. The Government's infrastructure bills of 2008 and 2012 (Swedish Government, 2008) emphasize that decision support should take into account the climate impact of infrastructure from a life cycle perspective. There is also the vision of Sweden having zero net emissions of greenhouse gases by 2050 (Swedish Government, 2014).

A life cycle perspective is needed to include the indirect emissions that occur as a result of producing the materials and fuel used in construction, as these emissions may be significant for the transport infrastructure system (Jonsson, 2007; Chester and Horvath, 2009). As described in the *Handbook on Life Cycle Assessment* by Guinée et al. (2002) and discussed by Harvey et al. (2016), Life Cycle Assessments (LCAs) can be categorized according to level of complexity, depending on the objective. These assessments may be benchmarking studies, studies with only a few impact indicators or selected life cycle stages, or full-scale LCA studies that include a life cycle impact assessment for a larger set of impact and resource indicators. LCAs can be applied on both a network level and a project level. Within the literature a wide range of LCAs for transport infrastructure have been published, differing in scope, system boundaries and complexity. There are studies that concern the entire transport system (Schlaupitz, 2008; Jonsson 2007), as well as studies that concern single projects or components of a project (i.e. Muench, 2010; Santero et al., 2010).

As suggested by Butt et al. (2015), in order to make appropriate methodological choices for an LCA, one should consider at which stage during the planning process the LCA results will be used. Choices made for early planning stages will influence energy use and climate impact during construction and maintenance. From an energy and climate perspective, there is a considerable difference between construction inside a tunnel, in a cutting, on high embankments and at ground level. The volume of earth movement and material use is affected by the choice of location and design. However, choices made during later planning stages will also affect energy use and climate impact. Specifically, these include choice of design, materials and suppliers. Miliutenko et al. (2014) have investigated the decision stages of road infrastructure planning in Sweden, Norway, Denmark and the Netherlands and suggested three main decision stages when a life cycle perspective may be applied. These are the choice of transport modality at the national level, choice of road corridor and construction type in a given project, and choice of specific construction design.

In order to achieve the goal of reduced emissions from transport infrastructure, there are several requirements to be considered: emissions need to be consistently attributed and reported; effective reduction measures need to be identified; and there need to be incentives that promote the implementation of those measures on both network and project levels, in both early and later stages. Furthermore, in order to achieve this in a way that is consistent and guarantees that all projects are evaluated on the same basis, the system boundaries, background data and other methodological choices need to be transparent and clearly defined. For this purpose the STA has developed the Klimatkalkyl climate calculation LCA model (STA, 2016a). This model has been implemented in the regular planning process and is also being used for formulating requirements in the procurement of contractors and consultants. The aim of this paper is to describe the model and its contribution to the implementation of a life cycle perspective on greenhouse gas emissions within the planning and procurement processes of Swedish transport infrastructure.

2 THE KLIMATKALKYL CLIMATE CALCULATION MODEL

2.1 *Model development*

The Klimatkalkyl climate calculation model has been developed to enable the calculation of potential climate impacts in terms of the Global Warming Potential (GWP) of greenhouse gas emissions and to account for energy use. The model covers construction as well as operation and maintenance of investments for roads and railways. In addition, the model has been developed for use both in the early planning stages when there is little project specific information available and in the later stages of a project. The model is designed so that climate calculations can be applied continuously to a certain project throughout the planning process. As the project proceeds, the results will gradually become more precise. The model is also designed to be user-friendly. No information input is required beyond what is available through the investment cost assessments. Furthermore, the model has been designed with the aim that it should apply to individual investments as well as to elements of investments, and that it should be possible to aggregate the results. Results for an entire transport plan can thus be obtained by adding together the individual results of the different investments within the plan.

The model has been developed by the environmental consultancy company WSP at the request of, and in consultation with, the STA. The first version of the model, Klimatkalkyl version 1.0, was developed in connection with the STA's preparations for its proposed 2013 National Transport Plan. This version of the model was used to estimate the greenhouse gas emissions derived from the construction of the investment elements specified in the plan (STA, 2015). The model has been updated annually since then; the latest version—version 4.0 – became operational in April 2016 (STA, 2016a).

2.2 *Model description*

The Klimatkalkyl model applies the basic principles of LCA in accordance with ISO 14044 (ISO, 2006a; 2006b). The model calculates energy use (primary energy) and greenhouse gas emissions (Global Warming Potential (GWP) quantified as emissions of CO_2 eq.) for an object or a measure by multiplying resource use by emission factors. The emission factors constitute background, or generic, LCA data that should be conservative average values representative for today's most common techniques. The LCA data should also be geographically representative. The model provides the option for constructors or designers to change the LCA data. The STA accepts LCA data based on Published Environmental Product Declarations (EPDs) in accordance with the standard for EPDs published as EN 15804 by the European Committee for Standardization under Technical Committee 350 (CEN, 2013). The emission factors include the energy use in—and emissions from—raw material extraction and the processing and transportation of energy resources and materials, as well as from the use (combustion) of the energy resources. The emission factors are reviewed and updated on a regular basis.

The model is a modular framework in which type measures contain a predefined set of components. Each component contains a predefined set of resources and each resource is linked to the generic LCA data provided as default. The model contains default resource templates for the expected use of resources in different type measures or components. Resource templates are based on earlier projects. Users of the model can choose either to use the default values for resource use and the generic LCA data, or to apply project-specific resource use and resource-specific LCA data. In the early planning stages the resource use and specific materials and suppliers are not yet known, and at these stages the default values are necessary, but as the level of detail increases toward the later planning stages, data that is more project specific can replace the default values. Resource templates also exist for the operation and maintenance of the most important type measures. Today, however, opportunities for using project-specific data for operation and maintenance are limited due to verification problems: while resource use and generic LCA data can be verified through project information and product-specific EPDs, future resource needs for operation and maintenance cannot be validated.

2.3 System boundaries

The Klimatkalkyl model considers the energy use (primary energy) and potential climate impact (emissions of CO_2 equivalents) of road and railway infrastructure. Construction, operation and maintenance are the life cycle stages included. An end-of-life stage for the construction as a whole is not included, as transport infrastructure is rarely demolished. Instead, the function of the infrastructure system is assumed to be kept constant by the ongoing replacement of components as they reach the end of their reference service life. The emissions from construction of each component, including material and processes, is divided by its reference service life and summed up as a total. This means that environmental impact is distributed forward in time throughout each components service life although the emissions occur during the limited time period of construction. Demolishing and waste management of each component is not included by default in the calculations, as it commonly contributes very little to the total emissions. However, demolition can be added manually by the user of the model. The components' reference service life are defined on the basis of previous experience.

The energy use and emissions of traffic are not considered in Klimatkalkyl; these are currently analyzed in other models. This is, however, one development potential of the model. As decisions regarding infrastructure influence future traffic, both on a project level (e.g. through roughness and rolling resistance) and on a network level (e.g. through the lengths of transport), there will be a future need to define the interface between Klimatkalkyl and existing models for the calculation of traffic emissions and energy use.

The model calculates emissions and energy use on the basis of current technology and material choices, and any marginal effects are not considered. This is in contrast with many road traffic analyses, which do consider future technology developments. However, Klimatkalkyl does allow for such aspects to be added manually for individual project calculations.

All types of transport generated within the projects that are specified as cost items in the original cost estimate of the project are included in the model. This means, for instance, that earth and rock moving within the project is included. Emissions from transport during raw material extraction and processing are included in the emission factors. However, transportation from production site to construction site is not included. These types of transport are generally assumed to represent a minor contribution to energy use and greenhouse gas emissions (Bothnia Line, 2010), although they may, in some cases, be significant. However, the model is being developed on an ongoing basis and the aim is to include resource templates for all types of transport.

Items included as ongoing operation and maintenance are winter road maintenance, pavement maintenance and tunnel operation (lighting, ventilation and pumping water). The model provides the opportunity to add lighting points separately. Winter road maintenance includes the use of salt and sand, as well as the energy used for spreading these and for snow clearing. Operation and maintenance of railways includes point machines, rail grinding, switch heating, heat and power to station buildings, power to signal systems, and tunnel operation (lighting, electronics, and frost avoidance for fire protection water).

The points mentioned above have been identified as significant contributors to energy use and climate impact, based on previous experience of the STA, certified Environmental Product Declarations (EPDs) and the STA's ongoing development within the field of life cycle cost assessments. On the basis of the above sources, a number of measures are assumed to be minor contributors to energy use and climate impact and these are not included in the model. Such operational measures include railway snow clearance, weeding, dust binding, inspection, sweeping, clearance of obstacles etc.

3 LIFE CYCLE THINKING IN PLANNING

3.1 *Implementation of Klimatkalkyl in the planning process*

Since 2015, the STA has regularly used the Klimatkalkyl climate calculation model to implement life cycle thinking for all new investments of greater than SEK 50 million and for projects included in Sweden's National Transport Plan. Climate calculations are performed from the early planning stage, when different solutions are compared on a network level, and throughout the planning process for the specific solution chosen, until the construction is finished and a climate declaration is delivered. The results are included in the overall impact assessment published for each project at certain predefined steps within the planning process, as prescribed by the guidelines of the STA (STA, 2015). At each of these steps, the cost calculations are revised and the climate calculation is updated on the basis of the cost calculations. Hence, climate calculations are systematically being performed and up-to-date, available as decision support throughout the planning process with a minimum of effort.

The climate calculations are used for decision support (i.e. what to build and how to build it in order to optimize climate performance) and for monitoring. Klimatkalkyl has been successfully used in major projects such as the Stockholm Bypass Project and the East Link Project—for example, to compare alternatives in Environmental Impact Assessments (EIAs) and identify emission hotspots. In the East Link Project, four different road corridors were compared, in terms of greenhouse gas emissions from the infrastructure. The one alternative with the most tunnels turned out to be the worst one due to the large amounts of concrete needed. In the Stockholm Bypass Project, concrete and steel were identified as the materials that contributed most to the greenhouse gas emissions. The project is now looking for steel suppliers with lower carbon footprint. Although the Klimatkalkyl climate calculation model is the only model prescribed by the official guidelines, it may also be combined with other, more detailed tools for the further identification of hot spots and resource-efficient design.

4 LIFE CYCLE THINKING IN PROCUREMENT

4.1 *Implementation of Klimatkalkyl in procurement*

In order to achieve the goal of the improved climate performance of the transport infrastructure, it is not enough to only implement life cycle thinking for decision support within the planning process. External actors such as contractors and material suppliers also need incentives for reducing greenhouse gas emissions within their processes. Therefore, the STA is now implementing climate requirements, based on life cycle climate calculations, in procurement (STA, 2016b).

Since February 2016, all projects greater than SEK 50 million that are planned to be finished in 2020 or later will be covered by the requirements. The requirements focus on achievements in terms of total decreased emissions rather than prescribing a specific technical solution, in order to achieve the most effective solutions and to stimulate innovations. Tenders are still being evaluated on the basis of price, but the contracts which have been generated include a requirement to reduce greenhouse gas emissions as compared against a predefined baseline. Consultants contracted by the STA at the planning stage or design stage are required to present measures for decreasing greenhouse gas emissions. Turnkey and construction contracts have quantitative requirements for reduced greenhouse gas emissions. Contractors' initial base-line for greenhouse gas emissions is set by the STA according to the Klimatkalkyl model. Compliance with the quantitative requirements is verified by a climate declaration within the same model. If the requested reduction in emissions is achieved, or even exceeded, contractors may be awarded a bonus. If the reduction is not achieved, other bonuses may also be suspended. By using the Klimatkalkyl model for all projects, for baseline setting and for verification by the final climate declaration, an equal process with similar system boundaries is ensured.

4.2 *Reduced greenhouse gas emissions*

The process of implementing the climate requirements and setting the level of reductions was carried out in close cooperation with consultants and contractors. During the first five-year period, on average a reduction of greenhouse gas emissions of 15% is asked for, as compared with the predefined baseline. The exact level of reduction to be required varies depending on the type of project. Reduced greenhouse gas emissions can be achieved either by changing the type or amount of material and fuel used, or by using material and fuel suppliers with a better climate performance in their production processes (e.g. products with a higher proportion of recycled material). In the former case, the default templates for resource use in the Klimatkalkyl model are replaced by project-specific values for the different predefined components, or new components are added if necessary. In the latter case, the generic LCA data provided in the model are replaced by resource-specific LCA data, verified through EPDs.

5 CONCLUDING DISCUSSION

5.1 *Current achievements*

In recent years, the Klimatkalkyl model has been used to produce life cycle perspective climate calculations for hundreds of projects. In 2013 the model was used to evaluate the 2014–2025 transport plan. The evaluation revealed that construction of the planned projects would result in emissions of 3.8 million tons of CO_2 eq (STA, 2014) (the model at that time included only the life cycle stages of raw material extraction, transport and processing, and construction). Since then, use of the model has been incorporated into regular practice for decision support, for identifying improvement measures, for monitoring purposes, and for defining initial baselines and verifying achievements within procurement. All larger projects are now accompanied by a climate calculation by which their climate performance and the effect of their improvement measures can be followed. Although the climate requirements are still a very new element of the procurement process, they are already having an effect in terms of the EPDs being generated. The EPDs are required for verifying the climate performance of contractors' solutions. The requirements are expected to stimulate the development of materials and fuels with better climate performance. There is an ongoing project to evaluate the responses to the procurement requirements among the actors concerned and to evaluate if the desired developments will be achieved.

5.2 Uncertainties

Uncertainties in the results from the Klimatkalkyl climate calculation model mainly derive from either the generic LCA data used, the default templates for resource use, or the project-specific data inserted by the user.

Regarding the generic LCA data, these values are based on published and reviewed sources (preferably EPDs) and are regularly revised and updated. As the number of published EPD is growing, the possibilities for improving the default values increases. The LCA data for steel, concrete, bitumen and diesel are of particularly great importance to the overall results.

Regarding the default templates for resource use, the possibilities for successively improvements of these templates will increase as climate declarations are performed, providing results regarding the resource use in completed projects. However, the variation between projects may largely depend on environmental aspects such as geological conditions, or specific design choices. Therefore the user is encouraged to routinely check the representativeness of the default values used and modify these when necessary.

The uncertainties that derive from the insertion of data by users are judged to be the most important. In the early planning stages in particular, knowledge of project components may be poor. Thus, when interpreting the results from the model, it is important to bear in mind that the results can never be more precise than the data used.

5.3 Further work

The Klimatkalkyl climate calculation model is constantly being improved to better support life cycle thinking in both planning and procurement. One example of needed improvement is the expansion of system boundaries so as to better address use and maintenance activities, the transport of materials from production to site, and demolition activities. The model may also be developed to include a broader range of environmental impact categories. A development of this kind is needed in order to address the high level of complexity of full LCA studies that can deliver a life cycle impact assessment for a large set of impact indicators (Harvey et al., 2016).

For use and maintenance activities, it is currently not possible for a user of the model to replace default templates with project-specific information. However, if future use and maintenance cannot be modified, there will be little incentive to design a construction with less intensive maintenance. Therefore, the development of such opportunities will be an important measure to avoid sub-optimization. The methodological challenge is to find a consistent, practical process for verifying project-specific data in future use and maintenance.

Furthermore, it is necessary to establish useful interfaces with other models—e.g. models which describe the environmental impact of traffic. Traffic emissions need to be considered when the model is to be used in the early planning stages—e.g. as decision support in the choice of road corridor, which today is a manual process involving the comparison of traffic calculations. With further development of the model, this process may become more effective and consistent. There is also a need to develop and improve the interface with more detail-oriented models that focus on subsystems, e.g. pavement design, so that the information obtained through the use of these models can be effectively utilized.

REFERENCES

Bothnia Line, 2010. *Environmental product declaration for railway track on the Bothnia Line.* [Online]. Available at: http://gryphon.environdec.com/data/files/6/7222/epd200.pdf, [Accessed 15 October 2016].

Butt, A.A., Toller, S. & Birgisson, B., 2015. *Life cycle assessment for the green procurement of roads: a way forward.* Journal of Cleaner Production, 90, 163:170.

CEN, 2013. *Sustainability of Construction Works—Environmental Product Declarations—Core Rules for the Product Category of Construction Products.* European Standard EN 15804:2012 + A1.

Chester, M.V. & Horvath, A., 2009. *Environmental assessment of passenger transportation should include infrastructure and supply chains.* Environmental Research Letters, 4, 024008.

Guinée, J.B., Gorrée, M., Heijungs, R., Huppes, G., Kleijn, R., Koning, A. de, Oers, L., van Wegener Sleeswijk, A., Suh, S., Udo de Haes, H.A., Bruijn, H., de Duin, R. & van Huijbregts, M.A.J., 2002. *Handbook on life cycle assessment. Operational guide to the ISO standards. I: LCA in perspective. IIa: Guide. IIb: Operational annex. III: Scientific background.* Dordrecht: Kluwer Academic Publishers.

Harvey, J., Meijer, J., Ozer, H., Al Quadi, I., Saboori, A., Kendall, A-., 2016. *U.S. Department of Transportation, Federal Highway Administration.* Pavement Life Cycle Assessment Framework. FHWA-HIF-16-014.

ISO 14040, 2006a. *Environmental Management—LCA principles and framework*, ISO 14040, International Organization for Standardization, Geneva.

ISO 14044, 2006b. *Environmental Management—LCA principles and framework*, ISO 14044, International Organization for Standardization, Geneva.

Jonsson, D.K., 2007. *Indirect energy associated with Swedish road transport.* European Journal for Transport Infrastructure, Res. 7, 183–200.

Miliutenko, S., Kluts, I., Lundberg, K., Toller, S., Brattebø, H., Birgisdóttir, H. & Potting, J., 2014. *Consideration of life cycle energy use and greenhouse gas emissions in road infrastructure planning processes: examples of Sweden, Norway, Denmark and the Netherlands.* Journal of Environmental Assessment Policy and Management, 16: 26.

Muench, S.T., 2010. *Roadway Construction Sustainability Impacts Review of Life Cycle Assessments, Transportation Research Record.* Journal of the Transportation Research Board, 2151: 36–45.

Santero, N., Masanet, E. & Horvath, A., 2010. *Life Cycle Assessments of Pavements: A Critical Review of Existing Literature and Research,* Illinois.

Schaupitz, H., 2008. *Energi och klimatkonsekvenser av moderne transportsystemer. Effekter ved byggning av høyhastighetsbaner i Norge* (in Norwegian). Oslo.

STA, 2014. *Planer för transportsystemet 2014–2015* (in Swedish). [Online]. Availabe at: https://trafikverket.ineko.se/Files/sv-SE/10799/RelatedFiles/2014_039_planer_for_transportsystemet_2014_2025_samlad_beskrivning_av_forslagen_till_nationell_plan_och_lansplaner.pdf., [Accessed 12 October 2016].

STA, 2015. *Klimatkalkyl—infrastrukturhållningens energianvändning och klimatpåverkan i ett livscykelperspektiv, Riktlinje TDOK 2015:0007* (in Swedish). [Online]. Available at: http://www.trafikverket.se/tjanster/system-och-verktyg/Prognos--och-analysverktyg/Klimatkalkyl/. [Accessed 12 October 2016].

STA, 2016a. *Klimatkalkyl version 4.0* (In Swedish). [Online]. Available at: http://www.trafikverket.se/klimatkalkyl. [Accessed 12 October 2016].

STA, 2016b. *Klimatkrav i planläggning, byggskede, underhåll och på teknisk godkänt järnvägsmateriel, Riktlinje TDOK 2015:0480* (in Swedish). [Online]. Available at: http://trvdokument.trafikverket.se/Versioner.aspx?spid=104&dokumentId=TDOK%202015%3a0480/.[Accessed 12 October 2016].

Swedish Government, 2014. *Kommite direktiv 2014:53* (in Swedish). [Online]. Available at: http://www.regeringen.se/contentassets/2c7a8ebcc6d847b9b188201869a401b7/klimatfardplan-2050--strategi-for-hur-visionen-att-sverige-ar-2050-inte-har-nagra-nettoutslapp-av-vaxthusgaser-ska-uppnas-dir–201453. [Accessed 10 October 2016].

Swedish Government, 2008. *Framtidens resor och transporter—infrastruktur för hållbar tillväxt* (in Swedish). [Online] Available at: http://www.regeringen.se/contentassets/4a7260246f324bb2adda1ef407d6b26b/framtidens-resor-och-transporter---infrastruktur-for-hallbar-tillvaxt-prop.-20080935. [Accessed 12 October 2016].

Emission-controlled pavement management scheduling

U.D. Tursun, R. Yang & I.L. Al-Qadi
University of Illinois Urbana–Champaign, USA

ABSTRACT: This paper describes an analytic approach that can be used to evaluate and propose rehabilitation schedules based on economic, performance, and environmental considerations for various types of pavements managed by the Illinois State Tollway Authority. A Mixed-Integer Nonlinear Program (MINLP) is formulated to model the agency's life-cycle cost and environmental impacts where the decision set consists of the maintenance overlay type and the thickness of the overlay proposed on a temporal scale over a planning horizon. The problem requires interaction of the integer and continuous variables that leads to MINLP formulation. Considering practical implications of the problem, the continuous variables are bounded into a finite and discrete set, while the integrality constraints are relaxed. The objective and constraints of the problem can be alternated to suit the needs of the agency, which may be interested in minimizing environmental impacts and restricting the cost to the agency, or vice versa, over the pavement life-cycle.

Keywords: Mixed-Integer Nonlinear Programming, Sustainability, Pavement Management

1 PAVEMENT MANAGEMENT SCHEDULING

1.1 *Introduction*

Although transportation agencies traditionally have considered cost as the main factor in deciding initial pavement design and maintenance schedules, interest in evaluating environmental impacts of transportation infrastructure has become common. In a 2011survey, it was found that 29 out of 35 responding state Departments of Transportation (DOTs) in the United States perform Life-Cycle Cost Analysis (LCCA) to evaluate alternative pavement designs for new construction and reconstruction projects (Hallin et al. 2011). Thus, both the initial cost of construction and the future cost of expected Maintenance and Rehabilitation (MR) schedules over the lifetime of the pavement are considered.

The parallel approach to LCCA for environmental consideration is Life-Cycle Assessment (LCA), which is not currently implemented by any US state DOT. Similar to LCCA, this methodology evaluates the environmental impacts incurred over the entire pavement life cycle from initial construction, MR, use, and disposal. If combined, LCCA and LCA result in quantitative cost and environmental measures, respectively, that can systematically inform an agency's decision-making process.

As LCCA is already an established component of pavement management, recent literature has investigated the application of LCA to pavement management and, specifically, how cost and environmental concerns can be balanced. Gosse et al. (2012) used a multi-objective genetic algorithm able to develop Pareto sets of maintenance plans in Virginia for various budgets, pavement performances, and emission levels. The scope of the work included only materials and construction, taking into account the deterioration of the pavement without user emissions or costs. Yu et al. (2013) further integrated LCA and LCCA for three types of overlay systems by using dynamic programming to consider both agency and user emissions, as well as costs from construction, work zone delay, and normal use of the pavement. Optimized strategies for each overlay system were developed by minimizing life-cycle cost, as well as environmental impacts (greenhouse gases, GHGs; and energy). Lidicker et al. (2012) in turn evaluated resurfacing policies based on minimizing life-cycle costs and emissions from both the user and the agency to find optimal overlay intervals (i.e. 15 years for minimum GHGs and 22 years for minimum cost).

Bryce et al. (2014) and Reger et al. (2014) applied optimization techniques at a multifacility level, incorporating pavement segments from a large network. Bryce et al. included various levels of MR treatment types, and also incorporated probabilistic distributions to account for uncertainties in the extent of treatment (e.g. thickness of overlays), transportation distances, and per unit environmental impact values themselves. A Pareto set for a network was given with respect to cost, energy, and pavement condition. Reger et al. minimized equivalent annual agency costs for a network while constraining GHGs. A Pareto set was formed and used to evaluate past and present MR policies in California. An updated study by Reger et al. (2015) also considered constraining the agency budget and minimizing emissions. In the network-level studies mentioned, delay effects from the work zone during construction was omitted.

Overall, a number of approaches have been implemented to integrate economic and environmental concerns into pavement management. However, this study extends the literature by including probabilistic consideration of changing traffic levels over time, updated inventory, and models specific to the targeted agency.

Transportation agencies have focused on material and process selection to reduce the environmental impacts of maintenance activity (Zapata & Gambatese 2005, Santero & Horvath 2009). Some of the applied environmental impact rating tools at the individual-project level are the Greenroads tool, the Infrastructure Voluntary Evaluation Sustainability Tool (INVEST) from the Federal Highway Administration (FHWA), and the Illinois Livable and Sustainable Transportation System (ILAST). Yet comprehensive, practical, and computationally tractable algorithms that would allow transportation agencies to add an environmental dimension to current Pavement Management Systems (PMSs) while reducing total economic impact and attaining performance targets are still not in place.

Many optimal decision problems in areas such as logistics, manufacturing, transportation, and the chemical and biological sciences involve both continuous and discrete decision variables over nonlinear system dynamics that require Mixed-Integer Nonlinear Programming (MINLP). The MINLP class of problems combines the combinatorial difficulty of optimizing over discrete variable sets, namely the Mixed-Integer Linear Programming (MILP), with the challenges of handling Nonlinear Programming (NLP). Nonlinear programming algorithms usually resort to certain convexity assumptions, leading to local optimization guaranteeing the global optimum. Without convexification, identifying a global optimum in the presence of multiple local optima is not guaranteed. Even when the objective function is convex, nonlinearities in the constraint set may give rise to local optima. Mixed-integer nonlinear programming entails optimization problems where the objective and/or constraints are nonlinear, with continuous and integer variables. MINLPs are a particularly challenging class of optimization problems. Even having only linear functions or merely continuous variables, which reduces MINLP into Mixed-Integer Linear Problems (MILPs) or Nonlinear Problems (NLPs), respectively, does not ease the computational intractability issues (Murty & Kabadi 1987, Garey & Johnson 1979). There have been efforts to solve a subclass of deterministic MINLPs, where the objective function and constraints are convex and upper-bounded (Gupta & Ravindran 1985, Quesada & Grossmann 1992a). When integer variables of MINLPs are relaxed, the feasible set is bound to be convex. Convex optimization on continuous variables has some strong advantages due to necessary and sufficient optimality conditions, duality theory, and reliable algorithms for reasonably large subclasses of these problems.

The literature has algorithms to generate and refine bounds on the optimal solution value of a convex MINLP with a finite number of constraints Bonami et al. (2012). A tree-search algorithm similar to a branch-and-bound algorithm was proposed by Dakin (1965) for convex MINLPs. Later, a generalized Benders decomposition by Geoffrion (1972) in which the parametrized subproblem of classical Benders decomposition was offered. Convex duality theory is used to derive the natural families of cuts corresponding to those in Benders' case. An outer-approximation algorithm was first presented by Duran & Grossmann (1986) for the same class of problems. Based on principles of decomposition, outer-approximation, and relaxation, the proposed algorithm consists of solving an alternating finite sequence of nonlinear programming subproblems and relaxed versions of a mixed-integer linear master program. The same algorithm was further improved by Fletcher & Leyffer (1994). An LP/NLP-based branch-and-bound algorithm in which the explicit solution of an MILP master problem is avoided at each major iteration in the framework of an outer-approximation algorithm was proposed by Quesada & Grossmann (1992b). Based on the algorithms proposed so far, the convex MINLP is broken into two pieces, MILP and NLP, which are solved separately.

The formulated MINLP problem that we solve is transformed into a pure integer problem where the continuous variables are bounded into a finite and discrete set. Also relaxing the integrality constraints leads to a convex formulation that can be solved with commercial solvers.

Firstly objective of this study is explained in section 1.2. In section 2, problem formulation and solution methodology including the details of Monte Carlo simulation is presented. And the paper is concluded by reporting the results of a demonstration case study on a mile section of jointed plain concrete pavement at 20-foot spacing (JPCP20) link. This study is the initial step of a comprehensive project. We discuss intended direction of research as the closing section of this paper.

1.2 Objective

A transportation agency typically utilizes a PMS to plan rehabilitation operations by identifying schedules based on pavement condition, subject to cost constraints. Current PMSs of the Illinois State Tollway Authority (Illinois Tollway) do not typically incorporate environmental considerations into the decision-making process to balance performance and environmental goals. This work seeks to develop a practical and computationally tractable algorithm that would allow Illinois Tollway to add an emission-control objective to the current goals related to cost and condition of pavement.

2 MIXED-INTEGER NONLINEAR FORMULATION OF EMISSION-CONTROLLED MAINTENANCE SCHEDULE

Our goal is to evaluate and propose maintenance schedules for various types of pavement for the Illinois State Tollway Authority, based on economic, performance, and environmental considerations. A mixed-integer nonlinear problem is formulated to model the life-cycle cost to the agency and the greenhouse gas emission due to both user and maintenance activities of a single pavement segment. There are two opposing motivations within each section of the objective function. While an increase in the International Roughness Index (IRI) value causes a linear surge in user emission, scheduling an overlay to lower the IRI value also leads to further emission due to construction activity. The decision set consists of maintenance type and thickness of overlay over the lifetime of the pavement. The problem requires interaction of integer and continuous variables within nonlinear equations leading to MINLP formulation. The inclusive mathematical formulation of the mixed-integer nonlinear problem is expressed as:

$$(P) \quad \min f(x,y)$$
$$\text{s.t.}$$
$$g(x,y) \leq 0$$
$$x \in X \subseteq \mathbb{Z}^p$$
$$y \in Y \subseteq \mathbb{R}^n,$$

where $f:(X,Y) \to \mathbb{R}, g:(X,Y) \to \mathbb{R}^m$ that are twice continuously differentiable functions. And m is the number of constraints. The number of integer variables is p, and the number of continuous variables is n. (P) is an NP-hard combinatorial problem Kannan & Monma (1978).

In general, nonconvex integer optimization problems are undecidable Jeroslow (1973). Jeroslow studied a class of integer programming problems under quadratical constraints, and he showed that no existing computing device can compute the optimum for all problems in this class. Yet (P) becomes decidable either by ensuring a compact feasible set or by assuming that the problem functions are convex.

There are two fundamental concepts underlying algorithms for solving decidable MINLPs:

- Relaxation
- Constraint enforcement

A relaxation is used to compute a lower bound on the optimal solution of (P). A relaxation is obtained by enlarging the feasible set of the MINLP. The "relaxed problem" of MINLP problem (P) is another optimization where a global optimum can be guaranteed and whose solution provides a lower bound on the optimal objective function value of (P). The relaxed problem is formulated as follows:

$$(R) \quad \min \overline{f}(\overline{x}, \overline{y})$$
$$\text{s.t.}$$
$$\overline{g}(\overline{x}, \overline{y}) \leq 0$$
$$\overline{x} \in \overline{X} \subseteq \mathbb{Z}^{\overline{p}}$$
$$\overline{y} \in \overline{Y} \subseteq \mathbb{R}^{\overline{n}},$$

where $\overline{X} \subseteq X$ and $Y \subseteq \overline{Y}$. Relaxing the integrality of the problem and enforcing lower and upper bounds on variables allow the search for a solution to terminate whenever the lower bound is larger than the current upper bound that can be obtained from any feasible point. Constraint enforcement refers to excluding solutions that are feasible to the relaxation but not to the original MINLP. Constraint enforcement may be accomplished by refining or tightening the relaxation, often by adding valid inequalities, or by branching. We should point out that (P) is its own relaxation, i.e. $\overline{f} \leq f$. The goal is to find a tight lower bound in the feasible set that is finite-valued, as shown in Figure 1. Based on the Weierstrass theorem, if a closed proper function $f : \mathbb{R}^n (-\infty, \infty)$ is coercive, then the set of minima of f over $\mathbb{R}^n$ is nonempty and compact Bertsekas et al. (2003).

Theorem 1. *If a function* $f : [a,b] \to (-\infty, \infty)$ *is lower semi-continuous,*

$$\liminf_{y \to x} f(y) \geq f(x)$$

for all $x \in [a,b]$, *then f is bounded below and attains its infimum.*

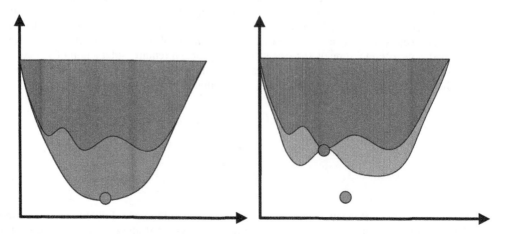

Figure 1. Constraint enforcement for global optimization.

3 EMISSION-CONTROLLED MAINTENANCE SCHEDULE FORMULATION

In light of the above theory, we present the problem we considered. The objective, as well as the constraints of the original model, including the relaxed form of the problem, is described as follows. A transportation agency typically utilizes pavement management systems to plan maintenance operations by identifying schedules based on the pavement condition, subject to cost constraints. Current PMSs of the Illinois State Tollway Authority do not typically incorporate environmental considerations into the decision-making process to balance performance and environmental goals. This work seeks to develop a practical and computationally tractable algorithm that would allow the Illinois State Tollway Authority to add an emission-control objective to the current goals—cost and condition of pavement. In this formulation, it is assumed that the agency is operating under a budget constraint, while tracking pavement condition through the IRI and ultimately minimizing greenhouse gas emissions due to maintenance and user activities. The environmental impact calculation from the construction of overlays takes into account both fixed and variable emissions, using results of statistical analysis presented by the Illinois Center for Transportation report on the tollway LCA tool for the Illinois State Toll Highway Authority Al-Qadi et al. (2014). There are two opposing motivations within each section of the objective function. Although an increase in the IRI value causes a linear surge in user emission, scheduling an overlay to lower the IRI value also leads to further emission due to construction activity.

The following forms of IRI progression and drop models were chosen based on work by Al-Qadi et al. (2014). IRI drop model was developed for major rehabilitation activities expected to change the smoothness of the existing pavement surface. Current literature (Wang et al. 2012, Irfan et al. 2008) indicates that the smoothness change of the existing pavement is related to the IRI value before major rehabilitation.

IRI Progression Model

$$IRI_{t+1} = IRI_t + a * d_t^b * ESAL_t^c \qquad (1)$$

IRI Drop Model

$$IRI_{t+1} = IRI_t - X_{t+1}(m * IRI_t + n) \qquad (2)$$

The MINLP formulation of the problem is presented as follows:

$$\min \sum_{i,t} d(i,t) * q_i * X(i,t) + p_i * X(i,t) + U(i,t) \qquad (3)$$

such that

$$IRI_{t+1} = IRI_t + \sum_i a_i * d(i,t)^b * ESAL_t^c * X(i,t)$$

$$+ \left[\sum_i m_i * \left(a_i * d(i,t)^b * ESAL_t^c * X(i,t) \right) + n_i \right] * X(i,t+1) \quad \forall t$$

$$\sum_{i,t} v_i * d(i,t) * X(i,t) + w_i * X(i,t) \leq B$$

$$\sum_i X(i,t) = 1 \quad \forall t$$

$$X(JPCP20, t+1) \leq X(JPCP20, t) \quad \forall t$$

$$X(HMA, t+1) \leq X(HMA, t) + X(HMA_{overlay}, t) \quad \forall t$$

$$X(SMA, t+1) \leq X(SMA, t) + X(SMA_{overlay}, t) \quad \forall t$$

$$d(HMA,t+1)*X(HMA,t+1) \geq \varepsilon - L*Y(HMA,t+1) \quad \forall t$$
$$d(HMA,t)*X(HMA,t) + d(HMA_{overlay},t)(HMA_{overlay},t) \leq M*(1-Y(HMA,t+1)) \quad \forall t$$
$$d(SMA,t+1)*X(SMA,t+1) \geq \varepsilon - L*Y(SMA,t+1) \quad \forall t$$
$$d(SMA,t)*X(SMA,t) + d(SMA_{overlay},t)(SMA_{overlay},t) \leq M*(1-Y(SMA,t+1)) \quad \forall t$$
$$d(JPCP20,1) = d_0$$
$$d(i,t) \leq u*X(i,t) \quad \forall i,t$$
$$d(i,t) \geq l*X(i,t) \quad \forall i,t$$

where

a_i, b, c: Coefficients of the IRI progression model based on pavement/overlay type
m_i, n_i: Coefficients of the IRI drop model based on type of maintenance activity
q_i: Variable emission proportional to thickness of pavement/overlay
p_i: Fixed emission due to maintenance activity
v_i: Variable cost proportional to thickness of pavement/overlay
w_i: Fixed cost due to maintenance activity
B: Total budget over lifetime of pavement
$U(i,t)$: User phase emission
$U(i,t) = f(IRI, traffic, speed)$
$M(i,t)$: Emission due to maintenance activity
$M(i,t) = d(i,t)*q_i*X(i,t) + p_i*X(i,t)$
IRI_t: International roughness index value at year, t
$X \in \{0,1\}$: Binary decision variable for state, i of pavement at year, t
$Y \in \{0,1\}$: Binary decision variable for conditional constraint of state, i of pavement at year, t
$d \in \mathbb{R}$: Thickness of pavement or overlay which is bounded as $l \leq d \leq u$
d_0: Original thickness of JCPC20
$i \in I$: State of pavement section
I = (JCPC20, HMA, SMA, HMA-overlay, SMA-overlay)
$t \in T$: Planning horizon
M: A sufficiently large upper bound
L: A sufficiently small lower bound
ε: Tolerance value
$ESAL$: Equivalent Single Axle Load for each time period, t

For model 3.3, three levels of decisions are defined: the schedule of maintenance activity, the state of the pavement, and the depth of the pavement/overlay. For time interval $t \in T$, pavement section jointed plain concrete pavement 20 ft. wide (JPCP20) can be in five separate states, which is indicated as a binary decision variable, $X(i,t), i \in I$ such that I = {JPCP2020, HMA-overlay, SMA-overlay, HMA, SMA}. JPCP20 stands for original state of the pavement that cannot be recovered once an overlay is constructed over it. The majority of the JPCP20 original depth within the network is 12 in. The overlay types are Hot Mix Asphalt (HMA) and Stone Mastic Asphalt (SMA). Pavement can also be in the state of either HMA or SMA where the overlay decision has already been made in previous time periods.

To model the 50 year lifetime of the pavement, our model starts from $t = 0, IRI_0 = 60$ that is the roughness index value for a brand new pavement. And also, we define an upper bound for the IRI value that acts as a threshold level. But it is also possible to initialize the model from a random IRI value at a random point in time t within the planning horizon T. The Monte Carlo approach that we used, which is explained in section 3.1, allows initialization parameter variation of IRI and forecasted traffic volume. The original pavement JPCP20 starts with the initial construction pavement thickness and continues to serve until it is replaced with one of the overlays, HMA-overlay or SMA-overlay. Although switching between overlay types is allowed, recovering to JPCP20 is not possible. Once an overlay decision; HMA-overlay or SMA-overlay, has been implemented at time period t, the decision set for the following time period is restricted either to staying at the current overlay or switching to the other overlay type. The depth of the pavement/overlay for time period t, $d(i,t)$ is a continuous decision variable with upper and lower bound values. The schedule of maintenance activity decision is embedded into the binary decision variable of $X(i,t)$. Once an overlay decision, HMA or SMA, has been made at time period t, and if keeping the aforementioned overlay has been decided for $\{t, t+1, t+2,...\}$, then the depth of the pavement stays the same for $\{d_t, d_{t+1}, d_{t+2},...\}$ until the next overlay decision is made. Also, the original depth of JCPC20, d_0 stays same until the first overlay decision is made.

The objective function of model 3.3 consists of $M(i,t)$ and $U(i,t)$ environmental impacts of variable and fixed emission due to maintenance activities in time period t and use of the roadway in time period t over the planning horizon for pavement/overlay i. The objective value is given in terms of tons of GHGs (greenhouse gases) per lane-mile. The environmental impacts for construction and use are calculated by considering all upstream and downstream materials and fuel consumption for construction and all upstream and downstream fuel usage from vehicles traversing the pavement for the use phase. An estimate of energy and emissions with respect to a reference speed and IRI values for use-phase framework for pavement LCAs is presented by Al-Qadi et al. (2014). The environmental impact for the use phase considers the additional fuel vehicles traversing the road consume due to additional road roughness. In this study, a model developed by Shakiba et al. (2016) is applied to consider the relationship between extra energy consumption and IRI, assuming an average vehicle distribution found on the Illinois Tollway.

The model uses the international roughness index as the performance criterion. The parameters m and n in equation 1 are related to the drop in IRI that occurs with a new HMA or SMA overlay. And the parameters a, b and c in equation 2 are associated with the thickness of the overlay or pavement and the Equivalent Single-Axle Load (ESAL) or traffic levels at time t. Thus, without intervention, the roughness, IRI_t of the pavement deteriorates following a convex accelerating trend, depending on the existing overlay thickness d_t, pavement/overlay type, and traffic level in terms of million $ESAL_t$. With intervention, the roughness, IRI_t is assumed to drop. If an overlay decision has not been made for the planning period, then increasing the IRI value leads to higher user-emission values. By contrast, if an overlay decision has been made, then the IRI value drops, while emission due to maintenance activity is incurred. The interaction between IRI and decision variables, $X(i,t)$ and d_t be seen in Figure 2.

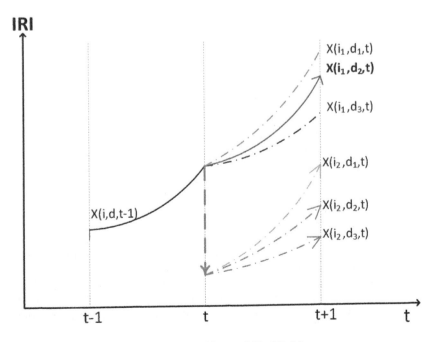

Figure 2. International roughness index vs. decision variable $X(i,d,t)$.

The problem stated in equation in 3 requires interaction of the integer and continuous variables that leads to MINLP formulation. Yet for practical purposes, the continuous variable, thickness of overlay $d \in \mathbb{R}$, can be chosen to be a finite discrete set with cardinality C. It can also be mounted into the binary variable as $X(i,d,t)$ which only increases the number of states of a pavement/overlay by C in a time period, t. For this particular case, we chose a set of three different thickness options that leads to 15 possible states for the overlay/pavement. The next step to transform the problem into a tractable form was relaxing the integrality of the formulation. We used the Branch-And-Reduce Optimization Navigator (BARON) solver within the General Algebraic Modeling System (GAMS) distribution 24.7.2 to solve the relaxed MINLP of problem 3.1 within acceptable execution time for the planning horizon of $T = 50$ years.

4 MONTE CARLO SIMULATION FOR PROJECTED TRAFFIC GROWTH

Monte Carlo simulation is a method for exploring the sensitivity of a mathematical model that is simulated in a loop by random parameter variation within statistical constraints. The results from the simulation are analyzed to determine the characteristics of the system in statistical terms.

The output measure from a run of the simulation model, where the number of independent runs of the model is n with the same initial conditions and with different streams of continuous uniform $u \sim U(a,b)$ random variates, then the output data $\{x_1,\ldots,x_n\}$ for each random variable can be treated as statistically independent observations of the random variable. When n increases, due to the central limit theorem, the distribution shape of the sample mean approaches a normal distribution, where

$$L \leq \mu \leq U$$

such that the upper and lower confidence limits $[L,U]$ are obtained from $U = \bar{x} + z_{\alpha/2} S_{\bar{x}}$ and $L = \bar{x} - z_{\alpha/2} S_{\bar{x}}$. $Z_{\alpha/2}$ is the value that gives $P(Z > z_{\alpha/2}) = \alpha/2$. The confidence interval on μ becomes $P(L \leq \mu \leq U) = 1 - \alpha$. The standard deviation of x is denoted as s; and the standard error of the mean, as $s_{\bar{x}}$ is obtained by as $s_{\bar{x}} = s/\sqrt{n}$. $x_{\bar{x}}$ is an estimate of the true mean, μ; and s is an estimate of the true standard deviation, σ. Based on the explanation above, we generated an expanded set of streams of annual traffic of $\{A_1,\ldots,A_{50}\}$ with a growing rate $2 \pm \mu\%$ that has an accompanying random noise of μ. We ran a relaxed form of problem 3 for each path of traffic stream and recorded the optimal solution set. By repeated random sampling, we created an extended set of traffic-growth paths to run the model.

5 RESULTS

Jointed Plain Concrete Pavement at 20-foot spacing (JPCP20) has been in use by the Illinois Tollway Authority. The JPCP20 pavement sections are maintained based on a predetermined maintenance and rehabilitation schedule. Current practice is SMA overlays of 3–4 inches at $t = 24, 36, 44$ years. Using the model formulation described in section 3, an alternative rehabilitation schedule of $t = 20 \pm 0.172, 37 \pm 0.121, 49 \pm 1.112$ years, 4-in. HMA overlays are recommended, with a 95% confidence interval in terms of minimizing emission while confirming performance and budget requirements. The resulting IRI performance curves is shown in Figure 3.

Also, using the same streams of annual traffic and the predetermined maintenance and rehabilitation schedule of Illinois Tollway Authority, we calculated the emission over the lifetime of JPCP20; and we observed a 7% reduction in greenhouse gas emission just by changing the set schedule for the pavement rehabilitation and the type of overlay. Also, we point out that the rehabilitation at $t = 49 \pm 1.112$ can be replaced with reconstruction of the existing pavement.

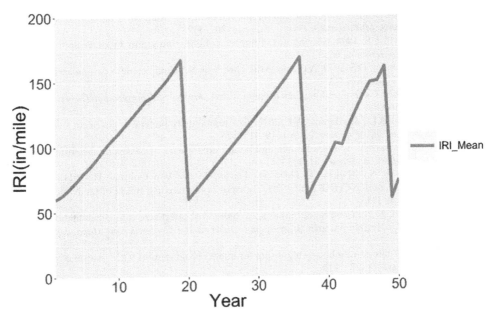

Figure 3. IRI time series as a result of proposed rehabilitation schedule.

6 CONCLUSIONS

This work is the preliminary step of creating an algorithm that allows agencies to determine the maintenance planning of individual links under budget and emission constraints while complying with performance criterion. We were able to set up an MINLP model of maintenance schedule, overlay type and overlay thickness over lifetime of a JCPC pavement. Although the MINLP model is not tractable, by using relaxation and outer approximation techniques, we were able to solve the problem with a commercial solver. The uncertainty accompanying traffic growth over the planning horizon was also remedied by randomly generating an extensive set of data paths.

As a result of this work we showed that by changing the schedule and type of overlay they currently use Illinois Tollway Authority can lower overall greenhouse gas emission while still staying within economic and performance requirements.

The next scope of this ongoing study is going to be creating a mathematical model for PMSs at network level that takes into account traffic interaction of links as well as user behavior.

REFERENCES

Al-Qadi, I., Di, W., Ziyadi, M., & Ozer, H. 2014. *Development of Rolling Resistance Models for Illinois State Toll Highway Authority Pavement LCA Tool*. Technical Report. Illinois Center for Transportation, University of Illinois Urbana-Champaign.

Bertsekas, D., Nedic, A., & Ozdaglar, A. 2003. *Convex Analysis and Optimization*. Cambridge, Massachusetts: Athena Scientific.

Bonami, P., Kilinc, M., & Linderoth, J. 2012. Algorithms and software for convex mixed integer nonlinear programs. In *Mixed Integer Nonlinear Programming*, 1–39. Springer New York.

Bryce, J., Katicha, S., Flintsch, G., Sivaneswaran, N., & Santos, J. 2014. Probabilistic life-cycle assessment as network-level evaluation tool for use and maintenance phases of pavements. *Transportation Research Record: Journal of the Transportation Research Board*: 44–53.

Dakin, R.J. 1965. A tree-search algorithm for mixed integer programming problems. *The Computer Journal* 8: 250–255.

Duran, M., & Grossmann, I. 1986. An outer-approximation algorithm for a class of mixed-integer nonlinear programs. *Mathematical Programming* 36: 307–339.

Fletcher, R., & Leyffer, S. 1994. Solving mixed integer nonlinear programs by outer approximation. *Mathematical Programming* 66: 327–349.

Garey, M.R., & Johnson, D.S. 1979. A Guide to the Theory of NP-Completeness. *Computers and intractability* 174. San Francisco, CA.

Geoffrion, A.M. 1972. Generalized benders decomposition. *Journal of Optimization Theory and Applications* 10: 237–260.

Gosse, C.A., Smith, B.L., & Clarens, A.F. 2012. Environmentally preferable pavement management systems. *Journal of Infrastructure Systems* 19: 315–325.

Gupta, O.K., & Ravindran, A. 1985. Branch and bound experiments in convex nonlinear integer programming. *Management Science* 31: 1533–1546.

Hallin, J., Sadasivam, S., Mallela, J., Hein, D., Darter, M., & Von Quintus, H. 2011. *Guide for pavement-type selection,* NCHRP report 703. Transportation Research Board of the National Academies, Washington, D.C.

Irfan, M., Khurshid, M.B., Labi, S., & Sinha, K.C. 2008. Cost-effectiveness of rehabilitation alternatives—the case for flexible pavements. In *Seventh International Conference on Managing Pavement Assets.*

Jeroslow, R. 1973. There cannot be any algorithm for integer programming with quadratic constraints. *Operations Research* 21: 221–224.

Kannan, R., & Monma, C.L. 1978. On the computational complexity of integer programming problems. In Optimization and Operations Research, 161–172. Springer. Springer Berlin Heidelberg. 161–172.

Lidicker, J., Sathaye, N., Madanat, S., & Horvath, A. 2012. Pavement resurfacing policy for minimization of life-cycle costs and greenhouse gas emissions. *Journal of Infrastructure Systems* 19: 129–137.

Murty, K.G., & Kabadi, S.N. 1987. Some np-complete problems in quadratic and nonlinear programming. *Mathematical Programming* 39: 117–129.

Quesada, I., & Grossmann, I.E. 1992a. An lp/nlp based branch and bound algorithm for convex minlp optimization problems. *Computers & Chemical Engineering* 16: 937–947.

Reger, D., Madanat, S., & Horvath, A. 2014. Economically and environmentally informed policy for road resurfacing: tradeoffs between costs and greenhouse gas emissions. *Environmental Research Letters* 9: 104020.

Reger, D., Madanat, S., & Horvath, A. 2015. The effect of agency budgets on minimizing greenhouse gas emissions from road rehabilitation policies. *Environmental Research Letters* 10: 114007.

Santero, N.J., & Horvath, A. 2009. Global warming potential of pavements. *Environmental Research Letters* 4: 034011.

Shakiba, M., Ozer, H., Ziyadi, M., & Al-Qadi, I.L. 2016. Mechanics based model for predicting Structure-induced Rolling Resistance (SRR) of the tire–pavement system. *Mechanics of Time-Dependent Materials*: 1–22.

Wang, T., Lee, I., Harvey, J., Kendall, A., Lee, E.B., & Kim, C. 2012. *UCPRC Life Cycle Assessment Methodology and Initial Case Studies for Energy Consumption and GHG Emissions for Pavement.* Technical Report. University of California Pavement Research Center.

Yu, B., Lu, Q., & Xu, J. 2013. An improved pavement maintenance optimization methodology: Integrating LCA and LCCA. *Transportation Research Part A: Policy and Practice* 55: 1–11.

Zapata, P., & Gambatese, J.A. 2005. Energy consumption of asphalt and reinforced concrete pavement materials and construction. *Journal of Infrastructure Systems* 11: 9–20.

Author index

Akbarian, M. 165
Alauddin Ahammed, M. 165
Al-Qadi, I.L. 133, 179, 289

Baliello, A. 189
Balzarini, D. 59
Braham, A.F. 103
Butt, A.A. 231

Cao, R. 199
Chatti, K. 59
Chebbi, W. 157
Chen, X. 211
Choy, D. 251
Cui, Q. 251

D'Angelo, G. 41
Dauvergne, M. 157
Dave, E. 241
Ding, W. 271
Dylla, H. 11

Finlayson, G. 165
Flintsch, G. 1
Flores, R. 1

Giacomello, G. 189
Goemans, C. 165
Gregory, J. 121

Harvey, J.T. 69, 231
Hassan, M. 261
Hettiwatte, M. 31

Inti, S. 145

Jones, D. 231
Jullien, A. 157

Keijzer, E. 1
Kirchain, R. 121
Kshirsagar, S. 31
Kulikowski, J. 89

Larsson, M. 281
Leng, Z. 199
Li, H. 69
Liu, A. 271
Liu, X. 251
Lu, Y. 271

Meil, J. 165
Mo, W. 241
Mukherjee, A. 11

Najm, H. 211
Neves, L.C. 51, 79
Noshadravan, A. 121

Osmani, F. 31
Ozer, H. 133, 179

Parry, T. 51, 79
Pasetto, M. 189
Pasquini, E. 189
Perrotta, F. 51
Praticò, F.G. 221
Presti, D.L. 41, 79

Qiao, Y. 241

Roesler, J.R. 111

Saboori, A. 69, 231
Santos, J. 1
Schwartz, C.W. 251
Sen, S. 111
Senadheera, S. 31
Senhaji, M.K. 179
Sharma, M. 145
Shu-Chien Hsu, M. 199
Soliman, N. 261
Strömberg, L. 23
Sullivan, S. 165
Swei, O. 121

Tandon, V. 145
Thyagarajan, S. 1
Toller, S. 281
Trupia, L. 51, 79
Tursun, U.D. 289

Valle, O. 241

Wang, H. 211
Wang, Y. 199
Wu, H. 271

Xu, X. 121

Yang, R. 289
Yazoghli-Marzouk, O. 157
Yu, H. 199

Zaabar, I. 59
Zhang, H.C. 31
Zhu, H. 271
Ziyadi, M. 133